Günther Rüdiger,
Leonid L. Kitchatinov, and
Rainer Hollerbach

Magnetic Processes in Astrophysics

Related Titles

McCarthy, D.D., Seidelmann, P.K.

Time – From Earth Rotation to Atomic Physics

2009
ISBN: 978-3-527-40780-4

Sidorenkov, N.S.

The Interaction Between Earth's Rotation and Geophysical Processes

2009
ISBN: 978-3-527-40875-7

Shore, S.N.

Astrophysical Hydrodynamics
An Introduction

2007
ISBN: 978-3-527-40669-2

Stahler, S.W., Palla, F.

The Formation of Stars

2004
ISBN: 978-3-527-40559-6

Foukal, P.V.

Solar Astrophysics

2004
ISBN: 978-3-527-40374-5

Spitzer, L.

Physical Processes in the Interstellar Medium

1998
ISBN: 978-0-471-29335-4

*Günther Rüdiger, Leonid L. Kitchatinov,
and Rainer Hollerbach*

Magnetic Processes in Astrophysics

Theory, Simulations, Experiments

Verlag GmbH & Co. KGaA

The Authors

Prof. Günther Rüdiger
Leibniz-Institut für Astrophysik Potsdam (AIP)
An der Sternwarte 16
14482 Potsdam
Germany

Dr. Leonid L. Kitchatinov
Institute for Solar-Terrestrial
Physiks
Lermontov st., 126 a
664033 Irkutsk
Russian Federation

Prof. Rainer Hollerbach
Institut für Geophysik
ETH Zürich
8092 Zürich
Switzerland

Cover Picture
The PROMISE facility for experimental studies of the helical and the azimuthal magnetorotational instability, constructed and operated at Helmholtz-Zentrum Dresden-Rossendorf.

All books published by **Wiley-VCH** are carefully produced. Nevertheless, authors, editors, and publisher do not warrant the information contained in these books, including this book, to be free of errors. Readers are advised to keep in mind that statements, data, illustrations, procedural details or other items may inadvertently be inaccurate.

Library of Congress Card No.:
applied for

British Library Cataloguing-in-Publication Data:
A catalogue record for this book is available from the British Library.

Bibliographic information published by the Deutsche Nationalbibliothek
The Deutsche Nationalbibliothek lists this publication in the Deutsche Nationalbibliografie; detailed bibliographic data are available on the Internet at http://dnb.d-nb.de.

© 2013 WILEY-VCH Verlag GmbH & Co. KGaA, Boschstr. 12, 69469 Weinheim, Germany

All rights reserved (including those of translation into other languages). No part of this book may be reproduced in any form – by photoprinting, microfilm, or any other means – nor transmitted or translated into a machine language without written permission from the publishers. Registered names, trademarks, etc. used in this book, even when not specifically marked as such, are not to be considered unprotected by law.

Print ISBN 978-3-527-41034-7
ePDF ISBN 978-3-527-64895-5
ePub ISBN 978-3-527-64894-8
mobi ISBN 978-3-527-64893-1
oBook ISBN 978-3-527-64892-4

Cover Design Grafik-Design Schulz, Fußgönheim
Typesetting le-tex publishing services GmbH, Leipzig
Printing and Binding Markono Print Media Pte Ltd, Singapore

Printed in Singapore
Printed on acid-free paper

Contents

Preface IX

1 Differential Rotation of Stars *1*
1.1 Solar Observations *2*
1.1.1 The Rotation Law *2*
1.1.2 Torsional Oscillations *5*
1.1.3 Meridional Flow *6*
1.2 Stellar Observations *9*
1.2.1 Rotational Evolution *9*
1.2.2 Differential Rotation *11*
1.3 The Reynolds Stress *13*
1.3.1 The Λ Effect *14*
1.3.1.1 Numerical Simulations *15*
1.3.1.2 Quasi-linear Theory of the Λ Effect *19*
1.3.2 Eddy Viscosities *22*
1.4 The Meridional Flow *24*
1.4.1 Origin of the Meridional Flow *26*
1.4.2 The Differential Temperature *28*
1.4.3 Advection-Dominated Solar Dynamo *32*
1.5 The Sun *35*
1.5.1 Sun without Λ Effect *38*
1.5.2 Sun without Baroclinic Flow *39*
1.5.3 Global Simulations *40*
1.6 Individual Stars *42*
1.6.1 Two MOST Stars *44*
1.6.2 Young Stars *46*
1.7 Dwarfs & Giants *50*
1.7.1 M Dwarfs *50*
1.7.2 F Stars *51*
1.7.3 Giants *55*
1.8 Differential Rotation along the Main Sequence *58*

2 Radiation Zones: Magnetic Stability and Rotation 63
2.1 The Watson Problem 65
2.1.1 The Stability Equations 65
2.1.2 2D Approximation 67
2.1.3 Stability Maps 69
2.2 The Magnetic Tachocline 72
2.2.1 A Planar Model 72
2.2.2 Magnetic Field Confinement by Meridional Flow 75
2.2.3 Tachocline Model in Spherical Geometry 79
2.3 Stability of Toroidal Fields 82
2.3.1 Equations 82
2.3.2 Nonexistence of 2D Magnetic Instabilities 85
2.3.3 No Diffusion 86
2.3.4 Growth Rates, Drift Rates and Radial Mixing 88
2.4 Stability of Thin Toroidal Field Belts 91
2.4.1 Rigid Rotation 92
2.4.2 Differential Rotation 93
2.4.3 High Fourier Modes 94
2.5 Helicity and Dynamo Action 94
2.5.1 Helicity and Alpha Effect 95
2.5.2 Dynamo Action 100
2.6 Ap Star Magnetism 103
2.7 The Shear–Hall Instability (SHI) 109

3 Quasi-linear Theory of Driven Turbulence 115
3.1 The Turbulence Pressure 116
3.2 The η-Tensor 124
3.2.1 Rotating Turbulence 124
3.2.2 Nonrotating Turbulence but Helical Background Fields 128
3.3 Kinetic Helicity and DIV-CURL Correlation 131
3.4 Cross-Helicity 134
3.4.1 Theory 135
3.4.2 Simulations and Observations 136
3.5 Shear Flow Electrodynamics 138
3.5.1 Hydrodynamic Stability of Shear Flow 138
3.5.2 The Magnetic-Diffusivity Tensor 140
3.5.3 Dynamos without Stratification 141
3.6 The Alpha Effect 143
3.6.1 Helical-driven Turbulence 143
3.6.2 Shear Flow 145
3.6.3 Shear-Dynamos with Turbulence-Stratification 149
3.6.4 Alpha Effect by Density Stratification 150
3.7 The Current Helicity 153

4	**The Galactic Dynamo** *157*	
4.1	Magnetic Fields of Galaxies *157*	
4.2	Interstellar Turbulence *161*	
4.2.1	Hydrostatic Equilibrium and Interstellar Turbulence *162*	
4.2.2	Alpha Effect by Supernova Explosions *165*	
4.2.3	The Advection Problem *168*	
4.3	Dynamo Models *170*	
4.3.1	Linear Models *171*	
4.3.2	Nonlinear Dynamo Models *173*	
4.4	Magnetic Instabilities *175*	
4.4.1	The Seed Field Problem *175*	
4.4.2	Magnetorotational Instability *176*	
4.4.3	Tayler Instability *180*	
5	**The Magnetorotational Instability (MRI)** *185*	
5.1	Taylor–Couette Flows *185*	
5.2	The Stratorotational Instability (SRI) *188*	
5.2.1	The Angular Momentum Transport *192*	
5.2.2	Electromotive Force by Magnetized SRI *195*	
5.3	The Standard Magnetorotational Instability (SMRI) *198*	
5.3.1	The Equations *200*	
5.3.1.1	The Rayleigh Limit *202*	
5.3.1.2	Pseudo-Kepler Rotation *203*	
5.3.2	Nonaxisymmetric Modes *204*	
5.3.3	Wave Numbers *206*	
5.3.4	Nonlinear Simulations *208*	
5.3.5	The Angular Momentum Transport *211*	
5.4	Diffusive Kepler Disks *214*	
5.5	MRI with Hall Effect *216*	
5.6	The Azimuthal MRI (AMRI) *218*	
5.6.1	The Equations *219*	
5.6.2	The Instability Map *223*	
5.6.3	Different Scalings with Pm *224*	
5.6.4	Nonlinear Results *224*	
5.6.5	The AMRI Experiment *228*	
5.7	Helical Magnetorotational Instability (HMRI) *231*	
5.7.1	From AMRI to HMRI *231*	
5.7.2	Nonaxisymmetric Modes for small Pm *236*	
5.7.3	Pseudo-Kepler Rotation *236*	
5.7.4	The Frequencies *237*	
5.8	Laboratory Experiment PROMISE *238*	
5.8.1	Experimental Results *240*	
5.8.2	Endplate Effects *242*	
5.8.3	PROMISE 2 *244*	

6 The Tayler Instability (TI) 247
- 6.1 Stationary Fluids *249*
- 6.2 Experiment GATE *254*
- 6.3 Rotating Fluids *256*
- 6.3.1 Rigid Rotation *257*
- 6.3.2 Differential Rotation *258*
- 6.3.3 Eddy Viscosity and Turbulent Diffusivity *262*
- 6.3.3.1 Eddy Viscosity *262*
- 6.3.3.2 Turbulent Diffusivity *263*
- 6.3.3.3 Mixing of Chemicals *265*
- 6.4 The Tayler Generator *267*
- 6.5 Helical Background Fields and Alpha Effect *272*
- 6.5.1 Helical Fields with Weak Axial Current *272*
- 6.5.2 Uniform Electric Current *275*
- 6.5.3 Alpha Effect *278*
- 6.5.3.1 The Helicities *278*
- 6.5.3.2 The Alpha Effect *280*
- 6.6 TI with Hall Effect *282*

7 Magnetic Spherical Couette Flow 287
- 7.1 Stewartson Layers *287*
- 7.2 Shercliff Layers *289*
- 7.3 Finite Re in an Axial Field *296*
- 7.3.1 Numerics *296*
- 7.3.2 The Maryland Experiment *302*
- 7.3.3 The Princeton Experiment *305*
- 7.4 The Grenoble DTS Experiment *307*
- 7.5 Other Waves and Instabilities *313*
- 7.5.1 Inertial Oscillations *313*
- 7.5.2 Torsional Oscillations *314*
- 7.5.3 Alfvén Waves *316*
- 7.5.4 The Magnetostrophic MRI *317*
- 7.6 Linear Combinations of Axial and Dipolar Fields *318*
- 7.7 Dynamo Action *321*

References *327*

Index *341*

Preface

In 2004 two of us (Rüdiger and Hollerbach) published a previous book entitled *The Magnetic Universe: Geophysical and Astrophysical Dynamo Theory*, describing the origin of magnetic fields in objects ranging in size from planets to galaxies. Ever since then, we have considered the possibility of writing a second edition, updating developments of the past decade. However, ultimately there were so many recent developments in areas not covered at all before that it ended up as a completely new book, with only minimal overlap (in parts of Chapters 1 and 4) with corresponding formulations in Rüdiger and Hollerbach (2004). In particular, the subjects of these two chapters (differential rotation theory, the galactic dynamo) have developed so rapidly since then that a new discussion was clearly necessary.

On the one hand, the successful asteroseismic space missions MOST, CoRoT and KEPLER ushered in a new era of knowledge of the internal stellar rotation laws. Stars exhibit much greater variety of turbulent convection zones and angular momentum than found in the Sun, so that it is now possible to develop the theory of the rotation of stars by means of the new data. It is also clear that only facts about stellar differential rotation allow us to understand the magnetic activity of the main-sequence stars.

On the other hand, the interstellar medium forms one of the most impressive realizations of driven cosmical turbulence under the influence of a (nonuniform) rotation, where both can be observed *in situ*. Observers find a strong stratification in the vertical direction of gas and turbulence, hence large-scale helicity should exist. In such cases mean-field dynamo theory predicts the instability of the solution $B = 0$ and consequently the existence of large-scale magnetic fields which can also be observed. The correctness of this scenario has been successfully probed with numerical simulations which are presented here as a convincing instance of the state of the modern dynamo theory.

One significant change though from Rüdiger and Hollerbach (2004) is the switch in emphasis from being primarily on dynamo theory more toward magnetic instabilities such as the magnetorotational instability or the Tayler instability. That is, instead of seeking to explain the origin of magnetic fields, we now take the existence of large-scale fields as given, and study the ways in which instabilities can destroy the large-scale structures again, and give rise to small-scale turbulence instead.

Another aspect that has changed significantly since 2004 is the increasing importance of liquid metal (and plasma) laboratory experiments, not only in attempts to create laboratory dynamos, but also involving externally imposed magnetic fields. In addition to the chapters on the magnetorotational and Tayler instabilities, the chapter on magnetic spherical Couette flow describes a number of new experiments. In all of these areas, the interplay between basic theory, detailed numerical simulations, and experiments has been particularly fruitful, with generally good agreement also between theory and experiment (perhaps "disappointingly" good, if one views unexpected experimental results as the ones most likely to further lead to fundamentally new insights). At any rate, we hope that this book will be of interest not just to astrophysicists but to fluid dynamicists more generally, or anyone else wanting to understand liquid metal experiments and the insights they can yield.

Numerous colleagues have contributed to this book, either directly or by general discussions over many years. GR particularly thanks Rainer Arlt, Detlef Elstner, Marcus Gellert, Andrea Hans, Manfred Küker and Manfred Schultz of the Leibniz-Institut für Astrophysik Potsdam for their substantial support with countless technical details. LLK and GR acknowledge the continuous encouragement of the Deutsche Forschungsgemeinschaft and the Alexander von Humboldt Foundation stimulating a number of the developments presented here. RH thanks Prof Andy Jackson of the Institute of Geophysics at ETH Zürich for the invitation to visit ETH (with funding by the European Research Council). The time away from regular duties was invaluable in finishing this project in time. Finally, among the vast MHD community we particularly thank Gunter Gerbeth, Thomas Gundrum, Martin Seilmayer, Frank Stefani and the entire group at the Helmholtz-Zentrum Dresden-Rossendorf for many intensive discussions and collaborations over the past decade, as presented in several of the chapters here, and which we hope may stimulate further developments in MHD laboratory astrophysics.

Potsdam *Günther Rüdiger*
Irkutsk *Leonid L. Kitchatinov*
Zürich *Rainer Hollerbach*
2013

1
Differential Rotation of Stars

Magnetic activity of solar-type stars is closely related to stellar rotation. The differential rotation participates in stellar dynamos by producing toroidal magnetic fields by rotational shear. Differential rotation and meridional flow can be understood in the context of mean-field hydrodynamics in stellar convection zones. Stratification in convection zones is so strong that the Schwarzschild criterion ($dS/dr < 0$, where S is the specific entropy) is fulfilled and the entire zone becomes turbulent. Due to the radial stratification the turbulence fields are themselves stratified with the radial preferred direction. Interaction of such a turbulence with an overall rotation leads to the formation of large-scale structure. Lebedinskii (1941), Wasiutynski (1946), Biermann (1951) and Kippenhahn (1963) were the first to find that differential rotation and meridional flow might be direct consequences of the rotating anisotropic turbulence. Details of the long history of this concept were presented by Rüdiger (1989, Chapter 2).

Whether a star is of solar-type is controlled by its structure. Stars of this type possess external (turbulent) convection zones. The solar convection zone only includes < 2% of the total mass ($M_\odot$) but it extends about 30% in radius. The outer convection zones in cooler stars become deeper as stellar mass decreases until for M stars the convection zone reaches down to the center. On the other hand, for A stars the outer convection zone becomes very thin, but an inner zone becomes convectively unstable. For B stars this inner convection zone reaches considerable dimensions.

The level of stellar activity depends strongly on spectral type. There is, however, the striking fact that the linear depth of the outer convection zone, at 200 000 km, does not vary too much among the solar-type stars. We shall see later how important the total thickness of a convection zone is for the formation of differential surface rotation.

It is certainly unrealistic to expect a solution of the complicated problem of stellar dynamos if the internal stellar rotation laws cannot be predicted or observed (by asteroseismology). Differential rotation is explained here as turbulence-induced with only a small magnetic contribution. Mean-field hydrodynamics provides a theoretical basis for differential rotation modeling, so that the models can be constructed with very little arbitrariness. Nevertheless, differential rotation of the Sun can be reproduced by computations very closely and the dependence of differential rotation on stellar parameters can be predicted.

1.1
Solar Observations

1.1.1
The Rotation Law

The rotation of the solar photosphere was measured using the Doppler shifts of photospheric spectral lines or tracking rotation of sunspots and various other tracers. Doppler measurements of Howard *et al.* (1983) and the classical work of Newton and Nunn (1951) on sunspot rotation are the well-known examples. Within a small percentage, all measurements yield similar results. Obtained by tracing bright coronal structures in SOHO images Wöhl *et al.* (2010) give

$$\Omega = 0.253 - 0.044 \sin^2 b - 0.013 \sin^4 b \quad [\text{rad/day}] \tag{1.1}$$

for the sidereal rotation rate, with $b = 90° - \theta$ as the heliographic latitude. The angular velocity of 0.25 rad/day leads to a frequency of 462 nHz at the equator. The observed equator–pole difference of the angular velocity, $\delta\Omega$, from (1.1) is 0.057 rad/day. We shall characterize the existence of differential rotation by the quantity $\delta\Omega = \Omega_{eq} - \Omega_{pole}$ rather than by the ratio

$$k = \frac{\Omega_{eq} - \Omega_{pole}}{\Omega_{eq}} \tag{1.2}$$

(here ≈ 0.23) because only $\nabla\Omega$ is relevant for the inducting action of differential rotation but not its normalized value k. With (1.2) we follow the notation of the seminal paper by Hall (1991) who derived from photometric stellar observations a relation $k \propto \Omega^{-0.85}$ (corresponding to the very flat relation $\delta\Omega \propto \Omega^{0.15}$ for rotating stars, see also Barnes *et al.* (2005)) which is rather close to the essentials presented in the theoretical part of this chapter.

Brown (1985) made the first attempt to infer how the latitudinal differential rotation varies with depth from rotational splitting of frequencies of global acoustic oscillations. Today the helioseismological inversions provide a detailed portrait of the internal solar rotation (Wilson, Burtonclay, and Li, 1997; Schou *et al.*, 1998). Figure 1.1 shows the distribution of rotation rate inside the Sun. Latitudinal differential rotation seen on the solar photosphere survives throughout the convection zone up to its base. Helioseismology detects the location of the inner boundary of the convection zone at $r_{in} = 0.713 R_\odot$ (Christensen-Dalsgaard, Gough, and Thompson, 1991; Basu and Antia, 1997). Latitudinal differential rotation at the inner boundary is reduced about twice compared with the surface (Charbonneau *et*

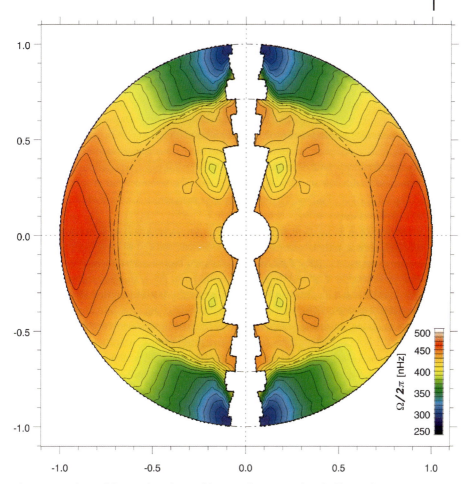

Figure 1.1 Isolines of the angular velocity of the Sun after Korzennik and Eff-Darwich (2011). The rotation of the polar and the near-center regions is difficult to measure. With permission of the authors.

al., 1999). A remarkable feature of Figure 1.1 is the sharp transition from differential to rigid rotation in a thin layer near the base of the convection zone. This layer, called after Spiegel and Zahn (1992) the solar "tachocline," extends not more than 4% in radius (Kosovichev, 1996; Antia, Basu, and Chitre, 1998). Its midpoint is at $(0.692 \pm 0.005)\, R_\odot$, and it is slightly prolate in shape (Charbonneau *et al.*, 1999). The tachocline is, therefore, located mainly if not totally beneath the base of the convection zone, in the uppermost radiative zone. Rotation beneath the tachocline is almost rigid at least down to $0.2\, R_\odot$ (Couvidat *et al.*, 2003; Korzennik and Eff-Darwich, 2011).

The main empirical features of the solar differential rotation can be summarized as follows (see Figure 1.1):

- an equatorial acceleration of about 23% at the surface,
- a near-surface shear layer with negative Ω-gradient in radius[1],
- a sharp transition to rigid rotation in a thin tachocline,
- a rigid rotation of the deeper radiative core.

The 'observed' phenomenon of the sharp transition layer between the outer domain of differential rotation and the inner domain of rigid-body rotation is hard to understand without the assumption of internal empirically unknown magnetic fields. We shall show in Section 2.2 that indeed fossil fields with amplitudes of only 1 mG are enough to explain not only the existence of the tachocline but also its small radial extension.

The present state of differential rotation may, however, differ from other epochs when magnetic activity of the Sun was different. Ribes and Nesme-Ribes (1993) used statistics of sunspot observations over the Maunder minimum at the Observatoire de Paris to find a rotation rate slower by about 2% at the equator and by about 6% at midlatitudes than at the present time. The differential rotation was thus stronger than today. The more magnetic the Sun, the faster and more rigidly its surface rotates. Balthasar, Vázquez, and Wöhl (1986), however, could not find similar results for a regular minimum. Also Arlt and Fröhlich (2012), who worked with data obtained from the drawings of Staudacher from the period from 1749 till 1799 did not find a significant difference to the present-day value of $\delta\Omega \simeq 0.050$ rad/day derived by Balthasar, Vázquez, and Wöhl (1986) from sunspot rotation. The reported average value of 0.048 indicates a slightly smaller value but this difference is not yet significant.

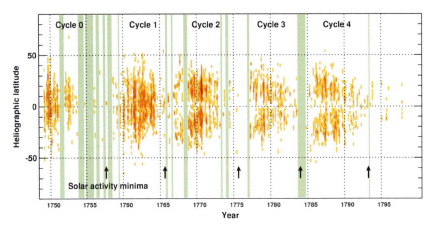

Figure 1.2 The butterfly diagram shortly after the Maunder minimum, as derived from the drawings of Staudacher between 1749 and 1799. Courtesy of R. Arlt.

---

[1] Young spots rotate faster by about 4% than the solar surface plasma, see (1.37).

The results are nevertheless highly interesting as they demonstrate the reliability of the data which also led to the construction of a butterfly diagram for the four cycles covered by the observations. The main question here is whether the dipolar parity which now dominates the solar activity already existed shortly after the Maunder minimum. This is certainly the case for the last two cycles shown in Figure 1.2 but it seems to be questionable for the older two cycles. For these cycles, which are closer to the Maunder minimum at least an overpopulation of near-equator sunspots is indicated by the data (Arlt, 2009).

1.1.2
Torsional Oscillations

As magnetic activity of the Sun varies with time, differential rotation may also be expected to be time-dependent. Variations of solar rotation law are indeed observed. Schrijver and Zwaan (2000), Stix (2002) and Thompson *et al.* (2003) presented detailed historical and data-based overviews of all phenomena concerning the temporal variations of the solar rotation law. As the magnetic force is quadratic in the magnetic field, the resulting flow is expected to vary with twice the frequency of the 22-year magnetic cycle. The 11-year torsional oscillations were first observed by Howard and LaBonte (1980).

Figure 1.3 shows the oscillation pattern. At a fixed latitude there is an oscillation of fast and slow rotation with an 11-year period. The whole pattern migrates at about 2 m/s toward the equator. The migration follows the equatorial drift of magnetic activity. Latitudinal shear of differential rotation is increased in the activity belt with faster than average rotation on the equatorial side of the belt and slower than average rotation on the polar side. Howe, Komm, and Hill (2002) showed by helioseismological inversions that the migrating torsional oscillation exists not

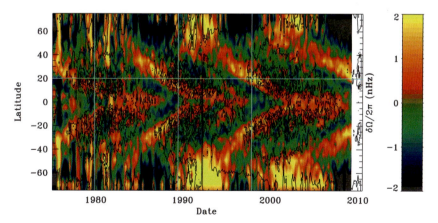

Figure 1.3 Torsional oscillations derived from Doppler shift measurements. The flow pattern follows the equatorial drift of magnetic activity. The flow at a given latitude oscillates with a period of about 11 years. Courtesy of Howe *et al.* (2011).

1 Differential Rotation of Stars

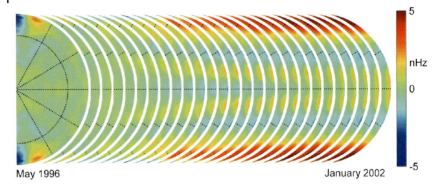

Figure 1.4 Helioseismology detects two branches of torsional oscillations migrating to the equator and to the poles from midlatitude and extending deep into the convection zone. From Vorontsov *et al.* (2002). Reprinted with permission from AAS.

only at the surface, but that it extends at least 60 000 km down into the convection zone.

An important question of dynamo theory is whether a poleward migrating branch is present at high latitudes. Schou (2001) and Vorontsov *et al.* (2002) reported the detection of such a branch of torsional oscillation for the rising phase of solar cycle 23 from helioseismological data between 1996 and 2002. Figure 1.4 shows the polar branch together with the low-latitude equatorial branch penetrating deep into the convection zone.

Close correlation with solar activity is indicative of a magnetic origin of torsional oscillations. Details of the mechanism producing the oscillations remain, however, uncertain. The oscillations may be produced by the global Lorentz force (Yoshimura, 1981; Schüssler, 1981; Rüdiger *et al.*, 1986) or the magnetic backreaction on small spatial scale of turbulence (Kitchatinov, 1990; Rüdiger and Kitchatinov, 1990). Meridional flow induced by entropy disturbances in the magnetic activity belt may also be relevant (Spruit, 2003; Cameron and Schüssler, 2012).

1.1.3
Meridional Flow

It was recognized since the work of Kippenhahn (1963) that differential rotation and meridional flow are closely related and it is not possible to correctly describe one if the other is not allowed for.

The relatively slow meridional circulation is difficult to measure. Ward (1965) noticed that the flow is problematic to define by the method of tracers: an inhomogeneity of tracer distribution over latitude together with latitudinal turbulent diffusion results in a false meridional flow. This is probably why early measurements using sunspots as tracers gave conflicting results. A more coherent picture is provided by using more uniformly distributed 'small magnetic features' as tracers. Komm, Howard, and Harvey (1993) found a meridional flow from the equator

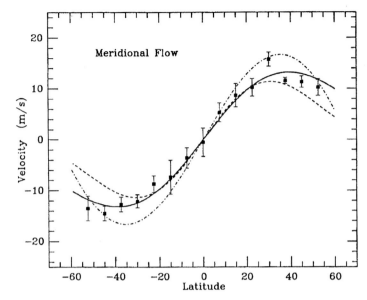

Figure 1.5 Full line and symbols show the meridional flow measured by using small magnetic features as tracers. Overplotted are Doppler measurements of Ulrich (1993) (dashed-dotted) and Snodgrass (1984) (dashed). From Komm, Howard, and Harvey (1993).

to pole with amplitude slightly above 10 m/s. The flow is shown in Figure 1.5. The flow pattern shows no hemispheric asymmetry and did not migrate in latitude during a solar cycle. The amplitude of the flow varies, however, over the activity cycle. Meridional velocity changes from below average during solar maximum to above average during solar minimum. The relative amplitude of the variation is about 25%. This picture is supported by recent measurements of Hathaway and Rightmire (2010) and by seismological sounding of the flow by Basu and Antia (2010).

Zhao and Kosovichev (2004) measured the meridional flow for seven Carrington rotations of years 1996–2002 covering the epoch from solar activity minimum to maximum. The measurements by time-distance helioseismology show the poleward flow decreasing with depth in the surface layer of 12 000 km. In addition to the dominating poleward flow of order 20 m/s, cells of weaker flow converging to the activity belts were found in both hemispheres. These cells migrated towards the equator following the migration of activity belts as the solar cycle evolved. These migrating cells may be a counterpart of torsional oscillations in the meridional flow.

Gizon and Rempel (2008) analyzed MDI data from 1996 to 2002. The resulting surface flow is poleward up to ±50° latitude. The flow velocity peaks at about 25°. The peak-to-peak variation in time is, at 7 m/s, rather large. There is no indication for a second (polar) cell of meridional circulation (Figure 1.6). The same is true for the flow at a depth of 60 000 km where the amplitude of the poleward flow is 6 m/s peaking again at 25°. Schad, Timmer, and Roth (2012) even reach $x = 0.8$, that is a

8 *1 Differential Rotation of Stars*

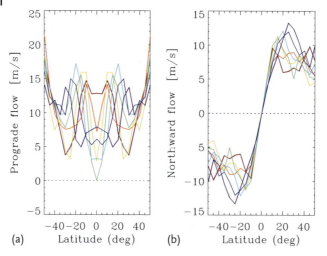

Figure 1.6 Helioseismology results (MDI) for the rotational velocity (a) and meridional flow (b) in their dependencies on latitude and time (from 1996 (blue) to 2002 (red)). The meridional flow is poleward with an amplitude of 10 m/s, which varies in time by 7 m/s. From Gizon and Rempel (2008).

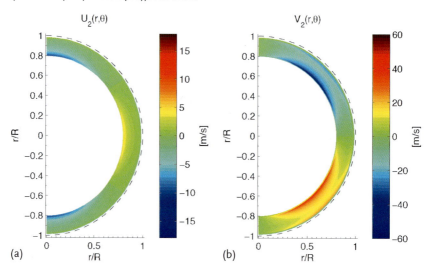

Figure 1.7 Helioseismology results for the radial (a) and the latitudinal (b) components of the meridional flow from the surface down to $x = 0.8$. The latitudinal flow is always poleward while the radial flow is upward at the equator and downward at the poles. Courtesy of M. Roth.

depth of 140 000 km (Figure 1.7). The main conclusion is that indeed only one cell with counterclockwise circulation exists in the solar convection zone, whose return flow exists below $x = 0.8$. This result is of basic significance for both the theory of differential rotation and the theory of the solar dynamo.

The question for the equatorward return flow deep in the convection zone has also been considered by Hathaway *et al.* (2003). From the sunspot data since 1874 they found an anticorrelation between the drift rate of the center of the butterfly diagram and the cycle length. The faster the drift of the butterfly diagram the shorter the cycles. With such statistics an amplitude of 1.2 m/s for the return flow velocity at the bottom of the convection zone has been estimated.

1.2 Stellar Observations

1.2.1 Rotational Evolution

Rotation of a star is an important parameter of hydromagnetic dynamos controlling the rate of magnetic field generation. The dynamo activity in turn decreases the rotation rate. Solar-type stars are observed to exhibit a steady decline in rotation rate between the ages from about 10^8 to 4.5×10^9 years (Skumanich, 1972).

The spin-down is commonly explained as follows (Kraft, 1967): magnetic activity of stars with external convection zones produce hot stellar coronae. Similar to the Sun, the hot coronae emanate stellar winds. The material making up the wind does not lose contact with the parent star after leaving its photosphere, but corotates with the star due to magnetic coupling to its surface. The extent of the coupling can be (very crudely) estimated by the Alfvén radius, R_A, where the wind velocity equals the Alfvén speed. As the angular momentum loss is proportional to R_A^2, magnetic activity enhances the rotational braking. Spindown of the main-sequence dwarfs closely obeys the Skumanich law,

$$P_{\rm rot} \propto \sqrt{t}, \tag{1.3}$$

relating rotation period $P_{\rm rot}$ of a star to its age t.

This law does not, however, apply to all stellar ages. Solar mass stars are born with rotation periods of about one week. Subsequently, these stars spin-up very quickly during contraction to the main-sequence to attain a rotation period of about one day or even shorter as ZAMS stars (Hartmann and Noyes, 1987). Close to the end of their main-sequence lives, stars seem to deviate from relation (1.3) as well. Figure 1.8 shows that the upper-left corner on the plot of $P_{\rm rot}$ vs. B-V color index is empty. This suggests that the dwarf stars are not decelerated beyond a maximum rotation period depending on the spectral type. The maximum period is larger for cooler stars.

The spin-down law (1.3) applies to the solar-type stars over a major part of their main-sequence life. Gray (1982) and Rengarajan (1984) forwarded the idea that the proportionality constant in the relation (1.3) is a single-valued function of stellar mass or other equivalent parameter. This idea eventually led to the development of gyrochronology establishing an empirical relation between age, rotation period and mass (Barnes, 2003, 2007, 2010; Collier Cameron *et al.*, 2009; Meibom, Mathieu,

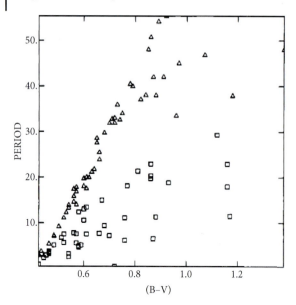

Figure 1.8 Plot of P_{rot} (in days) versus B-V color for main-sequence stars. The squares and triangles refer to young and old stars, respectively. From Rengarajan (1984). Reprinted by permission of AAS.

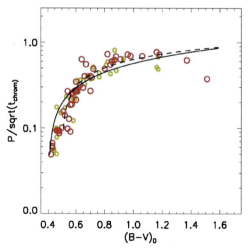

Figure 1.9 The ratio of P_{rot} (days) to square root of t (Myr) in dependence on B-V color for main-sequence stars of the Mount Wilson sample. Small and large circles show young and old stars, respectively. From Barnes (2007). Reprinted by permission of AAS.

and Stassun, 2009). The relation is illustrated by Figure 1.9 where the dependence of the ratio $P_{rot}/\sqrt{t}$ on B-V color is shown. Different functional expressions for the empirical relation have been suggested, for example

$$P_{\text{rot}} = a t^n (B - V - c)^b \tag{1.4}$$

in days with $n = 0.519 \pm 0.007$, $a = 0.773 \pm 0.011$, $b = 0.601 \pm 0.024$, $c = 0.4$, where t is measured in Myr (Barnes, 2007). The characteristic error of this relation when applied to gigayear-old stars from early M to late F is reported as within 20%.

1.2.2
Differential Rotation

Stellar differential rotation is measured mainly by the same method, which originally was used for the Sun, that is, by tracing the rotation of thermal or magnetic spots. As stars are typically point sources (only in very rare, exceptional cases can stellar surfaces be resolved), the methods are very sophisticated and demanding of observational data. The tracer method can be realized using high precision (space-based) photometry or high resolution spectroscopy with the Doppler imaging techniques. The differential rotation was also measured using shapes of spectral lines (Reiners and Schmitt, 2003a,b) and variations of Ca II H&K emissio (e.g., Donahue, Saar, and Baliunas, 1996). Doppler imaging (Khokhlova, 1975; Vogt and Penrod, 1983) provides detailed mapping of stellar surfaces but can be used only for young rapidly rotating stars because the projected rotation velocity $v \sin i$ should typically not be smaller than 15 km/s. Measurements of differential rotation by Doppler imaging were summarized by Barnes et al. (2005) to reveal strong dependence on spectral type: the hotter the star, the larger the pole–equator difference in rotation

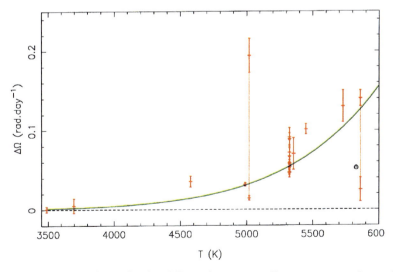

Figure 1.10 Dependence of surface differential rotation on effective temperature for rapidly rotating solar analogs by Doppler imaging (Barnes et al., 2005). Copyright © 2005 RAS.

rate. Observed differential rotation is usually fit by the $\sin^2 b$ profile

$$\Omega = \Omega_{eq} - \delta\Omega \sin^2 b . \qquad (1.5)$$

Figure 1.10 shows the pole–equator difference $\delta\Omega$ as a function of temperature for young solar-type stars. The largest differential rotation belongs to the hottest stars. Jeffers and Donati (2008) found the slightly premain-sequence G0 star HD 171488 (V889 Her) with $\delta\Omega \simeq 0.5$ rad/day exceeding all stars of Figure 1.10 in magnitude of its differential rotation (see Section 1.6.2).

The dependence of $\delta\Omega$ on the rotation rate is probably mild. The very rapidly rotating ($P_{rot} = 0.51$ day) solar analog AB Dor possesses almost the same differential rotation as the Sun (Donati and Collier Cameron, 1997). Figure 1.11 shows differential rotation of several stars close to the Sun by mass but rotating with different rates. Rotation of the stars with $P_{rot} < 2$ days of this figure was defined by Doppler imaging. For the slowest rotator in this sample, LQ Hya, large variations with time has been seen by Donati et al. (2003). Differential rotation of two moderate rotators, κ^1 Ceti and ε Eri, was measured using high precision photometry of the MOST mission (Croll et al., 2006; Walker et al., 2007). The stars of Figure 1.11 spanning almost two orders of magnitude in rotation rate show a very similar amount of differential rotation. This suggests that absolute value of differential rotation varies mainly with stellar surface temperature, variation with rotation rate being mild.

When common statistics for stars of different spectral types is used, however, an increase of differential rotation with rotation rate is usually found (cf. Donahue, Saar, and Baliunas (1996)). As can be seen from Figure 1.8 or the gyrochronology equation (1.4), slow rotators are mainly represented by K stars while G and F stars

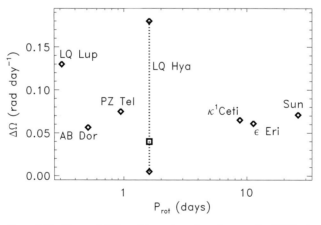

Figure 1.11 Observed differential rotation of stars close to the Sun by mass but rotating with different rates. LQ Lup is a premain-sequence star (Donati et al., 2000). AB Dor (Donati and Collier Cameron, 1997), PZ Tel (Barnes et al., 2000) and LQ Hya (Donati et al., 2003) are ZAMS stars. Flores-Soriano and Strassmeier (2013) give $\delta\Omega \simeq 0.04$ rad/day for LQ Hya. Differential rotation of these stars was measured by Doppler imaging. High-precision photometry of the MOST mission was used for older stars ε Eri (Croll et al., 2006) and κ^1 Ceti (Walker et al., 2007).

show much shorter rotation periods. The increase of differential rotation with rotation rate is found because rotation of cooler stars is more uniform (Figure 1.10).

The primary chromospherically active stars of the close RS CVn binary systems seem to show a more complicated behavior. Kővári et al. (2012) report for ζ And the characteristic value $\delta\Omega = 0.02$ rad/day. Another example is the highly active K2 giant II Peg with its mass of 0.8 $M_\odot$ and the rotation period of 6.72 days. Henry et al. (1995) report a very weak solar-type rotation law with $\delta\Omega \simeq 0.005$ rad/day while Siwak et al. (2010) derive the much higher value $\delta\Omega \simeq 0.023$ rad/day from MOST data. Roettenbacher et al. (2011) basically confirm the small value of Henry et al. (1995). Weber and Strassmeier (2005) yield with $\delta\Omega \simeq 0.04$ rad/day an even higher value. They also showed that in their sample single stars exhibit significantly higher values of $\delta\Omega$ than members of binaries (Figure 9 in Weber and Strassmeier (2005)). It is insofar interesting that Oláh et al. (2013) find the small but similar values $\delta\Omega \simeq 0.006$ rad/day for V2253 ($P_{\rm rot} = 21.55$ days) and $\delta\Omega \simeq 0.007$ rad/day for IT Com ($P_{\rm rot} = 65.1$ days).

For the very young T Tau stars the early investigations led to almost solid-body rotation. Rice and Strassmeier (1996) found only a ratio $\delta\Omega/\Omega \simeq 0.001$ for V410 Tau with its rotation period of 1.87 days. The equator–pole difference results in the positive but small value $\delta\Omega \simeq 0.0035$ rad/day. In a recent analysis of MOST data Siwak et al. (2011) confirmed the smallness of this value ($\delta\Omega \simeq 0.002$ rad/day) but for two other weak-line TTS the values $\delta\Omega \simeq 0.026$ rad/day and $\delta\Omega \simeq 0.045$ rad/day have been found which do not confirm the solid-body hypothesis.

1.3
The Reynolds Stress

The theory of differential rotation is mainly the theory of angular momentum transport. The angular momentum equation for a turbulent rotating fluid reads as

$$\frac{\partial}{\partial t}(\rho R^2 \Omega) = -\nabla \cdot \left\{ \rho R \langle u_\phi \boldsymbol{u} \rangle + \rho R^2 \Omega \, \boldsymbol{U}^{\rm m} - \frac{R}{\mu_0} \left[\bar{B}_\phi \bar{\boldsymbol{B}} + \langle b_\phi \boldsymbol{b} \rangle \right] \right\}, \quad (1.6)$$

where $R = r \sin\theta$ is the distance to the rotation axis, $\bar{\boldsymbol{U}}$ and $\bar{\boldsymbol{B}}$ are the mean flow and the mean magnetic field, $\boldsymbol{u}$ and $\boldsymbol{b}$ are their fluctuating parts, $\boldsymbol{U}^{\rm m}$ is the meridional flow, and angular brackets signify the averaging over an ensemble of realizations of turbulence. The vector in curly brackets of (1.6) is the angular momentum flux. The angular momentum can be transported by turbulence, global meridional flow and by magnetic stress.

Turbulence is well known to be capable of transporting momentum by the effect of turbulent viscosity. Turbulent mixing smooths out the mean velocity shear. This turbulent viscosity effect can only bring a star to the state of uniform rotation. It has been found, however, that rotating turbulence can transport angular momentum even in a state of rigid rotation. This nondiffusive transport, named the Λ effect (Rüdiger, 1989), is of key importance for understanding differential rotation of convective stars.

1.3.1
The Λ Effect

The pseudovector of angular velocity alone does not suffice to construct a polar vector of angular momentum flux. The turbulent fluid, therefore, has to possess a preferred direction for the Λ effect to emerge (Lebedinskii, 1941; Biermann, 1951). The preferred direction in stellar convection zones is provided by gravity.

The physical origin of the Λ effect is illustrated by Figure 1.12. The dashed arrows show the original motions and the solid arrows show the motions perturbed by the Coriolis force. A fluid particle, which moves originally in radius, attains azimuthal velocity, which can be estimated as $u_\phi \simeq -2\tau_{\text{corr}} \Omega\, u_r \sin\theta$, where τ_{corr} is the characteristic time of turbulent mixing. The product $u_r u_\phi \simeq -2\tau_{\text{corr}} \Omega\, u_r^2 \sin\theta$ is negative independently of whether the original radial motion is upward or downward. For an original azimuthal motion, radial velocity $u_r \simeq 2\tau_{\text{corr}} \Omega\, u_\phi \sin\theta$ is produced by the Coriolis force, and the product $u_r u_\phi \simeq 2\tau_{\text{corr}} \Omega\, u_\phi^2 \sin\theta$ is positive. On average, we have

$$\langle u_r u_\phi \rangle \simeq 2\tau_{\text{corr}} \Omega \left(\langle u_\phi^2 \rangle - \langle u_r^2 \rangle \right) \sin\theta ,$$

$$\langle u_\theta u_\phi \rangle \simeq 2\tau_{\text{corr}} \Omega \left(\langle u_\phi^2 \rangle - \langle u_\theta^2 \rangle \right) \cos\theta .$$

Anisotropy of turbulence with different intensity of radial and horizontal mixing is required for the net radial flux of angular momentum to arise. A latitudinal flux of angular momentum results if the two components of the horizontal turbulence intensities differ. It thus makes sense to introduce the anisotropy parameters $A_V = \langle u_\phi^2 - u_r^2 \rangle / u_{\text{rms}}^2$ and $A_H = \langle u_\phi^2 - u_\theta^2 \rangle / u_{\text{rms}}^2$ so that $Q_{r\phi} \propto A_V$ and $Q_{\theta\phi} \propto A_H$ results.

The above expressions for the cross-correlation involve the Coriolis number

$$\Omega^* = 2\tau_{\text{corr}} \Omega \tag{1.7}$$

as a key parameter of the differential rotation theory. This parameter measures the intensity of interaction between convection and rotation. Its value defines whether turbulent eddies are long-lived enough for rotation to significantly influence them. The Coriolis number is reciprocal to another commonly used Rossby number Ro =

Figure 1.12 Illustration of angular momentum transport by rotating turbulence. The direction of rotation is shown at the top. See text.

P_{rot}/τ_{corr}, that is, $\Omega^* = 4\pi \mathrm{Ro}^{-1}$. The Coriolis number depends on depth in a stellar convection zone. Figure 1.14a shows its depth profile for the Sun. Ω^* exceeds unity in the major part of the convection zone. This condition of strong interaction between convection and rotation, $\Omega^* > 1$, which is typical of solar-type stars, largely complicates the Λ effect theory. It means that the above linear estimates no longer apply, and a better theory should be fully nonlinear in rotation rate.

1.3.1.1 Numerical Simulations

Käpylä and Brandenburg (2008) simulated with the PENCIL CODE homogeneous but anisotropic turbulence by use of variable anisotropic forcing functions in boxes with 256^3 grid points. In all simulations the radial velocity fluctuations dominated the other components so that always $A_V < 0$. Without rotation both the horizontal velocity intensities are equal. The global rotation suppresses the vertical turbulence and increases the horizontal rms values. By this influence the (negative) A_V is reduced and a small but positive A_H results so that a negative radial angular momentum flux and a positive horizontal angular momentum flux can be expected. Because of the rotational isotropizing of the turbulence a sufficiently rapid rotation should remarkably quench the two components of the angular momentum flux (radial and latitudinal). Figure 1.13a shows the calculated off-diagonal elements of the Reynolds stress tensor. Indeed, the resulting $Q_{r\phi}$ is negative while the result-

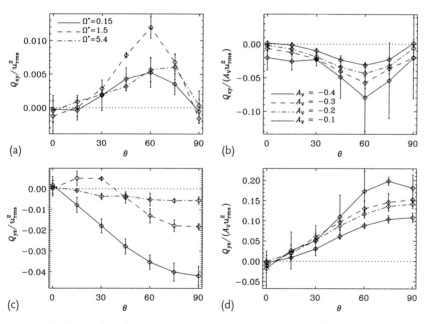

Figure 1.13 The Λ effect of turbulence due to anisotropic forcing. The numerical values are also the cross-correlation coefficients. (a,c) The off-diagonal elements of the normalized Reynolds tensor Q_{ij}/u_{rms}^2 vs. the colatitude. (a,b) $Q_{\theta\phi}$ and (c,d) $Q_{r\phi}$. (b,d) The same but for $Q_{\theta\phi}/(A_V u_{rms}^2)$ and $Q_{r\phi}/(A_V u_{rms}^2)$. The Coriolis number is fixed. From Käpylä and Brandenburg (2008). Reproduced with permission © ESO.

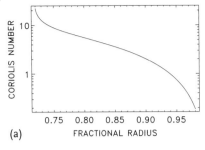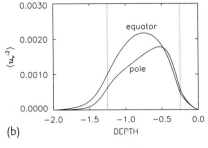

Figure 1.14 Depth profile of the Coriolis number (1.7) in the solar convection zone according to the solar structure model by Stix and Skaley (1990) (a). Shown also is the radial turbulence intensity for Ta $= 10^6$ (b). The flow amplitudes are normalized with the sound velocity at the surface of the unstable domain.

ing $Q_{\theta\phi}$ is positive (but small). That the radial transport of the angular momentum is indeed due to the radial anisotropy of the turbulence can be demonstrated with Figure 1.13b. There the ratio $Q_{r\phi}/A_V$ is shown as almost independent of the numerical value of the turbulence field considered.

The anisotropy in solar and stellar convection zones can be studied in detail with 3D simulations of thermal convection (Pulkkinen et al., 1993; Chan, 2001; Käpylä, Korpi, and Tuominen, 2004; Rüdiger, Egorov, and Ziegler, 2005). In the following representation of numerical simulations with the Nirvana Code the Λ effect is renormalized in accordance to

$$\Lambda_V^* = \frac{\Lambda_V \Omega}{c_{ac}^2}, \quad \Lambda_H^* = \frac{\Lambda_H \Omega}{c_{ac}^2} \tag{1.8}$$

so that

$$V = \Lambda_V^* \frac{c_{ac}^2}{\nu_T \Omega}, \quad H = \Lambda_H^* \frac{c_{ac}^2}{\nu_T \Omega}. \tag{1.9}$$

For the Sun the value of $c_{ac}^2/(\nu_T \Omega)$ is of order 10^4. In order to get V and/or H of order unity the simulations for Λ^* should yield rather small values. In the calculations Ra $= 3 \times 10^5$ and Pr $= 0.1$, while Ta varies from 10^4 to 10^6 which may represent the realization of slow rotation and fast rotation. The value Ta $= 10^6$ does not describe the real solar convection but it seems to be large enough to reveal the relation between anisotropy and angular momentum transport in rotating convection. All velocities are normalized with the speed of sound (c_{ac}) at the upper layer of the unstable domain. Its solar value is about 6.4 km/s. In Figure 1.14 the maximum velocity in the box is given with 0.044, which would mean about 300 m/s for the solar case. This value corresponds to the velocity amplitudes which are characteristic in the bulk of the solar convection zone. Figure 1.14b also reveals that for Ta $= 10^6$ the turbulence intensity $\langle u_r^2 \rangle$ grows from the pole to the equator. This unexpected result for rapid rotation is also present in the simulations of convection in boxes with weak density stratification (Chan, 2001; Käpylä and Brandenburg, 2008; Snellman et al., 2009).

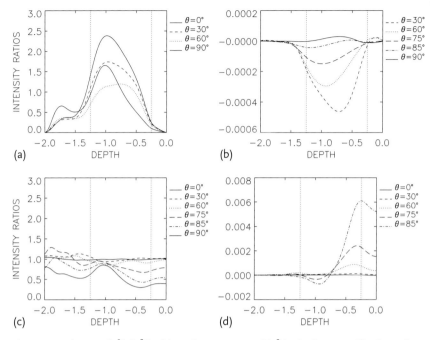

Figure 1.15 The ratio $\langle u_r^2\rangle/\langle u_\phi^2\rangle$ of the turbulence intensities for rotating turbulence fields (a), the function Λ_V^*. Ta = 10^6 (b), and the intensity ratio $\langle u_\theta^2\rangle/\langle u_\phi^2\rangle$ (c). Note the dominance of $\langle u_\phi^2\rangle$. The function Λ_H^* is basically positive (d). Ta = 10^6. The instability domain is located between the two vertical lines in each plot. From Rüdiger et al. (2005a).

The anisotropy between vertical and azimuthal turbulence intensities is shown in Figure 1.15a. Without rotation the turbulence is vertically dominated except in the top layer. This is also true for the lower overshoot zone between the unstable and the stable layer ('tachocline'). We therefore expect the occurrence of negative Λ_V in the bulk of the convection zone.

For Ta = 10^4 the resulting Λ_V is small. Already for Ta = 10^6, however, the results are very clear (Figure 1.15b). The function Λ_V^* is zero at the equator and is negative in both hemispheres (see Chan, 2001). The vanishing of Λ_V^* at the equator is not trivial and requires formulations such as $Q_{r\phi} \propto \cos^{2l}\theta \sin\theta$ with $l > 0$. Note also that the Λ_V^* vanishes in the upper overshoot layer while it remains basically negative in the tachocline layer.

The amplitude of Λ_V^* for Ta = 10^6 is 4×10^{-4} which leads to $V \simeq 10$. The behavior of the anisotropy parameter $\langle u_r^2\rangle/\langle u_\phi^2\rangle$ for rapid rotation is of particular interest. At the poles the influence of the rotation is rather small. It is much stronger in the equatorial region. There, we find a tendency of return-to-isotropy as a consequence of the Taylor–Proudman theorem. A possible vanishing of the radial angular momentum transport at the equator might be due to this phenomenon.

The horizontal angular momentum transport is based on the anisotropy in the turbulence field between both the horizontal components which only exists for rotating stars (Figure 1.15c). It is therefore not surprising that in contrast to the radial angular momentum transport, for rapid rotation (Ta = 10^6) only, a remarkable effect exists. The anisotropy in the turbulence field between $\langle u_\phi^2 \rangle$ and $\langle u_\theta^2 \rangle$ exists mainly in the equatorial region. There $\langle u_\phi^2 \rangle$ dominates, leading to positive cross-correlations close to the equator, that is $\Lambda_H^* > 0$ (Figure 1.15d).

The amplitude of $Q_{\theta\phi}$ is greater than the amplitude of $Q_{r\phi}$, and is positive in the upper half of the convection zone. The amplitude of H, therefore, exceeds the amplitude of V by a factor of about 10 – similar to Chan's results obtained with a completely different code. In the lower half of the convective domain it is much smaller and negative and also highly concentrated at the equator.

The situation in the lower overshoot region is also of interest. Note that the turbulence in both (rotating) overshoot regions is horizontal rather than vertical, that is $A_V > 0$. At the top of the convection box the situation is more complicated. For slow rotation (Ta = 10^5) there is no H but a negative V which only depends slightly on the latitude. The same is true for the negative slope of the outer solar rotation law in the supergranulation layer (see Figure 1.1). For faster rotation (Ta = 10^6) one finds the opposite. V goes to zero and H is a positive and large number (in the equatorial region). This is a basic problem existing in the entire box.

Motivated by this problem, and in order to avoid possible numerical artifacts of box simulations, Käpylä et al. (2011) designed global simulations in a 'wedge' geometry defined by $0.65 \leq x \leq 1$, $15° \leq \theta \leq 165°$ and $0 \leq \phi \leq 90°$. For the thermal stratification a piecewise polytropic setup is used with the logarithmic temperature gradient $\nabla = \partial \log T / \partial \log P = 1/(n+1)$ which describes an unstable domain for $n < 1.5$. This convection zone has been sandwiched by two stable overshoot regions. The radial and latitudinal boundaries are taken to be impenetrable and stress-free, and the heat-fluxes are suppressed through the latitudinal boundaries. The simulations were performed with the PENCIL CODE code in spherical coordinates (for details see Mitra et al., 2009).

The results have been obtained with fixed low Mach number but with a free Coriolis number of order unity. The latter is insofar important as the wanted off-diagonal elements of the Reynolds stress tensor only exist for sufficiently rapid rotation. Figure 1.16 presents the resulting radial flux of angular momentum $Q_{r\phi}$ normalized with the turbulence intensity $u_{\rm rms} = \sqrt{\langle u_r^2 + u_\theta^2 \rangle}$. Indeed, there is almost no signal for the two lowest Ω^* (Figure 1.16a,b). For the intermediate rotation rates symmetric profiles with respect to the equator appear with predominantly negative signs confirming the basic result of the above presented box simulations. For the fastest rotation with $\Omega^* \simeq 6$, $Q_{r\phi}$ becomes smaller and even positive.

As it should, the signals for the latitudinal flux of angular momentum are antisymmetric with respect to the equator with positive values at the northern hemisphere so that the angular momentum is transported from the poles to the equator (Figure 1.17). For the fastest rotation with $\Omega^* \simeq 6$ the cross-correlation values at the top and bottom of the convection zone are rotationally quenched. The striking maxima very close to the equator which are characteristic for the box simulations (see also Hupfer, Käpylä, and Stix, 2005, 2006) no longer appear in the global simulations.

1.3 The Reynolds Stress | 19

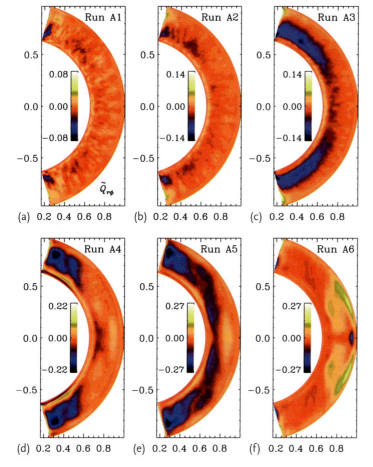

Figure 1.16 The radial Reynolds stress $Q_{r\phi}/u_{\mathrm{rms}}^2$ for $\Omega^* = 0.13$ (a), 0.25 (b), 0.50 (c), 0.94 (d), 2.56 (e), 6.09 (f). For intermediate rotation rates the correlation coefficient grows with the Coriolis number. Note the clear equatorial symmetry of the signals. From Käpylä et al. (2011). Reproduced with permission © ESO.

1.3.1.2 Quasi-linear Theory of the Λ Effect

An analytical theory of the Λ effect for density-stratified convection zones was performed by Kitchatinov and Rüdiger (1993, 2005). The derivations show that the nondiffusive part, Q_{ij}^{Λ}, of the velocity correlation tensor $Q_{ij} \equiv \langle u_i(\boldsymbol{x}, t) u_j(\boldsymbol{x}, t) \rangle$ has the structure

$$Q_{ij}^{\Lambda} = \nu_{\mathrm{T}} \left(\frac{\ell_{\mathrm{corr}}}{H_\rho} \right)^2 \Omega_k g_l \left[V(\Omega^*) \left(g_i \epsilon_{jkl} + g_j \epsilon_{ikl} \right) \right. $$
$$\left. - H(\Omega^*) \frac{\boldsymbol{g} \cdot \boldsymbol{\Omega}}{\Omega^2} \left(\Omega_i \epsilon_{jkl} + \Omega_j \epsilon_{ikl} \right) \right], \quad (1.10)$$

where repeated subscripts signify summation, ν_{T} is the eddy viscosity, ℓ_{corr} is the correlation length, $H_\rho = -\mathrm{d}r/\mathrm{d}\log\rho$ is the density scale height, V and H are di-

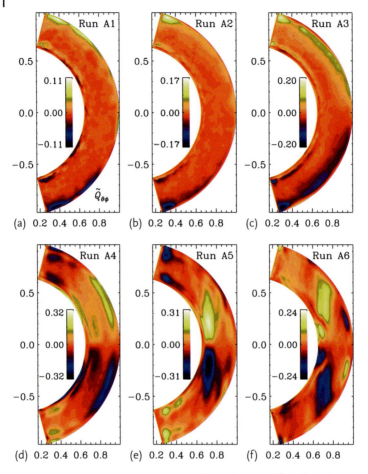

Figure 1.17 The same as in Figure 1.16 but for the horizontal flux of angular momentum $Q_{\theta\phi}$ which for not too slow rotation is directed to the equator (a–f). Reproduced with permission © ESO.

mensionless functions of the Coriolis number. The functions V and H consist of two parts including the effects of stratification and turbulence anisotropy. The radial unit vector is $\mathbf{g}$. As they must, the expressions (1.10) are odd in the basic rotation vector $\mathbf{\Omega}$ so that they are antisymmetric under the transformation $\Omega \to -\Omega$. This is exactly required for the azimuthal cross-correlation $Q_{r\phi}$ and $Q_{\theta\phi}$ which are responsible for the turbulent angular momentum transport.

The stratification in convection zones is known to be close to adiabatic. Anisotropy of stellar convection is less certain though. In the case of rapid rotation ($\Omega^* \gg 1$), the contribution of stratification dominates and there is no arbitrariness in specifying the Λ effect. Figure 1.14, however, shows that near-surface layers are in a state of slow rotation with $\Omega^* < 1$. The resulting uncertainty in the Λ effect can be excluded by using the anisotropy resulting in 3D numerical sim-

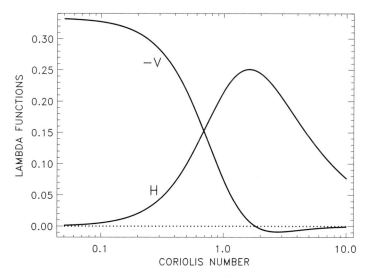

Figure 1.18 The functions $V(\Omega^*)$ and $H(\Omega^*)$ of the Λ effect of (1.10) and (1.11) after Kitchatinov and Rüdiger (2005).

ulations. After this is done, the Λ functions V and H assume the form shown in Figure 1.18.

The cross-correlations

$$Q^\Lambda_{r\phi} = \nu_T \left(\frac{\ell_{\text{corr}}}{H_\rho}\right)^2 \Omega \sin\theta \left[V(\Omega^*) - H(\Omega^*)\cos^2\theta\right],$$

$$Q^\Lambda_{\theta\phi} = \nu_T \left(\frac{\ell_{\text{corr}}}{H_\rho}\right)^2 \Omega \sin^2\theta \cos\theta\, H(\Omega^*) \quad (1.11)$$

of the fluctuating velocities in spherical coordinates (r, θ, ϕ) are proportional to the angular momentum fluxes in radius $(Q_{r\phi})$ and co-latitude $(Q_{\theta\phi})$. The $V(\Omega^*)$ function contributes to the radial flux of (1.11) only. Therefore, the part of (1.10), which includes V, represents radial flux of angular momentum. It can be shown that the angular momentum fluxes of (1.10) and (1.11) are the superposition of two nonorthogonal fluxes: $V(\Omega^*)$ is the normalized flux in radius, and $-H(\Omega^*)\cos\theta$ is normalized angular momentum flux along the rotation axis.

In the case of small Ω^*, the (negative) vertical flux dominates, $|V| \gg H$, and the Λ effect transports angular momentum downward. This is the reason for the existence of the radial near-surface shear seen in helioseismological inversions of Figure 1.1. In the opposite case of rapid rotation, $\Omega^* \gg 1$, the vertical flux is small, $H \gg |V|$, and the angular momentum is transported parallel to the rotation axis towards the equatorial plane. This picture basically agrees with the 3D numerical simulations presented above.

1.3.2
Eddy Viscosities

The correlation tensor of fluctuating velocities also includes the viscous part, Q_{ij}^ν, in line with the Λ effect[2]

$$Q_{ij} = Q_{ij}^\Lambda + Q_{ij}^\nu, \quad \text{with} \quad Q_{ij}^\nu = -\mathcal{N}_{ijkl}\frac{\partial \bar{U}_k}{\partial x_l}, \qquad (1.12)$$

where $\mathcal{N}_{ijkl}$ is the eddy viscosity tensor. The reason to write the viscosity as a tensor is the effect of rotation. Turbulent mixing becomes anisotropic under influence of rotation so that the relevant eddy viscosity depends on the orientation of the mean velocity and of the direction in which the velocity varies, relative to the rotation axis. Quasi-linear theory of turbulent transport provides the following expression for the viscosity tensor for rotating fluids

$$\begin{aligned}\mathcal{N}_{ijkl} = {} & \nu_1(\delta_{ik}\delta_{jl} + \delta_{jk}\delta_{il}) \\ & + \nu_2\left(\delta_{il}\frac{\Omega_j\Omega_k}{\Omega^2} + \delta_{jl}\frac{\Omega_i\Omega_k}{\Omega^2} + \delta_{ik}\frac{\Omega_j\Omega_l}{\Omega^2} + \delta_{jk}\frac{\Omega_i\Omega_l}{\Omega^2} + \delta_{kl}\frac{\Omega_i\Omega_j}{\Omega_2}\right) \\ & + \nu_3\delta_{ij}\delta_{kl} - \nu_4\delta_{ij}\frac{\Omega_k\Omega_l}{\Omega^2} + \nu_5\frac{\Omega_i\Omega_j\Omega_k\Omega_l}{\Omega^4}.\end{aligned} \qquad (1.13)$$

Five coefficients of the viscosity tensor (1.13) depend on the rotation rate,

$$\nu_n = \nu_T \phi_n(\Omega^*) \qquad (1.14)$$

with $n = 1, 2, \ldots, 5$. Figure 1.19 shows the viscosity quenching functions. In the case of slow rotation, $\Omega^* \ll 1$, only the coefficients ν_1 and ν_3 remain finite, while all other coefficients vanish and the viscosity becomes isotropic. In the opposite limit of rapid rotation, $\Omega^* \gg 1$, all the coefficients decrease in inverse proportion to Ω^*. The decrease is caused by rotational suppression of turbulence.

The coefficient ν_T of (1.14) (same as in (1.10) and (1.11) for the Λ effect) is the isotropic turbulent viscosity for a nonrotating fluid, that is, this is the viscosity which would take place under actual sources of turbulence, but if there were no rotation. This coefficient, therefore, is not measurable. The mixing-length expression for nonrotating fluids,

$$\nu_T = -\frac{\tau_{\text{corr}}\ell_{\text{corr}}^2 g}{15 C_p}\frac{\partial S}{\partial r}, \qquad (1.15)$$

can, however, be used to express the viscosity in terms of the entropy gradient; g is gravity, C_p is the specific heat at constant pressure, and $\ell_{\text{corr}} = \alpha_{\text{MLT}} H_p$ is the mixing length.

---

[2] We omit the turbulent pressure, which is not significant for the problem of differential rotation.

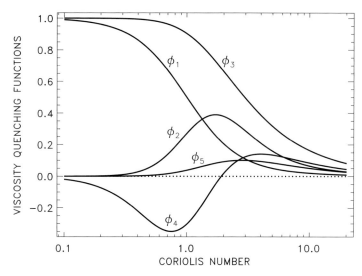

Figure 1.19 Quenching functions of the eddy viscosities of (1.14) after quasi-linear theory of turbulent transport in rotating fluids (Kitchatinov, Pipin, and Rüdiger, 1994).

Equations (1.12) and (1.13) lead to the following expressions for the viscous fluxes of angular momentum

$$Q^v_{r\phi} = -\nu_1 r \sin\theta \frac{\partial\Omega}{\partial r} - \nu_2 \sin\theta \cos\theta \left(r\cos\theta \frac{\partial\Omega}{\partial r} - \sin\theta \frac{\partial\Omega}{\partial\theta} \right),$$

$$Q^v_{\theta\phi} = -\nu_1 \sin\theta \frac{\partial\Omega}{\partial\theta} - \nu_2 \sin^2\theta \left(\sin\theta \frac{\partial\Omega}{\partial\theta} - r\cos\theta \frac{\partial\Omega}{\partial r} \right). \quad (1.16)$$

The first terms in the RHS of these equations are the same as in the case of isotropic viscosity. The expressions in brackets are proportional to the angular velocity gradient along the rotation axis. Therefore, ν_2 is the viscosity excess for the direction along the rotation axis compared to the viscosity ν_1, which applies to the direction normal to this axis.

The viscous fluxes of (1.16) are proportional to spatial derivatives of angular velocity. These fluxes increase with inhomogeneity of rotation. The Λ effect of (1.11), which produces this inhomogeneity, depends on rotation rate but not on the rotational shear. A steady state of differential rotation can to some extent be understood as a balance between the Λ effect and eddy viscosities. This picture is, however, very approximate and not complete because it does not allow for the angular momentum transport by the meridional flow.

1.4
The Meridional Flow

To understand the differential rotation induced by the meridional flow alone, we consider a simplified problem where the meridional flow is supposed as given, the eddy viscosity ν_T is isotropic, and the Λ effect in the turbulence-induced Reynolds stress is ignored (see Balbus, Latter, and Weiss, 2012). In this case, the stationary equation for the angular momentum (1.6) reduces to

$$\nabla \cdot \left(\rho R^2 \Omega\, U^m - \rho R^2 \nu_T \nabla \Omega\right) = 0 . \tag{1.17}$$

As a further simplification, we assume a one-cell meridional flow as sketched in Figure 1.20.

This simplified problem has one controlling parameter of the Reynolds number $\mathrm{Re} = U_0 R_*/\nu_T$, where U_0 is the amplitude of the meridional flow and R_* the star's radius.

It might be expected that conservation of angular momentum will result in an 'antisolar' rotation with angular velocity increasing with latitude. This is the case, however, only for large Reynolds number $\mathrm{Re} \gg 1$. According to Kippenhahn (1963), Steenbeck and Krause (1965) and Köhler (1969) a slow circulation with equatorward flow at the top of the convection zone can produce equatorial acceleration. Hence, any theory of the solar rotation law which only works with meridional flow as the nondiffusive angular momentum transporter leads to a circulation with *downflows* in the equatorial region. If large-scale upflows are observed there the equatorial acceleration must have a different origin. Note, however, that antisolar rotation can also be produced by use of a magnetic setup (see Kitchatinov and Rüdiger, 2004, and references therein).

This result for small Reynolds numbers can be understood as follows. In a steady state the material flux through the conical surface $(r_i - r_e)$ of constant θ in Fig-

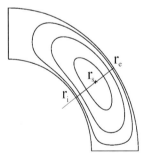

Figure 1.20 Streamlines of meridional flow in a spherical layer with inner radius r_{in} and outer radius r_e. The term r_s is the radius of the stagnation point where the meridional velocity changes its sign. For low Reynolds number the direction of angular momentum transport by the meridional flow coincides with the direction of the surface flow (see text).

ure 1.20 is zero, that is

$$\int_{r_i}^{r_e} \rho U_\theta^m r \, dr = 0 \,. \tag{1.18}$$

The angular momentum flux, t_θ, produced by the meridional flow, through the same surface reads as

$$t_\theta = 2\pi \sin^3 \theta \int_{r_i}^{r_e} \rho U_\theta^m \Omega r^3 \, dr \,. \tag{1.19}$$

Differential rotation in the case of small Reynolds number is also small. Therefore, we can put Ω in (1.19) outside of the integral sign. On doing so and using (1.18), the angular momentum flux (1.19) can be written as

$$t_\theta = 2\pi \sin^3 \theta \, \Omega \int_{r_i}^{r_e} \rho U_\theta^m \left(r^2 - r_s^2 \right) r \, dr \,, \tag{1.20}$$

where r_s is the radius of the stagnation point where $U_\theta^m = 0$ (Figure 1.20); note that U_r^m is not necessarily zero at this point. The expression under the integral in the right-hand side of (1.20) is sign-definite. It has the same sign as U_θ^m on the top. Therefore, the directions of meridional flux of angular momentum and surface meridional flow coincide. In the case of small Reynolds number, the surface meridional flow points in the direction of increasing angular velocity. This is the rule only for the case where the time of viscous smoothing of the angular velocity distribution is short compared to the meridional flow circulation time.

The situation is different for large Reynolds numbers, $\mathrm{Re} \gg 1$. Then the second term in the RHS of (1.17) can be neglected to write this equation as

$$\frac{1}{\sin^2 \theta} \frac{\partial \psi}{\partial r} \frac{\partial (\sin^2 \theta \, \Omega)}{\partial \theta} - \frac{1}{r^2} \frac{\partial \psi}{\partial \theta} \frac{\partial (r^2 \Omega)}{\partial r} = 0 \,, \tag{1.21}$$

where ψ is the stream function of the meridional flow

$$U^m = \frac{1}{\rho} \mathrm{curl} \left(\frac{\psi \, e_\phi}{r \sin \theta} \right) \,, \tag{1.22}$$

where e_ϕ is the azimuthal unit vector. The solution of (1.21), $\Omega(r, \theta) = F(\psi)/r^2 \sin^2 \theta$, can be written in terms of an arbitrary function $F(\psi)$. The solution describes antisolar rotation with specific angular momentum constant along the streamlines of meridional flow. The solution does not apply to close vicinities of the rotation axis and the equatorial plane, where the circulation time cannot be small compared to the viscous timescale so that the viscosity cannot be neglected.

The change of character of the differential rotation with Reynolds number is shown in Figure 1.21. The latitudinal differential rotation depends on the sense of

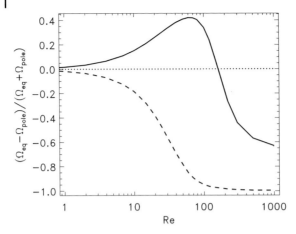

Figure 1.21 Latitudinal differential rotation produced by the meridional flow as a function of the Reynolds number. Clockwise flow (solid), counterclockwise flow (dashed). Without Λ effect positive k are only possible for clockwise flow.

meridional flow for Re $\lesssim$ 1, while a fast flow always produces antisolar rotation for Re $\gg$ 1 independent of its orientation. The transition from slow to fast circulation regimes occurs at Reynolds numbers of about 100. The Sun with meridional flow of order 10 m/s and eddy viscosity $\nu_T \simeq 5 \times 10^{12}$ cm^2/s belongs to the case of moderate Reynolds numbers, Re $\sim$ 10.

If the helioseismology finds for the solar meridional circulation a one-cell counterclockwise pattern which rises (sinks) in the equatorial (polar) region then the observed equatorial acceleration *must* be a result of the angular momentum transport by Reynolds stress (in form of the Λ effect) – and cannot be a result of the angular momentum transport of the (say) baroclinic flow. With the empirical findings of Section 1.1.3 concerning the meridional circulation geometry it is not possible to explain the solar rotation law as due to the action of baroclinic flows.

1.4.1
Origin of the Meridional Flow

Centrifugal and buoyancy forces are the most essential drivers of meridional flow in stellar convection zones. The buoyancy force results from a slight dependence of mean temperature on latitude. Both forces are strong: each of them alone would drive a meridional flow of several hundreds of meters per second on the Sun (Durney, 1996). The forces, however, almost balance each other and the meridional flow results from the slight imbalance between them.

The meridional flow equation can be obtained as azimuthal component of the 'curled' (1.6). The resulting steady equation reads as

$$\mathcal{D}(U^m) = \sin\theta \, r \frac{\partial \Omega^2}{\partial z} - \frac{g}{C_p r} \frac{\partial S}{\partial \theta}, \qquad (1.23)$$

where $\partial/\partial z = \cos\theta\,\partial/\partial r - r^{-1}\sin\theta\,\partial/\partial\theta$ is the spatial derivative along the rotation axis. The most essential terms only are kept in (1.23). The left side of this equation accounts for the meridional flow braking by the eddy viscosity,

$$\mathcal{D}(\boldsymbol{U}^{\mathrm{m}}) \equiv -\epsilon_{\phi ij}\frac{\partial}{\partial x_i}\left[\frac{1}{\rho}\frac{\partial}{\partial x_k}\left(\rho\mathcal{N}_{jkln}\frac{\partial U_l^{\mathrm{m}}}{\partial x_n}\right)\right], \qquad (1.24)$$

where the first subscript on the right-hand side indicates the azimuthal component in spherical coordinates.

If (1.23) is normalized by dividing it by v_{T}^2/R_*^4, its left side scales as the Reynolds number. This number is of order 10 for the Sun. Each term on the right-hand side of (1.23) is much larger. The first term scales as the Taylor number Ta $= 4\Omega^2 R_*^4/v_{\mathrm{T}}^2$. We have Ta $\sim 10^7$ for the Sun. The second term on the right-hand side of (1.23) is estimated by the Grashof number

$$\mathrm{Gr} = \frac{gR_*^3}{v_{\mathrm{T}}^2}\frac{\delta T}{T}, \qquad (1.25)$$

where $\delta T = T_{\mathrm{pole}} - T_{\mathrm{eq}}$ is the 'differential temperature.' The ratio $\delta T/T$ varies moderately with depth in stellar convection zones. We can take this ratio for the surface of the Sun to find that the Grashof number is about 10^7 for each kelvin of the surface differential temperature, Gr $\sim 10^7 \delta T_{\mathrm{surf}}$.

Such a large difference in characteristic values of the left and right-hand sides of (1.23) shows that this equation might be satisfied by the two terms of the RHS balancing each other. This state is known as the 'thermal wind balance.' Meridional flow arises from slight deviations from the balance. It is remarkable that the flow in turn maintains the balance. Any considerable deviation from the balance would produce a large meridional flow which reacts back on the distributions of angular velocity and entropy to re-establish the balance. A successful theory must, therefore, keep the meridional flow in both the angular momentum and entropy equations. A 'self-influence' of the meridional flow is, however, of secondary importance and the term nonlinear in U^{m} is dropped in (1.23). It is small compared to the term nonlinear in the angular velocity.

If angular velocity decreases with distance from the equatorial plane ($\mathrm{d}\Omega/\mathrm{d}z \lesssim 0$), the (nonconservative) centrifugal force is largest near the equator. The resulting torque drives meridional circulation with poleward flow at the top of the convection zone. From the early model of Köhler (1969, 1970) to the formulations of Glatzmaier (1985), Gilman and Miller (1986), Brandenburg et al. (1990), Miesch et al. (2000) and Brun and Toomre (2002), the inclusion of a meridional flow always led to the 'Taylor-number puzzle': cylindrical isorotation contours are found, independently of the Λ effect applied. The isolines of Ω in the solar convection zone, however, are *not* cylindrical (Figure 1.1). The above estimations show that positive differential temperature with the poles warmer than the equator by about 1 K suffices to produce a considerable deviation from the cylindrical rotation law.

The impact on the maintenance of differential rotation can be seen from the equation for the meridional flow. For fast rotation (large Taylor number) the

Reynolds stress and nonlinear terms can be neglected and (1.23) is reduced to

$$2r\sin\theta\, \Omega \frac{\partial \Omega}{\partial z} \simeq \frac{g}{rT}\frac{\partial \delta T}{\partial \theta} . \tag{1.26}$$

It follows that a gradient of the angular velocity along the axis of rotation is needed to balance a horizontal temperature gradient. A disk-shaped rotation pattern, therefore, results in the polar area (only) due to the action of the baroclinic flow. For warmer poles (δT sinks equatorwards) the z-gradient of Ω must be negative as observed. The meridional flow without the baroclinic component only yields $d\Omega/dz \simeq 0$ (see Figure 1.32). Hence, the empirical finding of negative $d\Omega/dz$ at the northern polar axis proves the existence of baroclinic flows in the solar convection zone.

As we shall demonstrate by means of Figure 1.30, the superrotation beneath the solar equator is a direct indication for the action of the Λ effect in the convection zone. All models without Λ effect but with a clockwise (equatorward at the surface) circulation lead to $d\Omega/dr \lesssim 0$ in the bulk of the convection zone beneath close to the equator. Such clockwise flows are able to accelerate the equator relative to the poles but they cannot produce the superrotation beneath the equator.

1.4.2
The Differential Temperature

Heat transport in a stellar convection zone is governed by the mean entropy equation,

$$\rho T \frac{\partial S}{\partial t} + \rho T \boldsymbol{u} \cdot \nabla S = -\nabla \cdot \left(\boldsymbol{F}^{\mathrm{conv}} + \boldsymbol{F}^{\mathrm{rad}} \right) + \varepsilon , \tag{1.27}$$

where ε is the source function, $\boldsymbol{F}^{\mathrm{rad}}$ and $\boldsymbol{F}^{\mathrm{conv}}$ are the radiative and convective heat-fluxes[3].

The differential temperature is currently understood as an outcome of rotational influence on convective heat transport. The tensor χ_{ij} of the eddy thermal diffusion that controls the convective heat-flux,

$$F_i^{\mathrm{conv}} = -\rho T \chi_{ij} \frac{\partial S}{\partial r_j} , \tag{1.28}$$

includes the rotationally induced anisotropy and quenching

$$\chi_{ij} = \chi_T \left[\phi(\Omega^*)\delta_{ij} + \phi_\parallel(\Omega^*) \frac{\Omega_i \Omega_j}{\Omega^2} \right] . \tag{1.29}$$

The diffusivity quenching functions are shown in Figure 1.22, and the background diffusivity χ_T can be expressed in terms of the entropy gradient,

$$\chi_T = -\frac{\tau_{\mathrm{corr}} \ell_{\mathrm{corr}}^2 g}{12 C_p} \frac{\partial S}{\partial r} , \tag{1.30}$$

similar to the (1.15) for the eddy viscosity.

3) For details see Rüdiger and Hollerbach (2004).

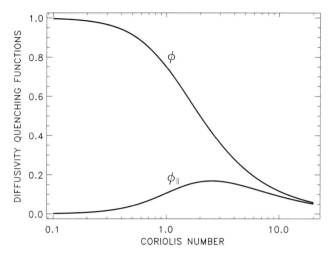

Figure 1.22 Quenching functions of the eddy thermal diffusivity of (1.29) after quasi-linear theory of turbulent transport in rotating fluids (Kitchatinov, Pipin, and Rüdiger, 1994).

Weiss (1965) suggested that the eddy thermal diffusivity in rotating fluids depends on latitude. Equation (1.30) allows for the dependence:

$$\chi_{rr} = \chi_T \left[\phi(\Omega^*) + \phi_\|(\Omega^*) \cos^2 \theta \right] . \qquad (1.31)$$

The diffusivity increases with latitude to imply that the poles are warmer than the equator. A differential temperature of this sense is required to explain the equatorial acceleration of Figure 1.1 (Gilman, 1986; Durney, 1987). However, the latitude-dependent heat transport does not help to resolve the Taylor-number puzzle. Much more significant is the anisotropy of the eddy heat transport. The diffusivity tensor (1.29) possesses the finite off-diagonal component

$$\chi_{\theta r} = -\chi_T \phi_\|(\Omega^*) \cos \theta \sin \theta . \qquad (1.32)$$

This means that even if the mean entropy varies only with radius, the convective heat-flux deviates from radial direction to poles, $F_\theta^{conv} \cos \theta < 0$. The differential temperature resulting from the convective heat-flux in latitude is sufficiently large (~ 1 K) to resolve the Taylor-number puzzle.

There were many attempts to observe the differential temperature on the Sun. Recent observations of Rast, Ortiz, and Meisner (2008) suggest that the differential temperature on the solar photosphere is indeed positive, $\delta T \simeq 2.5$ K.

We always find a higher temperature at high latitudes than at the equator. This is a consequence of the tilt of the convective heat transport vector towards the axis of rotation caused by the Coriolis force. In spherical polar coordinates the components of the heat-flux read as

$$F_r = -\rho T \chi_{rr} \frac{\partial S}{\partial r} - \rho T \chi_{r\theta} \frac{\partial S}{\partial \theta} , \qquad (1.33)$$

$$F_\theta = -\rho T \chi_{\theta r} \frac{\partial S}{\partial r} - \rho T \chi_{\theta\theta} \frac{\partial S}{\partial \theta} \,. \tag{1.34}$$

The first term in the horizontal component precludes a purely radial stratification. Any variation of the specific entropy with the radius will cause a horizontal heat-flux and thus build up a horizontal gradient. This case is profoundly different from that of a latitude-dependent but still purely radial heat-flux. The latter would have a much smaller impact and would result in a much weaker differential rotation.

Figure 1.23a shows the depth profile of the correlation $\langle u_r \vartheta \rangle$ with the temperature fluctuation $\vartheta = T - \bar{T}$ for various latitudes which is positive (negative) in the convection zone (overshoot layer). More important is that the values differ between poles and equator. The pole–equator difference in the radial heat-flux slightly depends on the radius. Except for the top layer, the eddy heat-flux at the equator exceeds the eddy heat-flux at the poles. In the top layer, however, where the turbulence is horizontally dominated, the polar heat-flux dominates the equatorial one. Rieutord et al. (1994), Käpylä, Korpi, and Tuominen (2004) and Hupfer, Käpylä, and Stix (2005) reported similar results. The crossover happens where the vertically dominated turbulence changes to a horizontally dominated turbulence. The radial heat-flux directly reflects the rotation-influenced turbulence intensity $\langle u_r^2 \rangle$. It is shown in Figure 1.14 that the $\langle u_r^2 \rangle$ at the equator exceeds the value at the poles. A similar crossover does not exist for the latitudinal eddy heat-flux $\langle u_\theta \vartheta \rangle$ plotted in Figure 1.23b at the poles and the equator. Due to the action of the Coriolis force the heat flows towards the pole in the convection zone and towards the equator in the lower overshoot region.

The key question regarding the importance of the anisotropic heat transport has been attacked by Rempel (2005, 2011). He showed that it is also possible to produce the typical large-scale flow pattern in the solar convection zone without anisotropic heat transport. In this approach the warm poles which are necessary to violate the Taylor–Proudman theorem are produced by the inclusion of a stably stratified sublayer ('tachocline') below the convection zone with low viscosity and a rigid-rotation lower boundary condition. The latter condition together with the differential rotation within the convection zone produced by the Λ effect leads to negative values of

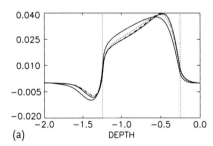
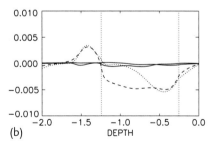

Figure 1.23 The correlations $\langle u \vartheta \rangle$ for different colatitudes in box simulations after horizontally and time-averaging vs. depth, Ta = 10^6. (a) $\langle u_r \vartheta \rangle$. (b) $\langle u_\theta \vartheta \rangle$. Pole (solid line); $\theta = 30°$ (dashed line); $\theta = 60°$ (dotted line); equator (triple-dot-dashed). The convective domain is between the two vertical lines. From Rüdiger et al. (2005a). Reproduced with permission © ESO.

dΩ^2/dz in the higher latitudes of the tachocline. With the low viscosity given there by (1.15) a negative latitudinal gradient of the entropy immediately results leading to warmer poles below the convection zone[4]. The screening by the convection zone of the entropy (temperature) differences produced in the tachocline layer, however, is not weak. Screening factors of order 5×10^{-2} are typical for convection zones with their high thermal conductivity (Spruit, 1977; Stix, 1981; Rüdiger, 1989). Figure 1.24b demonstrates the meaning of the screening problem for the presented 'tachocline models' of the solar differential rotation.

The main differences of these models and those with anisotropic turbulent heat transport are (i) that the Rempel model needs strong pole–equator differences of

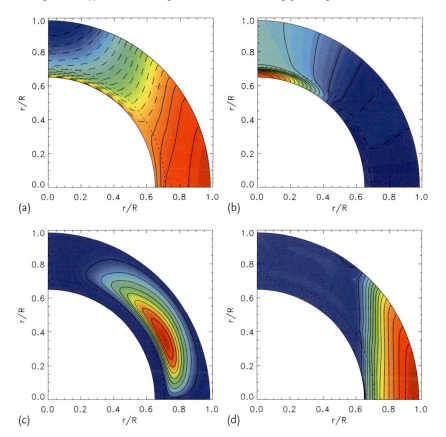

Figure 1.24 Contour plots of differential rotation (a,d), entropy profile (b) and streamlines of the meridional flow (c) in a model without anisotropic heat transport. Complete suppression of the latitudinal temperature differences leads to the Taylor–Proudman state for the rotation law (d). Note the strong screening of the entropy (temperature) differences of the tachocline by the convection zone. From Rempel (2011).

---

4) This can also be due to the gravity darkening of rapidly rotating stars, see Rüdiger and Küker (2002).

the temperature at the bottom of the convection zone and (ii) that stars without a tachocline (i.e., without internal fossil magnetic fields) must have rotation laws in their convection zone which strongly differ from the solar one (see Figure 1.24d).

1.4.3
Advection-Dominated Solar Dynamo

Traditionally, mean-field models of the solar dynamo explain the generation of the solar magnetic field with an interplay between the differential rotation, which winds up the poloidal field and thus generates the toroidal field, and the α effect, which generates a poloidal field from the toroidal field. The $\alpha\Omega$-dynamo, as this mechanism is called, explains the sunspot cycle and (more or less) the cycle time but fails to produce the observed butterfly diagram if the helioseismologically derived rotation law in the convection zone is applied ('dynamo dilemma' Parker, 1992).

This shortcoming of the $\alpha\Omega$-dynamo can be overcome through the inclusion of the meridional flow, U^m, if the eddy diffusivity η_T is so small that the magnetic Reynolds number $\text{Rm} = U^m R/\eta_T$ reaches values of the order 10^3. Depending on the location of the dynamo wave, one expects a strong modification of the dynamo. This possibility has been the subject of intense numerical investigation (Choudhuri, Schüssler, and Dikpati, 1995; Dikpati and Charbonneau, 1999; Dikpati and Gilman, 2001; Küker, Rüdiger, and Schultz, 2001; Bonanno et al., 2002) where it has been shown that solutions with large magnetic Reynolds numbers provide the correct cycle period and butterfly diagram for a positive (negative) α effect in the northern (southern) hemisphere. The new problem with the models was that it was no longer trivial to find the solutions with the correct dipolar parity as the fundamental mode which can only be overcome if the α effect is concentrated at the bottom of the convection zone (see Bonanno et al., 2002).

Slow meridional circulations will only modify the known $\alpha\Omega$-dynamo solutions (Roberts and Stix, 1972). This influence can be expected to be just a modification if its characteristic timescale exceeds the cycle time of about 11 yr. One finds $U^m \simeq 2\,\text{m/s}$ as a critical value for the flow velocity. If the flow is faster its influence might be much more than a modification and one can hope that the direction of the dynamo wave can indeed be reversed. There is indeed no continuous transition from the simple $\alpha\Omega$-dynamo to the advection-dominated dynamo solutions. Without circulation the butterfly diagram of course becomes 'wrong,' with a poleward migration of the toroidal magnetic belts. One can also understand that the counterclockwise flow (poleward at the surface) rapidly destroys the dynamo, in this case already for an amplitude of (say) 20 m/s. For such fast flows the dynamo action ceases. As a consequence of the small value of η_T, the cycle times are longer than the solar cycle time for the majority of the models.

The observed rotation pattern of the solar convection zone is characterized by a strong subrotation in the polar region (cf. Figure 1.1). For magnetic fields with dipolar geometry a strong toroidal field will always be induced at polar latitudes unless the eddy diffusivity is increased there. Strong polar activity is not observed, however, on the Sun. All recent dynamo models with the observed rotation law are faced with this problem, even in the case that the α effect has been strongly reduced in the polar region by the relation $\alpha \propto \sin^2 \theta \cos \theta$.

A characteristic model of an advection-dominated dynamo has recently been presented by Bonanno (2012). A realistic rotation law of the convection zone is used. The α effect is positive (negative) in the northern (southern) hemisphere, with two different latitudinal profiles where the second one artificially suppresses the α effect at the poles (see Figure 1.25). Additionally, the α effect is concentrated at the base of the convection zone. This 'overshoot dynamo' works with an eddy diffusivity of 4.66×10^{11} cm^2/s through the bulk of the convection zone. The magnetic Reynolds number of the (rather fast) flow is about 400. At the base of the convection zone the eddy diffusivity is reduced by two order of magnitudes, the radial component of the circulation vanishes there (so that the requirement of a small penetration depth is fulfilled) and the stagnation point of the circulation is deep inside the convection zone ($x = 0.8$).

Without circulation the toroidal field is generated in midlatitudes where the latitudinal shear $\partial \Omega / \partial \theta$ peaks. Then the toroidal belts mainly drift in the radial direction rather than in latitude (Köhler, 1973). Figure 1.25 shows the results. Toroidal fields only exist in midlatitudes even in the case that the α effect is concentrated towards the equator by putting its polar value to zero. A counterclockwise circulation, however, is able to generate the equatorward migration of the toroidal field which is observed in form of the solar butterfly diagram. Figure 1.26 displays the butterfly diagram of the model in the flux-dominated regime. It also shows the preference of the dipolar parity, that is the antisymmetry of the field components with respect to the equator. In the original paper by Bonanno (2012) it is shown that for circulation speeds exceeding 10 m/s indeed the dipolar parity dominates the quadrupolar parity of the dynamo-maintained magnetic fields. There is indeed no monotonic transition from the low to the high magnetic Reynolds number regimes, which are separated by a Reynolds number of 100 (see above).

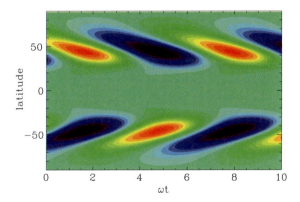

Figure 1.25 Butterfly diagram for an $\alpha \Omega$-dynamo model with positive α and without meridional circulation. $\alpha \propto \cos \theta$. Solid and dashed lines represent the radial surface field. The butterfly diagrams do not reproduce the observations. Courtesy of A. Bonanno.

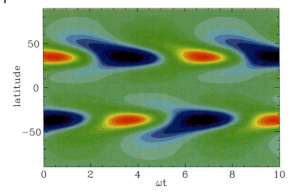

Figure 1.26 The same as in Figure 1.25 but with the induction by a counterclockwise meridional flow with Rm = 400. The ratio of the cycle time to the rotation period results as 300. Courtesy of A. Bonanno.

In the fast flow model the ratio of the rotation rate to the cycle frequency is about 300, that is 300 rotations happen within the cycle time of 22 years, which is very close to the observations.

There are many studies about the advection dominated solar dynamo which works with positive α effect mainly at the bottom of the convection zone (see Charbonneau, 2005, and many references therein). It needs a high magnetic Prandtl number there which is hard to explain. The alternative overshoot or (what is more or less the same) tachocline dynamo which works with negative α effect in the overshoot region has its basic problem with the minor thickness of this layer of only 7000 km (Basu, 1997) which unavoidably leads to a multistriped butterfly diagram (Rüdiger and Brandenburg, 1995).

A drastically different concept has been developed by Brandenburg (2005a) who considers the negative shear of the solar rotation law beneath the surface together with positive α effect as the main components of the solar dynamo. As demonstrated above this negative shear is the immediate result of the interaction of *slow* rotation and stratified turbulence. For the Sun it exists for $x > 0.95$, that is its thickness is 35 000 km. It should be broader for slow rotators and it should be even narrower for fast rotators like F stars (see Figure 1.47).

Also numerical models simulating rotating convection in spherical domains lead to oscillating dynamos. Käpylä, Mantere, and Brandenburg (2012) added the magnetic equations to their model of differential rotation (see Figure 1.16) and obtained cyclic magnetic activity with butterfly diagrams with (for the first time) equatorward migration of the toroidal large-scale fields[5].

---

5) The toroidal field strength does not strongly exceed the poloidal field strength and also a meridional flow does not exist.

1.5 The Sun

In the following we discuss mean-field models of differential rotation and confront them with observations and 3D simulations. The models are based on a uniform theoretical basis: everything the modeling needs, including the Λ effect, the eddy viscosities and thermal diffusivities, have been derived within the same approximations of turbulence theory. As a result, there is almost no freedom in formulating and tuning the mean-field models. The models reproduce the solar helioseismologically derived internal rotation and observations of many individual stars rather closely.

The models solve the three steady hydrodynamic mean-field equations for the angular velocity, meridional flow and the entropy in a stellar convection zone. The latest models are described by Küker, Rüdiger, and Kitchatinov (2011) and Kitchatinov and Olemskoy (2011). Figure 1.27 shows the computed internal rotation of the Sun in comparison to the observations. The theoretical figure includes the tachocline region and deeper radiation zone just for completeness of the picture. The tachocline was computed with a separate model discussed in Chapter 2. The tachocline model uses the results of the computation of the differential rotation in the convection zone as a boundary condition but does not influence that computation in any way. Figure 1.27 is also quite similar to the helioseismological rotation law of Figure 1.1.

Quasi-linear theory, on which the mean-field model is based, may not be very precise. This raises the question of how sensitive are the results to variations of the model parameters. With the exception of one parameter – the anisotropy of thermal diffusivity – the sensitivity is quite low; for example, varying the Λ effect changes the results only a little. Decreasing the Λ effect by 50% decreases the

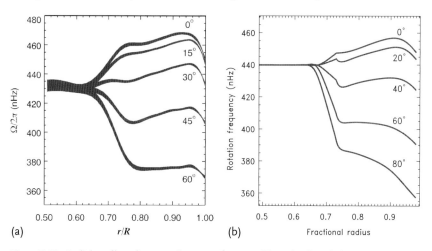

Figure 1.27 Radial profiles of rotation frequency for several latitudes from helioseismology (National Solar Observatory) (a). Theoretical depth profiles of the rotation rate for several latitudes (b). From Kitchatinov and Rüdiger (2005).

pole–equator difference $\delta\Omega$ by about 10% (from 30 to 27% in relative magnitude of $\delta\Omega$). It is remarkable that increasing Λ effect by 50% also *decreases* $\delta\Omega$ by the same 10%. This is the effect of the meridional flow: the flow keeps the thermal wind balance by restricting variations of $\delta\Omega$. The model is also moderately sensitive to an increase of the eddy viscosity or thermal diffusivity. A considerable reduction of the diffusivities is not possible because that would make the model unstable to thermal convection (Tuominen et al., 1994) and thus inconsistent.

The only parameter to which the model is quite sensitive is the anisotropy of thermal diffusion. The anisotropy can be varied by introducing the parameter C_χ into the anisotropic part of the diffusivity tensor (1.29):

$$\chi_{ij} = \chi_T \left[\phi(\Omega^*)\delta_{ij} + C_\chi \phi_\|(\Omega^*) \frac{\Omega_i \Omega_j}{\Omega^2} \right]. \tag{1.35}$$

The best agreement with the results of helioseismology is obtained with $C_\chi = 1.5$. Figure 1.27 and all the results of Section 1.6 were obtained with this value. C_χ is the only adjustable parameter of the model.

Figure 1.28 shows the meridional flow for the same model as Figure 1.27. The flow velocity decreases with depth near the surface (see Zhao and Kosovichev, 2004). The flow is relatively slow in the bulk of the convection zone but increases toward the boundaries. This structure of the meridional flow can be interpreted as follows. The thermal wind balance discussed in Section 1.4.1 is not compatible with the stress-free boundary conditions[6]. Thin boundary layers are formed where the balance is violated (Durney, 1989). Meridional flow results from the deviations from the thermal wind balance. The flow concentrates in the boundary layers where the deviations are most pronounced. This interpretation is supported by Figure 1.29 showing the thermal wind balance in the bulk of the convection

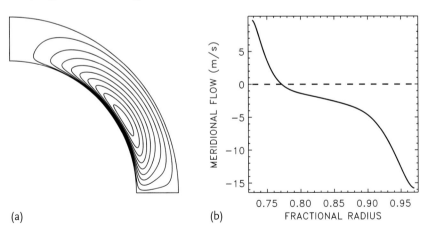

Figure 1.28 Streamlines of meridional flow (a) and radial profile of meridional velocity (b) for 45° latitude for the same model as Figure 1.27. Negative velocity means poleward flow.

6) The conditions of zero surface stress, $Q_{r\phi} = Q_{r\theta} = 0$, guarantee that the global circulation in the convection zone is not imposed by prescribed surface forces.

zone and its violation near the boundaries. The boundary layers were also found in 3D simulations of Brun and Toomre (2002) and Brown et al. (2008).

The boundary layers in the slowly rotating Sun are relatively thick. We shall see that the layers are much thinner in rapidly rotating stars. The meridional velocity at the bottom of the convection zone is not small compared to that of the surface (Figure 1.28). Note, however, that the flow decreases rapidly with depth beneath the convection zone (Gilman and Miesch, 2004, see Section 2.2.2).

Our model includes two effects that are capable of maintaining differential rotation, the Λ effect and the baroclinic flow as due to a horizontal temperature gradient. This can be seen when the baroclinic term in (1.23) is rewritten in terms of the temperature,

$$\frac{1}{\rho^2}(\nabla \rho \times \nabla P)_\phi \approx -\frac{g}{rT}\frac{\partial \delta T}{\partial \theta} . \tag{1.36}$$

The term can be a powerful driver of meridional flows, which in turn can drive differential rotation. In our theory the latitudinal temperature profile is due to the anisotropic heat-conductivity tensor in the presence of rotation. The poles are always slightly warmer than the equator but with an unobservable temperature excess. A positive pole–equator temperature difference will drive a *clockwise* flow at the northern hemisphere, that is from pole to equator. It transports angular momentum toward the equator and, therefore, leads to an accelerated rotation at low latitudes. A baroclinic flow by warmer poles can therefore maintain differential rotation with a solar-type surface rotation.

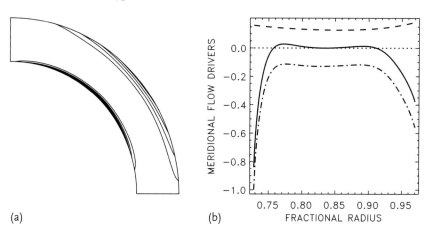

Figure 1.29 The thermal wind balance is fulfilled in the bulk of the convection zone but not close to the boundaries. Isolines of the LHS of (1.23) (a). The baroclinic (dashed), centrifugal (dash-dotted) terms of (1.23) and their sum (solid) for a latitude of 45° in the same model as Figures 1.27 and 1.28 (b). Normalized values are plotted. The dotted line shows the zero level.

1.5.1
Sun without Λ Effect

For an illustration we first repeat our computation for the Sun with the Λ-terms canceled within the Reynolds stress. The resulting rotation and flow patterns are shown in Figure 1.30 (and Figure 1.31). The rotation is indeed of the solar-type but with $\delta \Omega = 0.04$ rad/day weaker than observed, and the contours are distinctly disk-shaped at the poles and even in the midlatitudes. The equator region rotates almost rigidly, that is $\partial \Omega / \partial r \lesssim 0$. The typical superrotation of the deep convection zone beneath the equator does not occur. The reason is simple: if a fast meridional circulation transports the angular momentum with a pattern symmetric with

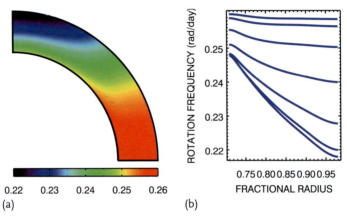

Figure 1.30 Solar rotation law computed without Λ effect. The contours of the angular velocity are nearly disk-shaped (a). Note the steep negative gradient of the angular velocity at the poles and the almost rigid rotation beneath the equator (b). From Küker, Rüdiger, and Kitchatinov (2011). Reproduced with permission © ESO.

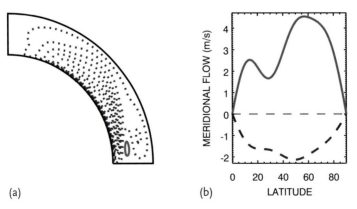

Figure 1.31 The same as in Figure 1.30 but for the meridional flow (a). The meridional circulation flows equatorwards at the top of the convection zone (solid line) and flows polewards at the bottom of the convection zone (dashed line) (b). Reproduced with permission © ESO.

respect to the equator then the angular momentum becomes uniform in the equatorial region except the boundary layers. Hence, $r^2 \Omega \simeq$ const independent of the flow direction, but in contradiction to the observation. Without the Λ effect a radial superrotation beneath the equator can only be due to a counterclockwise flow which is sufficiently slow.

With amplitudes of 4.7 m/s at the top and 2.1 m/s at the bottom of the convection zone the baroclinic clockwise meridional flow is weaker than the observed flow and it goes in the 'wrong' direction. Induced by a meridional flow the observed equatorial acceleration can only be produced by an equatorwards directed transport of angular momentum, that is by an equatorwards directed meridional flow. As such a flow has never been observed at the solar surface the solar rotation law cannot be explained by a baroclinic flow.

1.5.2
Sun without Baroclinic Flow

Next the baroclinic term in the equations will be ignored while keeping the Λ effect which maintains the differential rotation together with the meridional flow caused by the former through the first term on the RHS of (1.23). The flow, however, acts back on the differential rotation and reduces it.

Figure 1.32 shows the resulting rotation pattern. The rotation law is of solar-type, that is the rotation period is shortest at the equator and longest at the poles. With $\delta \Omega = 0.014$ rad/day the surface rotation is much more rigid than observed. The contour plot shows a clear cylinder-shaped pattern in the bulk of the convection zone while at the top and bottom boundaries the pattern deviates from the cylinder geometry and the rotation rate falls off with increasing radius at all latitudes (Figure 1.32). The unavoidable dominance of the cylinder geometry of the Ω-isolines

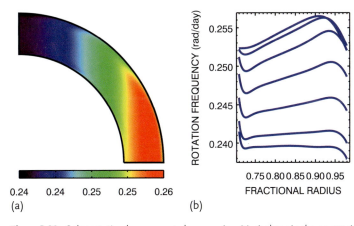

Figure 1.32 Solar rotation law computed without baroclinic terms. The Ω-isolines have cylindrical geometry (Taylor–Proudman theorem) (a). No shear along the polar axis but (positive) shear in the equatorial region. It shows superrotation while the polar axis rotates almost rigidly (b). Reproduced with permission © ESO.

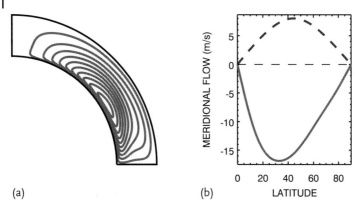

Figure 1.33 The same as in Figure 1.32 but for the meridional flow of the northern hemisphere (a). The flow is of solar-type, that is it goes polewards at the top (solid line) of the convection zone and equatorwards at its bottom (dashed) (b). Reproduced with permission © ESO.

due to the Taylor–Proudman theorem ($d\Omega/dz \simeq 0$) has been called the Taylor-number puzzle. Note that the negative shear along the polar axis which is typical for the solar rotation law and which is also produced by baroclinic flows (cf. Figure 1.30) does not exist here.

The meridional flow has a very similar geometry as in the full model but it is faster with amplitudes of 17 and 8 m/s at the top and bottom. It is *counterclockwise* in the northern hemisphere, that is it flows polewards (equatorwards) at the top (bottom) of the convection zone (Figure 1.33). This is a direct consequence of the rotation law in Figure 1.32 which by the action of the given Λ effect in both boundary layers possesses shear domains with $d\Omega/dz < 0$. It could easily have the opposite sign for another sign of the Λ effect.

Very fast rotation will also produce more cylinder-shaped rotation patterns as the baroclinic term is overwhelmed by the centrifugal term and the system approaches the Taylor–Proudman state.

1.5.3
Global Simulations

It makes sense to compare the above solutions to the results of the 3D hydrodynamic simulations with the ASH CODE for solving the anelastic equations (Miesch et al., 2008; Matt et al., 2011) and with the PENCIL CODE for fully compressible fluids (Käpylä et al., 2011). As the density stratification strongly influences the formation of the Λ effect, it is clear how important the use of the anelastic equations is. It is striking, however, that the Taylor-number puzzle in such simulations must be solved by adopting an artificial temperature gradient at the bottom of the convection zone (with warm poles, see Rempel, 2005, 2007). The simulations do *not* provide sufficient anisotropic heat transport in order to produce the warm poles consistently (Figure 1.34).

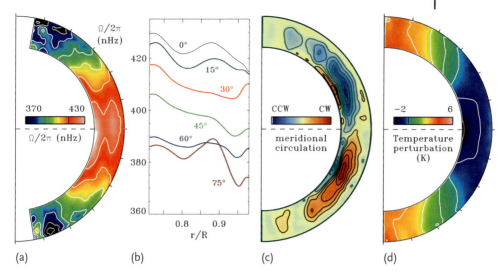

Figure 1.34 Differential rotation (a,b), meridional flow (c) and temperature profile (d, imposed from below) in a rotating convection zone. All quantities are averaged over longitude. The equator is accelerated as observed. The meridional circulation in the northern hemisphere is counterclockwise with polewards directed 20 m/s as observed. The observed strong negative shear along the polar axis does not exist. From Miesch et al. (2008). Reproduced by permission of AAS.

Matt et al. (2011) with the same code even provide models for the flow system in the convection zones of K7 to G0 stars of the main-sequence. Their masses are 0.5, 0.7, 0.9 and 1.1 solar masses, always rotating with the solar value of the angular velocity. The thickness of the convection zones hardly varies within the sample considered. No temperature gradient is imposed at the base of the convection zone. Among the many new results, the finding that the turnover time of the convective cells in the various stellar models decreases for increasing masses is of high importance for the formation of differential rotation and meridional flow. Hence, as the angular velocity is one and the same for all stars, the Coriolis number (1.7) sinks for more massive stars which, therefore, are only slow rotators. Then the Λ effect becomes so small that for the 1.1 $M_\odot$ star the meridional circulation is the only transporter of the angular momentum. As due to the antisolar temperature gradient in latitude it flows counterclockwise in the northern hemisphere (see the outer shell in Figure 1.35) the resulting rotation law *must* be antisolar, that is with faster polar vortices.

The lighter stars are faster rotators so that the Λ effect becomes stronger, transporting the angular momentum towards the equator. The consequence is that despite a counterclockwise meridional circulation the rotation law is of the solar-type, that is with accelerated equator. This can only be a result of the existence of the Λ effect. If a not too fast one-cell meridional circulation rises at the equator so that it flows counterclockwise then – without Reynolds stress – the rotation law must be of antisolar-type (see, however, Figure 1.36).

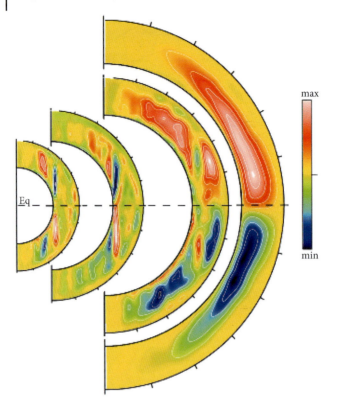

Figure 1.35 The streamlines of the meridional flows for the K- and G-type stars with 0.5 (a), 0.7 (b), 0.9 (c), and 1.1 $M_\odot$ (d). Blue: clockwise flow, red: counterclockwise flow. The data are temporally and azimuthally averaged. Courtesy of S.P. Matt.

1.6
Individual Stars

There are two main difficulties in applying the differential rotation models to individual stars. First, the stellar structure parameters like mass, radius, or convection zone depth, which are necessary to model stellar rotation, are often not known. Second, accurate measurements of differential rotation are difficult and thus rare. This is especially relevant to the stars with moderate rotation rates where the Doppler imaging technique cannot be applied.

So far no single main-sequence star has been found showing antisolar rotation, that is a polar cap rotating with a shorter period than the stellar equator (see Strassmeier, 2004; Weber, Strassmeier, and Washuettl, 2005). Stellar differential rotation is usually characterized by the surface shear, $\delta\Omega = \Omega_{eq} - \Omega_{pole}$, where Ω_{eq} and Ω_{pole} are the rotation rates at the equator and the poles, respectively. The surface shear is related to the lapping time, $t_{lap} = 2\pi/\delta\Omega$. For stars, a surface rotation law of the form $\Omega = \Omega_{eq}(1 - k\cos^2\theta)$ is usually assumed, where θ is the colatitude.

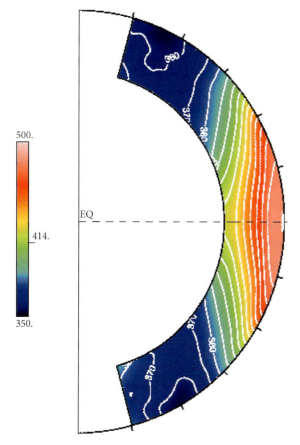

Figure 1.36 The contour lines of the rotation law of the model with 0.9 $M_\odot$ by Matt et al. (2011). The data are temporally and azimuthally averaged. The solar-type rotation law of this model with its counterclockwise flow (see Figure 1.35) can serve as numerical proof of the existence of the Λ effect. Courtesy of S. Matt.

With that type of rotation law we have with the definition (1.2) $\delta\Omega = 2\pi|k|/P_{\rm rot}$, where $P_{\rm rot}$ is the rotation period at the equator. While simplified rotation laws of this form are widely used for fitting observation data it should be noted that in the rotation law derived from the rotation of the solar plasma at the surface

$$\Omega = 0.245 - 0.026 \cos^2\theta - 0.045 \cos^4\theta \quad \text{[rad/day]} \tag{1.37}$$

the $\cos^4\theta$ term exceeds the $\cos^2\theta$ term. Only by use of the associated Legendre polynomials do the corresponding coefficients of higher orders monotonically decrease, that is

$$\Omega = 0.23\left(1 - 0.033\frac{P_3^1}{\sin\theta} - 0.0043\frac{P_5^1}{\sin\theta}\right) \quad \text{[rad/day]}. \tag{1.38}$$

Rotation laws derived from the observation of sunspots have k = 0.23, which is somewhat smaller than the equator–pole difference from the Doppler shifts and

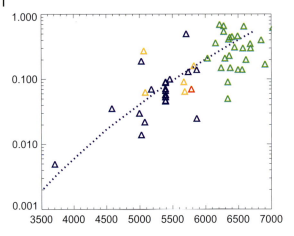

Figure 1.37 Observed surface shear values $\delta\Omega$ in rad/day of main-sequence stars from class M (left) to F (right). The red triangle gives the Sun while the green color represents the measurements of F stars (see text). The yellow symbols stand for the photometric results with MOST and CoRoT data while Doppler imaging led to the blue triangles. Courtesy of M. Küker.

corresponds to $\delta\Omega = 0.057$ rad/day and a lapping time of 135 days. From the rotation law (1.37) we find $\delta\Omega = 0.07$ rad/day and $t_{\rm lap} = 90$ days.

It might be enlightening to start the theoretical discussion of the empirical results for the shear $\delta\Omega$ of stellar rotation laws with a general overview summarized in the Figure 1.37.

1.6.1
Two MOST Stars

The differential rotation of two moderately rotating ($P_{\rm rot} \sim 10$ day) stars – ε Eri and κ^1 Ceti – have been measured using high-precision photometry of the MOST mission (Croll et al., 2006; Walker et al., 2007). Fortunately, parameters of ε Eri are known relatively well (Soderblom and Däppen, 1989). The parameters can be closely reproduced by the structure model of a 0.8 $M_\odot$ star with metallicity $Z = 0.01$ and an age of 1 Gyr. The parameters of κ^1 Ceti are less certain. Those used by Rucinski et al. (2004) and Walker et al. (2007) in differential rotation measurements can be roughly reproduced by the structure model of a 1 $M_\odot$ star with $Z = 0.02$ at

Table 1.1 Input parameters of the differential rotation models for two moderate rotators.

Star	$M/M_\odot$	$R/R_\odot$	$L/L_\odot$	Z	Age (Gyr)	$P_{\rm rot}$ (days)
ε Eri	0.8	0.724	0.337	0.01	1	11
κ^1 Ceti	1.0	0.907	0.758	0.02	0.6	9

the age of about 600 Myr. The parameters of the MOST-stars used in simulations of their differential rotation are listed in Table 1.1.

Figures 1.38 and 1.39 show the modeled differential rotation and meridional flow of ε Eri. The relative magnitude of the surface differential rotation can be estimated with the simple expression $\delta\Omega \simeq 0.063$ rad/day, or equivalently

$$k \simeq \frac{P_{eq}}{100 \text{ day}} \tag{1.39}$$

by means of a simplified model for solar-type main-sequence stars (Kitchatinov and Rüdiger, 1999). The model gives $k = 0.127$ for ε Eri, close to the observational value $k = 0.11$ by Croll et al. (2006). The agreement for κ^1 Ceti is not so close: $k = 0.13$ is

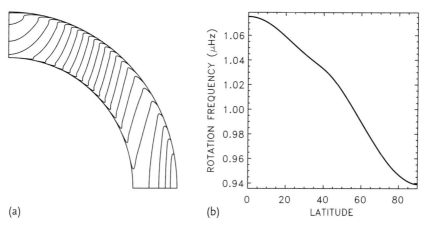

Figure 1.38 Angular velocity isolines (a) and surface profile of the rotation frequency ν (b) for the differential rotation model of ε Eri.

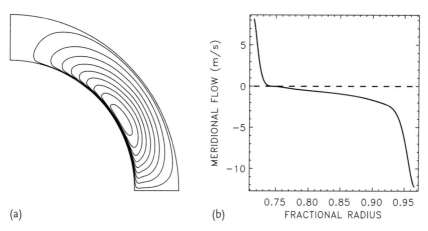

Figure 1.39 Meridional flow of ε Eri from the same model as Figure 1.38. Streamlines are shown (a), and the depth profile of the meridional velocity for a latitude of 45° (b). Negative velocity means poleward flow.

the computed value and k = 0.09 is the empirical result (Walker et al., 2007). The difference may be partly due to uncertainty in parameters of κ^1 Ceti. The simulated rotation laws for both moderate rotators are quite similar. The dependence of the rotation rate on latitude is not as smooth as for the solar model. There is a 'peculiarity' in the surface profile located around the latitude where the angular velocity contour tangential to the inner boundary at the equator reaches the surface. The peculiarity is typical of the rotation laws computed for stars rotating considerably faster than the Sun. This means that the often used approximation (1.5) may not be equally accurate for stars as it is for the Sun.

The boundary layers are thinner in Figures 1.38 and 1.39 compared to the solar model. The layers' thickness is estimated by the Ekman scale, $\sqrt{\nu_T/2\Omega}$, which decreases with rotation rate. Computations show that the thermal wind balance is satisfied very closely everywhere except for the boundary layers. Meridional flow of Figure 1.39 is concentrated in the boundary layers. The concentration becomes more pronounced with faster rotation.

1.6.2
Young Stars

The most frequent targets for observations of differential rotation by the method of Doppler imaging were the two young dwarfs AB Dor and LQ Hya. The parameters used in simulations of differential rotation of these two rapid rotators are listed in Table 1.2. These parameters help to reproduce closely the observational structure parameters of AB Dor given by Donati and Collier Cameron (1997), Ortega et al. (2007) and Guirado, Marti-Vidal, and Marcaide (2008), and LQ Hya (Kővári et al., 2004).

Figures 1.40 and 1.41 show the simulated rotation law and meridional flow of AB Dor. The computed differential rotation measure k = 4.37×10^{-3} is very close to the observational value of 4.5×10^{-3} (Donati and Collier Cameron, 1997). The peculiarity in the surface profile of rotation rate discussed above is even more pronounced in Figure 1.40 compared to the moderate rotation case of Figure 1.38. The profile can only be roughly approximated by a $\cos^2 \theta$-law.

Observational estimates of the differential rotation of slower rotating LQ Hya have a wide spread. The recent result of Flores-Soriano and Strassmeier (2013), however, with $\delta\Omega = 0.04$ rad/day for LQ Hya nearly perfectly fits into the general trend (Figure 1.11).

Table 1.2 Input parameters of the differential rotation models for rapidly rotating stars.

Star	$M/M_\odot$	$R/R_\odot$	$L/L_\odot$	Z	Age (Myr)	P_{rot} (days)
AB Dor	0.9	0.803	0.438	0.02	70	0.514
LQ Hya	0.77	0.698	0.273	0.01	100	1.6

The meridional flow of AB Dor of Figure 1.41 shows an extreme concentration in the boundary layers. The flow consists of two near-boundary jets linked by a very slow circulation in between. There is an increasing belief that the meridional flow is important for solar (Choudhuri, Schüssler, and Dikpati, 1995; Choudhuri, 2008) and stellar (Jouve, Brown, and Brun, 2010) dynamos. It is, however, hard to imagine that the boundary-layer flow of Figure 1.41 can be significant for magnetic field transport. The modeling predicts that the meridional circulation changes from distributed flow (Figure 1.28) to the near-boundary jets (Figure 1.41) with increasing rotation rate. This change of meridional flow structure may cause a change in dynamo regime. This may be the reason for the two separate branches for fast and

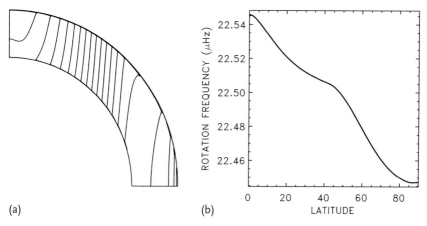

Figure 1.40 Angular velocity isolines (a) and surface profile of the rotation frequency (b) for the differential rotation model of rapidly rotating star AB Dor. From Kitchatinov and Olemskoy (2011). Copyright © 2011 RAS.

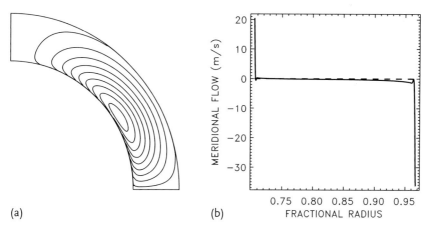

Figure 1.41 Meridional flow of AB Dor from the same model as Figure 1.40. Streamlines are shown (a), and the depth profile of the meridional velocity for a latitude of 45° (b). The meridional flow is highly concentrated in thin boundary layers. Copyright © 2011 RAS.

slow rotators in the observed dependence of dynamo-cycle period on the stellar rotation rate found by Saar and Brandenburg (1999).

Isorotational surfaces are cylinder-shaped near the equator and disk-shaped near the axis of rotation in all mean-field models (cf. Figures 1.27, 1.38 and 1.40). Isorotational surfaces are normal to the equatorial plane due to the symmetry of angular velocity distribution about the equator. As the distribution should be regular at the poles, isorotational surfaces are also normal to the rotation axis. It is thus an elementary rule that isorotational surfaces generally are cylinder-shaped near the equator and disk-shaped close to the poles.

CoRoT-2a is a young Sun, that is, it has solar mass but is much younger with an age of 0.5 Gyr. The star has been observed by the CoRoT satellite and found to have a planet with an orbital period of 1.743 days (Alonso et al., 2008). Besides planetary transits, the light curve shows periodic variation that is most easily explained by a rotating, spotted surface with basic rotation period of 4.5 days. Spot modeling using circular spots finds an excellent fit assuming three spots and a solar-type surface differential rotation of about 8%, that is with a shear of 0.11 rad/day (Fröhlich et al., 2009). Lanza et al. (2009) for the same star only reported a differential rotation of about 1%.

For two more young solar-type stars (KIC 7985370, KIC 7765135) Fröhlich et al. (2012) report the rather high values of 0.18 rad/day after use spot system models with at least seven spots (KIC 7985370) and nine spots (KIC 7765135) to provide a satisfactory fit to the data.

Our model star is based on a model from an evolutionary track for the Sun. The age is 0.5 Gyr, the radius 0.9 solar radii, and the luminosity is only 75% of the solar value. The bottom of the convection zone is at a fractional radius $x = 0.73$. Figure 1.42 shows the resulting rotation pattern. The surface differential rotation of 0.09 rad/day well reproduces the value observed by Fröhlich et al. (2009). The rotation pattern is more cylindrical than that of the Sun, but less than that of our

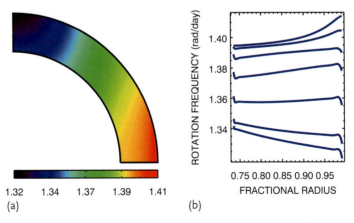

Figure 1.42 Rotation pattern of CoRoT-2a (a), a young Sun rotating with $P = 4.5$ days (b). Reproduced with permission © ESO.

fast-rotating Sun. Similarly, the boundary layers are more pronounced than in the Sun but less than for the fast-rotating Sun. The flow pattern is similar to the solar pattern. The amplitude is slightly larger with a maximum value of 18.6 m/s at the top and 10.6 m/s at the bottom of the convection zone.

The G dwarf R58 (HD 307938) rotates with a period of only 0.56 day. The two different values of 0.025 and 0.14 rad/day have been reported for its surface differential rotation (Marsden *et al.*, 2005). LQ Lup is a post-T Tauri star with an effective temperature of 5750 K, a rotation period of 0.31 day, and a mass of 1.16 $M_\odot$. The equator–pole difference of Ω is 0.12 rad/day.

For HD 171488 (V889 Her) Strassmeier *et al.* (2003) found a rotation period of 1.34 day, a mass of 1.06 $M_\odot$. Marsden *et al.* (2006) reported a rotation period of 1.31 day. At 0.4–0.5 rad/day the equator–pole difference in Ω is exceptionally large. The k-value of 0.10 misses the k = 0.013 predicted by (1.39) by a factor of 7.5. The corresponding factors for R58 and LQ Lup are much smaller. As the stars have similar radii and effective temperatures, only the differences in the internal structure may influence the surface rotation patterns. Two models with different depths of the convection zone are thus considered. The rotation period is the 1.33 day period of HD 171488. Both model stars are 30 Myr old.

The first model has a mass of 1.11 $M_\odot$. The value of 1.6 for the mixing-length parameter lead to a radius of 1.14 $R_\odot$. The bottom of the convection zone is located at $x_{in} = 0.77$ and has a temperature of 1.46×10^6 K. The rotation law is shown in Figure 1.43a. The isocontours are cylinder-shaped with the surface shear of 0.11 rad/day. The radial profiles show superrotation beneath the equator and subrotation along the rotation axis. The meridional flow is of solar-type, that is it flows at the surface flow towards the poles and the return flow is located at the bottom of the convection zone. The amplitudes are 23 and 14 m/s, respectively. The model has a pole–equator temperature difference of 24 K at the top and 124 K at the bottom.

A second model is defined by imposing the lower boundary at $x_{in} = 0.88$ (Küker, Rüdiger, and Kitchatinov, 2011). Now the theoretical surface shear with 0.5 rad/day is much stronger and close to the observations. For shallow convection zones the radial shear becomes weak below the equator but the latitudinal shear is increased. The early-G star HD 141943 may be a similar example of this phenomenon. Its rotation period is 2.18 day and the surface shear is reported as about 0.4 rad/day (Marsden *et al.*, 2011). The relative thickness of the convection zone of HD 141943 is only 0.16.

1 Differential Rotation of Stars

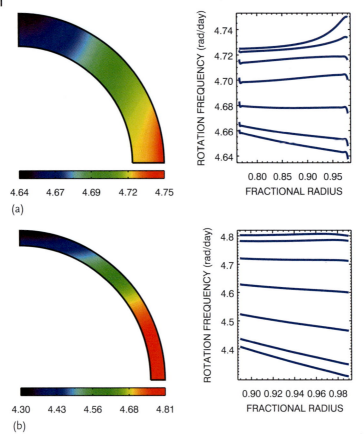

Figure 1.43 Internal rotation of young G dwarfs with rotation period of 1.33 days. Deep convection zone, $\delta\Omega = 0.11$ rad/day at the surface (a). Shallow convection zone, $\delta\Omega = 0.5$ rad/day at the surface (b). Reproduced with permission © ESO.

1.7
Dwarfs & Giants

1.7.1
M Dwarfs

Fully convective stellar structure models with less than 0.35 $M_\odot$ by Chabrier and Baraffe (1997) are used. The reference star has 0.1 $M_\odot$, 0.12 $R_\odot$, and a central density of 400 g/cm^3. From mixing-length theory the convection velocity results as typically 2 m/s throughout the whole star. Note that the density is two orders of magnitude larger but the convection velocities are much smaller than the solar values. The convective turnover times of about 100 days result in Coriolis numbers larger than unity, especially for young stars with rotation periods of several days (Proxima Cen: 31.5 day, Lalande 21185: 47 day). As a slow rotator the M dwarf

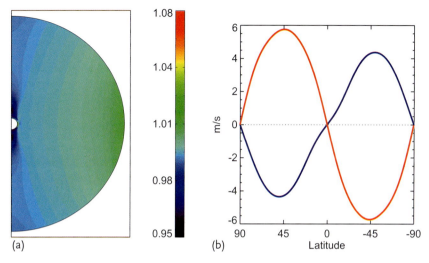

Figure 1.44 Contour plot of the internal rotation rate of a M dwarf (a). The flow amplitude at the top (red) and the bottom of the convection zone (black) (b). After Küker and Rüdiger (2008).

rotation period must exceed 1 year. Hence, the majority of M dwarfs represents rapid rotators (with consequences also for their dynamos, see Chabrier and Küker, 2006).

Figure 1.44b shows the meridional flow of an M dwarf rotating with a period of 5 days. As in the solar convection zone, there is only one flow cell per hemisphere with the surface flow directed towards the poles. The flow amplitude at the surface is about 6 m/s. Interestingly, the cross travel time of the meridional flow from equator to pole is about 11 years, as for the Sun. A small core with a radius of 5% of the stellar radius had to be introduced for numerical reasons. The contour plot for the rotation rate (Figure 1.44a) shows only little variation except close to the rotation axis which is probably caused by the boundary condition. A fit to a $\cos^2 \theta$ law yields k = 0.028 and $\delta \Omega \simeq 0.035$ rad/day. The lapping time is thus 180 days. Despite the fact that the rotation profile appears nearly rigid, the total latitudinal shear $\delta \Omega$ is about half the solar value. The rotation pattern is much closer to the Taylor–Proudman state, with Ω increasing with radius in the bulk of the convection zone.

1.7.2
F Stars

For sufficiently rapid rotators surface differential rotation can be detected from the broadening of spectral lines. Reiners and Schmitt (2003a,b) carried out measurements for F stars with moderate and short rotation periods. They found differential rotation to be much more common for stars with moderate rotation rates than for very rapid rotators. The main result is that the latitudinal shear reaches maximum values up to 1 rad/day for rotation periods of two and three days while 0.2 rad/day

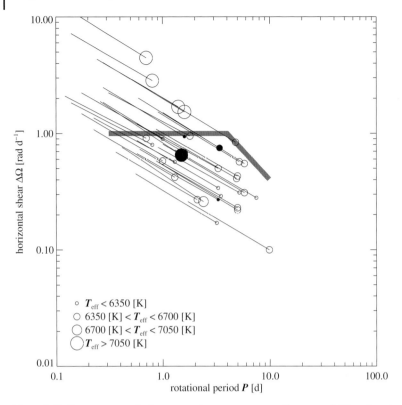

Figure 1.45 The spectroscopically measured surface shear $\delta\Omega$ of F stars vs. estimated projected rotation periods detected by profile analysis. The symbol size scales the effective temperature. The gray solid line gives the upper envelope to F-type rotators. Courtesy of M. Ammler-von Eiff and A. Reiners.

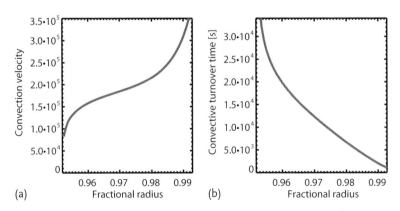

Figure 1.46 Convective velocity in cm/s (a) and turnover time (b) in the shallow convection zone of a main-sequence star of 1.5 $M_\odot$. From Küker and Rüdiger (2007).

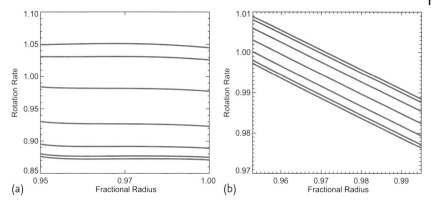

Figure 1.47 Normalized rotation rate as function of the fractional stellar radius for the 0°, 15°, 30°, 45°, 60°, 75°, and 90° latitude (from top to bottom in the diagrams) for rotation periods of 1 day (a) and 27 days (b). Copyright © 2011 RAS.

are typical at $P_{rot} = 10$ days. For faster rotation ($P_{rot} < 2$ days) the shear is hardly reduced (Ammler-von Eiff and Reiners, 2012, see Figure 1.45).

The convection zones are rather thin for early-type F stars and the typical timescales of the convection cells are more than one order of magnitude shorter than those of the Sun (Figure 1.46b). The maximal surface shear of the F stars exceeds the solar one by more than a factor of ten. Our model star with 1.4 $M_\odot$ represents the upper end of the lower main-sequence in the context of differential rotation and stellar activity. With a convection zone depth of about 54 500 km ($x_{in} = 0.95$) and a surface gravity roughly equal to the solar value, the main difference from the Sun is the higher luminosity (4.4 times the solar luminosity). The greater luminosity enforces a larger convective heat-flux. To ensure the necessary heat transport the convection velocity must be larger than in the solar convection zone (Figure 1.46a). The density stratification is steeper than in the solar convection zone, leading to smaller values of the density and pressure scale heights and hence to smaller values of the mixing length.

In Figure 1.47 the rotation rate is plotted vs. radius for latitudes at 0° (equator), 15°, 30°, 45°, 60°, 75°, and 90° (poles). The equator rotates faster than the poles, but the amplitude of the relative shear varies with the rotation rate. The model with the fastest rotation yields strong differential rotation. Note that for rapid rotation the lines are practically horizontal so that only latitudinal shear exists. For slow rotation there is only a (small) value of radial shear. For rapid rotation one finds the numerical value of the normalized latitudinal shear k $\simeq$ 0.17 at the surface. For a rotation period of 1 day that means $\delta \Omega \simeq 1.05$ rad/day which is much higher than that of the Sun. There is always a broad range of nearly constant shear around the maximum. It is thus indeed possible to explain the strong surface values of the latitudinal shear for rapidly rotating stars with rather thin outer convection zones. The possible existence of a characteristic maximum of $\delta \Omega$, however, is discussed in more detail in Section 1.8.

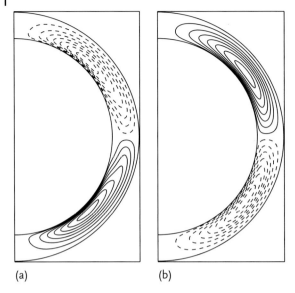

Figure 1.48 The same as in Figure 1.47 but for the streamlines of the meridional circulation for rotation periods of 1 day (a) and 27 days (b). Solid (dashed) lines refer to clockwise (counterclockwise) circulations. The radial scale is increased. Note that for slow and fast rotation the flow direction reverses. Reproduced with permission © ESO.

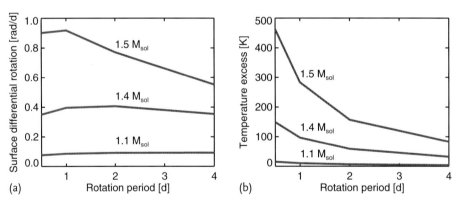

Figure 1.49 The increase of the surface shear $\delta\Omega$ (a) and the pole–equator temperature difference (b) for faster rotation and hotter stars. The largest pole–equator difference of temperature for the most massive star with a rotation period of 1 day is 280 K. Courtesy of M. Küker.

Figure 1.48 shows the meridional flow patterns. For $P_{\rm rot} = 1$ day there is only one flow cell per hemisphere directed towards the poles in the upper part of the convection zone and towards the equator at the bottom. For a rotation period of 27 days the flow is oppositely directed (clockwise). The change of the flow direction at the bottom of the convection zone should have dramatic consequences for the advection-dominated stellar dynamo.

The maximum flow speed at the top (bottom) of the convection zone is with 144 m/s (103 m/s) again one order of magnitude faster than the solar circulation. The same is true for the traveling time of the flow from the equator to the poles. Advection-dominated $\alpha\Omega$-dynamos should have rather short activity cycles.

Figure 1.49 summarizes the results for the pole–equator differences of the surface profiles of angular velocity and temperature for upper main-sequence stars. The value of $\delta\Omega$ strongly grows with the stellar mass so that the observed values of about 1 rad/day (see Figure 1.45) can be explained. The corresponding temperature differences δT reach values up to 280 K which is basically smaller than the 1000 K given by Augustson et al. (2012) for their F star models. Such large values might easily be a consequence of their artificial formulation of the heat-flux tensor which is not allowed to dissipate the heat of the axisymmetric modes in the latitudinal direction.

1.7.3
Giants

Another type of convective stars to which the differential rotation theory can be applied are the red giants. Observations revealed several striking regularities concerning rotation and magnetic activity of the evolved stars. The sharp decrease of characteristic rotation rates with spectral type known for mid-F main-sequence dwarfs is also observed for early-G normal giants of luminosity class III. For later spectral types, the rotation rate is uniquely defined by the spectral type, $U_{eq} = 7.31 - 0.417\,\text{Sp}\,[\text{km/s}]$ (Gray, 1989a) where Sp is the numerical spectral type equal 3 for G3, 4 for G4 up to 10 for K0. The giants with characteristic radii of the order of 10 $R_\odot$, therefore, rotate with periods of some years. Similar to the Sun, magnetic activity is confined to near-equator belts (Gray, 1989b). Long-term variations of magnetic activity similar to the solar cycle were also observed (Donahue, Saar, and Baliunas, 1996). Red giants seem to be more solar-like than the rapidly rotating main-sequence dwarfs.

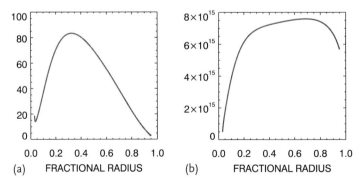

Figure 1.50 The convection zone data used for the model of the rotation law of Arcturus. Convective turnover time in days leading to a Coriolis number smaller than unity (a). The eddy viscosity in cm^2/s as computed with the standard expression (1.15) (b).

Preliminary rotation calculations for red giants predicted the relative magnitude of the surface differential rotation k ≥ 0.3 exceeding the solar value. Models for giants with deep convection zones also predicted substantial increase of rotation rate with depth. The giants were found to be slow rotators with Coriolis number $\Omega^* < 10$. Characteristic times of turbulent diffusion, $R_*^2/\nu_T \simeq 10$ years, are similar to the Sun.

Arcturus is a relatively close K2 III star. Gray and Brown (2006) inferred a two-year rotation period from high-resolution spectroscopic data of two decades of observations of this star. The two-year period has also been found in the Ca II emission of Arcturus which has been monitored by the Mount Wilson H+K Project since 1984 (Brown et al., 2008). The data also seem to indicate an activity cycle with a period of about 14 years. One finds the convective turnover times to be on the order of several months and the eddy viscosity proves to be two or three orders of magnitudes larger than the solar values (Figure 1.50).

Figure 1.51 illustrates the differential rotation model of Arcturus. The surface rotation is of solar-type with $\delta\Omega = 8 \times 10^{-3}$ rad/day. This is only marginally less

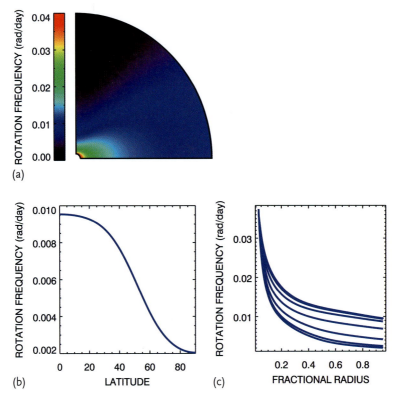

Figure 1.51 The differential rotation of Arcturus. Contour plot of the internal rotation law (a). Surface profile of rotation rate (b). Rotation rate as function of depth for the equator, latitudes of 15°, 30°, 45°, 60°, 75° and poles, from top to bottom (c). After Küker and Rüdiger (2011).

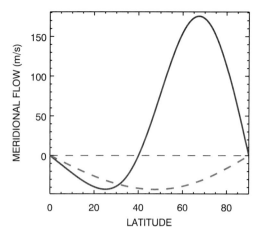

Figure 1.52 The flow amplitude at the top (solid) and the bottom of the convection zone (dashed) for one solar mass.

than the rotation rate of the equator, that is, k is close to one. While the latitudinal shear dominates in the outer layers of the convection zone, the rotation is almost uniform in latitude but increases strongly with depth near the bottom. The inner nonconvective core is very small: it extends to about 3% in radius only. In the whole convection zone, the rotation rate decreases with increasing radius. The decrease is about four times at the equator and the meridional flow at the surface reaches values up to 170 m/s (Figure 1.52). The results for HD 12545 (XX Tri, K0 III, $M = 2\,M_\odot$) demonstrates how important the parameter $\delta\Omega$ is. Künstler, Carroll, and Strassmeier (2013) report for this red giant a rotation period of 24 days with a solar-type differential rotation of k = 0.023, so that the observed $\delta\Omega \simeq 6 \times 10^{-3}$ rad/day is very close to the result for Arcturus. The same holds for the bright giant (G5) variable star FK Coma with its rapid rotation with 2.4 days period which has been reported as a differential rotator with $\delta\Omega \lesssim 8 \times 10^{-3}$ rad/day (Hackman et al., 2013).

Recent asteroseismological soundings of Beck et al. (2012) indicate that the steep increase of rotation rate with depth may be typical of giants with extended convective envelopes. The K giant KIC 8366239 rotates about ten times faster than Arcturus so that the calculated rotation law should differ from that given in Figure 1.51b in the sense that the surface shear is expected to be much stronger ($\Omega^* \simeq 15$). On the other hand, the fast rotation of the core may form the remnant of a much steeper rotation profile originating during the core-collapse during the formation of the red giant without internal angular momentum transport. Compared with the rotation of the new-born core of the red giant the observed rotation rate is much smaller. Eggenberger, Montalbán, and Miglio (2012) compute the viscosity which could generate the necessary spin-down of the core and obtain the value of 3×10^4 cm^2/s which exceeds the generally accepted value by a few orders of magnitude. Also the KEPLER mission data for the early red giant KIC 7341231 lead to a fast-rotating core. Deheuvels et al. (2012) report its rotation five times faster

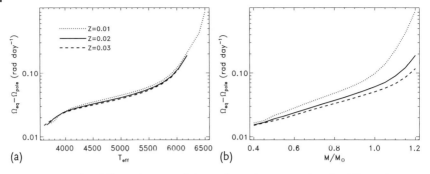

Figure 1.53 Surface differential rotation $\delta\Omega$ computed for ZAMS stars with $P_{rot} = 1$ day for different metallicities. The dependence on surface temperature (a). Mass-dependence of the shear (b). The metallicity dependence almost disappears when the differential rotation is considered as a function of T_{eff}. From Kitchatinov and Olemskoy (2011). Copyright © 2011 RAS.

than the rotation of the stellar envelope – which again is too slow to be consistent with the actual stellar evolution codes.

1.8
Differential Rotation along the Main Sequence

Figure 1.53 shows the dependence of differential rotation computed for ZAMS stars on the effective temperature T_{eff}. The dependence agrees with differential rotation measurements by Doppler imaging (Figure 1.10). Both observations and theory find the differential rotation increasing with temperature. The increase steepens for F stars compared to K and M dwarfs.

Figure 1.53 also illustrates the temperature scaling: differential rotation values computed for stars of given P_{rot} but various masses (M) and metallicities (Z) fall on a common line when plotted as a function of temperature. The temperature scaling, therefore, reduces the number of independent input parameters of the differential rotation models[7].

Further reduction is achieved by using gyrochronology discussed in Section 1.2.1. We apply the relation (3.3) of Barnes (2007) to fix the rotation period of a star of given age and mass.

Figure 1.54 shows the resulting horizontal shear at the surface as a function of the rotation period for a sample of stellar masses. Lines of this plot were computed for a certain initial mass but not for a given stellar structure. The structure changes as a star ages and rotation period increases. Note that the dependence of differential rotation on the rotation rate is rather mild. Figure 1.54 shows much stronger dependencies on stellar mass, with more massive and hotter stars having larger differential rotation. Note that Jeffers and Donati (2008) report $\delta\Omega = 0.5$ rad/day

---

7) All further results of this section were obtained with metallicity $Z = 0.02$.

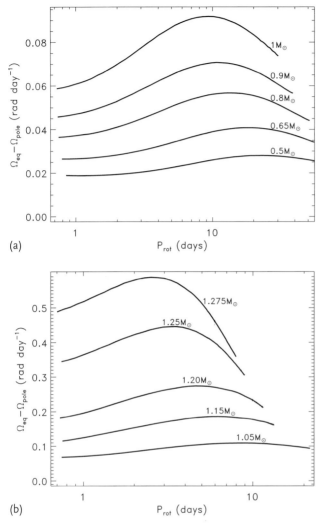

Figure 1.54 Surface differential rotation as a function of the rotation period for stars less (a) and more (b) massive than the Sun.

for the early G0 dwarf HD 171488 while Reiners (2006) argues for $\delta\Omega \leq 1$ rad/day for the hotter F stars.

For a given mass, the amount of the surface differential rotation of Figure 1.54 varies by about 30% as the rotation period changes. The surface differential rotation increases initially up to a maximum but then decreases with P_{rot}. Maxima on all lines are positioned at the values of the Coriolis number (1.7) between 10 and 20. The Coriolis number is estimated for the middle of the convection zone. Stars of smaller mass host slower convection with larger τ_{corr} and attain their largest differential rotation at smaller rotation rates. The increase of differential rotation with angular velocity for not too rapid rotation was also found in 3D simulations of Brown *et al.* (2008). The character of differential rotation changes between the cases of slow ($\Omega^* < 10$) and fast ($\Omega^* > 20$) rotation. Slow rotators have smooth distributions of angular velocity on the surfaces and inside their convection zones similar to the Sun (Figure 1.27). Isorotation surfaces are far from cylinders. Rapid rotators have thin boundary layers near the top and bottom of their convection zones (Figure 1.38). Their isorotation surfaces are much closer to cylinders.

The meridional flow patterns also differ for slow and fast rotation (cf. Figures 1.28 and 1.41). Violation of the thermal wind balance in the boundary layers results in relatively large values of the sources of meridional flow written on the right-hand side of (1.23). Accordingly, the flow attains its largest velocities close to the boundaries. The boundary layers in slow rotators are relatively thick. The flow is smoothly distributed over the entire thickness of the convection zone in this case. In rapid rotators, however, the flow is confined in thin layers near the boundaries. This agrees with 3D simulations of Brown *et al.* (2008, 2010, 2011) who found a decrease of the meridional flow energy with rotation rate.

The near-boundary flows are significant. The surface flow is potentially observable. The flow near the bottom may be important for dynamos. Figure 1.55 shows how the near-boundary flows vary with stellar mass and rotation rate. The bottom flow is smaller but not much smaller compared to the surface. Similar to differential rotation, the flow amplitude increases with stellar mass and it increases steadily with rotation rate. The flow in the bulk of convection zone, however, has the opposite tendency to decrease with rotation rate.

The meridional flow is produced, in particular, by the effect of the thermal wind due to the latitudinal temperature dependence. Figure 1.56 shows the depth pro-

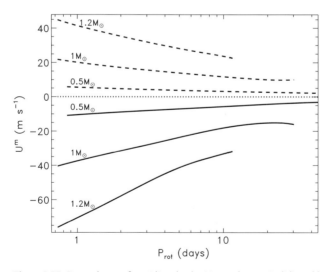

Figure 1.55 Dependence of meridional velocities at the top (solid) and bottom (dashed) of the convection zone on the rotation period for stars of different mass. The velocities are taken at a latitude of 45°. Negative values mean poleward flow.

files of the differential temperature. The differential temperature is produced mainly by anisotropy of the thermal diffusivity tensor (1.29). The anisotropy is induced by rotation and increases with rotation rate. Accordingly, the differential temperature in faster rotating star of Figure 1.56 is larger.

Figures 1.53 and 1.54 suggest that the external convection zones of F stars can possess strong differential rotation with pole–equator lap times less than 10 days. The question arises whether a strong differential rotation implies above-normal dynamo activity. The efficiency of differential rotation in generating magnetic fields can be estimated by the dynamo number

$$C_\Omega = \frac{\delta \Omega \, D^2}{\eta_T}, \tag{1.40}$$

where D is the convection zone thickness and η_T is the turbulent magnetic diffusivity. Figure 1.57 shows contours of C_Ω on the $T_{\text{eff}} - P_{\text{rot}}$ plane from the same computations as Figure 1.54. It is estimated for the middle of the convection zone

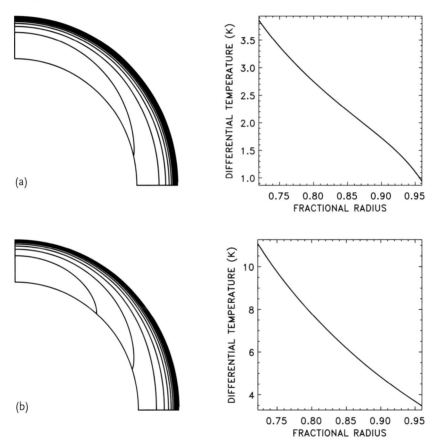

Figure 1.56 Entropy isolines and pole–equator temperature difference for 1 $M_\odot$ stars rotating with periods of 30 days (a) and 3 days (b).

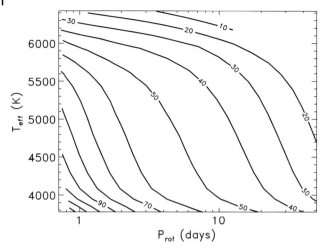

Figure 1.57 Contours of the magnetic dynamo number (1.40) in the $T_{\text{eff}} - P_{\text{rot}}$ plane, from the same computations as Figure 1.54. The dynamo number decreases with either P_{rot} or T_{eff}.

using the isotropic part of the eddy diffusivity $\eta_T = \chi_T \phi(\Omega^*)$ that coincides with the isotropic part of the thermal eddy conductivity (1.30).

For a given rotation period the C_Ω in Figure 1.57 *decreases* with T_{eff}. Strong differential rotation of F stars is not efficient at producing toroidal magnetic fields. This is in accordance with the idea of Durney and Latour (1978) that the convective dynamos cease to operate at about spectral type F6. The C_Ω increases steadily with decreasing temperature. This is because the convection slows down in low-mass stars to decrease eddy diffusion. The decline of magnetic eddy diffusion overpowers the decrease of differential rotation to produce larger C_Ω in the redder stars. This is in contrast to the common belief that the small differential rotation of M stars cannot be important for dynamos and that magnetic fields of these stars are generated by the α^2 mechanism. The α^2 dynamos produce nonaxisymmetric global fields (Rüdiger and Elstner, 1994; Chabrier and Küker, 2006). However, observations favor an axial symmetry of the global magnetic structure of M dwarfs (Donati et al., 2006). The increase of C_Ω-dynamo number with decreasing temperature may also be the explanation for why the low-mass stars are spinning down to smaller rotation rates (Figure 1.8).

2
Radiation Zones: Magnetic Stability and Rotation

The rotational state of the solar radiation zone revealed by helioseismology is challenging to the theory. This is not only because of the thin tachocline seen in Figure 1.1 in the upper radiation zone. The almost uniform rotation in deeper regions is also surprising. Before the emergence of helioseismology, it was believed very probable that the deep solar interior rotates much faster than the photosphere.

Solar-type stars at the beginning of their main-sequence life rotate fast, $P_{\rm rot} \simeq$ 1 day. It is very likely that the infant Sun was a fast rotator as well. It arrived at the present day state of slow rotation, $P_{\rm rot} \simeq 25$ day, by losing angular momentum to a magnetically coupled wind. The braking torque is applied to the solar surface. The eddy diffusion time in the convection zone, ~ 10 years, is relatively short. The entire convection zone, therefore, reacts almost instantaneously to the angular momentum loss, which proceeds on evolutionary timescales. However, there is no clear reason for a coupling between the base of the convection zone and the deep radiative core. It might be supposed that the solar core rotates with the primordial period of the order of 1 day (Dicke, 1970). Helioseismology, however, shows that rotation in the radiation zone is almost uniform with radius. An efficient coupling should, therefore, exist between the base of the convection zone and the deep solar interior.

Stellar observations show that the characteristic time of rotational coupling between core and envelope in solar-type stars is of the order of 100 Myr, and decreases with stellar mass (Hartmann and Noyes, 1987; Denissenkov, 2010). The origin of this coupling is problematic. The microscopic viscosity, $\nu \sim 10\,{\rm cm}^2/{\rm s}$, is too small to provide it. A larger turbulent viscosity could be produced by a hypothetical turbulence, but this possibility is ruled out by the observed lithium abundances.

Abundances of light elements like lithium, beryllium and boron can be used to constrain the intensity of mixing in stellar radiation zones. Because these elements are destroyed in proton capturing reactions at relatively low temperatures, they can survive only in convection zones and upper layers of radiation cores of solar-mass stars. Figure 2.1 shows the depth profiles of the light elements in the Sun expected if they are altered by nuclear reactions only (Barnes, Charbonneau, and MacGregor, 1999). The stepwise profiles of Figure 2.1 are caused by extreme temperature sensitivity of the light element burning. According to the graph, lithium vanishes at depths below $0.65\,R_\odot$ where temperature exceeds 2.7×10^6 K. Beryllium and

Magnetic Processes in Astrophysics, First Edition. G. Rüdiger, L.L. Kitchatinov, and R. Hollerbach.
© 2013 WILEY-VCH Verlag GmbH & Co. KGaA. Published 2013 by WILEY-VCH Verlag GmbH & Co. KGaA.

2 Radiation Zones: Magnetic Stability and Rotation

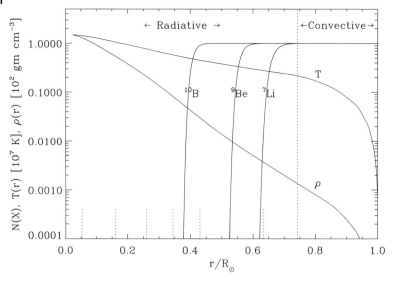

Figure 2.1 Radial profiles of the expected abundances of Li, Be and B normalized to their primordial contents for the present-day solar model, in the absence of radial mixing in the radiation zone. The extreme temperature sensitivity of the nuclear reaction responsible for the light element depletion results in a step-like radial dependencies of the abundances. Courtesy of P. Charbonneau.

boron are burned-out below the radii of 0.54 and 0.4 $R_\odot$ where temperature exceeds 3.5×10^6 and 5×10^6 K, respectively.

Figure 2.1 illustrates the hypothetical case of no mixing at all in the radiation zone. A finite intensity of mixing results in a depletion of the light elements everywhere inside a star, not only in the deep radiation core. Observations of the light elements content therefore impose constraints on the mixing intensity.

The amounts of beryllium and boron are difficult to measure. Determinations of their abundances indicate that these elements are depleted in the Sun by at most a factor of two (or perhaps not at all), while for lithium a factor of about 160 is almost certain (Anders and Grevesse, 1989; Balachandran and Bell, 1998; Asplund et al., 2009). This means that the radial turbulent diffusivity in the upper radiation zone, averaged over the Sun's past, cannot exceed 10^3 cm^2/s. A turbulent viscosity of that magnitude is much too small to provide uniform rotation in the solar radiation zone.

Any core-envelope coupling of rotation should proceed without radial mixing of material (Zahn, 1992, 1993). This leads naturally to the idea of magnetic coupling (Spruit, 1987; Charbonneau and MacGregor, 1992, 1993). Radiation zones of solar-type stars can possess primordial magnetic fields. Ferraro's law (Ferraro, 1937) demands that the angular velocity be uniform along the field lines in a state of steady rotation. We know that the solar tachocline can also be explained by an effect of magnetic fields (Rüdiger and Kitchatinov, 1997; MacGregor and Charbonneau, 1999). Magnetized radiation zones, however, are prone to magnetic instabilities (Tayler, 1957; Vandakurov, 1972; Tayler, 1973; Wright, 1973). The problems of radiation zones, that is their rotation laws and their inner mixing of chemicals, and their hydromagnetic stability are, therefore, closely related.

2.1
The Watson Problem

The solar tachocline can be thought of as a result of an instability. If the latitudinal differential rotation in the radiation zone is unstable, the instability can destroy the differential rotation, thus limiting its extension beneath the base of the convection zone. The tachocline was not yet known, however, when Watson (1981) first addressed the stability problem with the idea that whatever origin the solar differential rotation has, its observed value may be marginal for the instability. Later on, we shall see that the instability does not exist in the limit of vanishing buoyancy frequency, $N \to 0$, so that differential rotation is stable in the convection zone of almost adiabatic stratification. The instability may, however, switch on in the region of penetrative convection close below the base of the convection zone.

The question of the tachocline stability is important for dynamo theory. It would be difficult to conceive that the site of the solar dynamo is beneath the convection zone if the tachocline is stable. The solution of the lithium problem is also strongly influenced by the stability/instability characteristics of the tachocline. The possible transmission of rotational energy into other types of motion by means of the instability may be relevant to various astrophysical processes. Excitation of r-modes of global oscillations in differentially rotating neutron stars is considered as a possible source of detectable gravitation waves.

2.1.1
The Stability Equations

Stability of the *latitudinal* differential rotation is analyzed, neglecting deviations of stellar structure from spherical symmetry. Radial displacements in stably stratified radiation zones are opposed by buoyancy and should thus be small. This means that the radial scale of the disturbances can be assumed small compared with the radius of a star. We shall see in Section 2.3 that the most unstable modes do indeed have short radial scales. The present stability analysis will be local in radius but global in horizontal dimensions. The dependencies on time, longitude ϕ, and radius r is taken in the form of the Fourier modes, $\exp(-i\omega t + im\phi + ikr)$. As the radial displacements are small, the velocity field can be assumed to be divergence-free, $\mathrm{div}\, \boldsymbol{u} = 0$. This assumption filters out all acoustic oscillations.

Different versions of (physically identical) linear stability equations can be found in the literature (cf. Cally, 2003; Miesch and Gilman, 2004; Gilman, Dikpati, and Miesch, 2007), also including magnetic fields. The nonmagnetic version of these equations is used, written in terms of the normalized disturbances of entropy (S) and scalar potentials for toroidal (W) and poloidal (V) flows. The disturbances in physical units follow from their normalized values by the relations

$$s = -\frac{iC_p N^2}{gk} S, \quad P_u = \frac{\Omega_0 r^2}{k} V, \quad T_u = \Omega_0 r^2 W, \quad (2.1)$$

where Ω_0 is the angular velocity at the equator, and N is the buoyancy (Brunt–Väisälä) frequency. The velocity field can be recovered from the potentials of poloidal (P_u) and toroidal (T_u) flows,

$$\boldsymbol{u} = \frac{1}{r}\left[\frac{1}{r}\hat{L}P_u, -\left(\frac{1}{\sin\theta}\frac{\partial T_u}{\partial\phi} + \frac{\partial^2 P_u}{\partial r\partial\theta}\right), \frac{\partial T_u}{\partial\theta} - \frac{1}{\sin\theta}\frac{\partial^2 P_u}{\partial r\partial\phi}\right] \quad (2.2)$$

in spherical coordinates (r, θ, ϕ) where

$$\hat{L} = \frac{1}{\sin\theta}\frac{\partial}{\partial\theta}\sin\theta\frac{\partial}{\partial\theta} + \frac{1}{\sin^2\theta}\frac{\partial^2}{\partial\phi^2} \quad (2.3)$$

is the angular part of the Laplacian operator (Chandrasekhar, 1961).

The system of three equations for linear stability includes the equation for the toroidal flow,

$$(\hat{\omega} - m\hat{\Omega})(\hat{L}W) = -m\frac{\partial^2\left[(1-\mu^2)\hat{\Omega}\right]}{\partial\mu^2} W$$
$$+ \frac{\partial\left[(1-\mu^2)\hat{\Omega}\right]}{\partial\mu}(\hat{L}V) + \frac{\partial^2\left[(1-\mu^2)\hat{\Omega}\right]}{\partial\mu^2}(1-\mu^2)\frac{\partial V}{\partial\mu}, \quad (2.4)$$

the equation for the poloidal flow,

$$(\hat{\omega} - m\hat{\Omega})(\hat{L}V) = -\hat{\lambda}^2(\hat{L}S) + 2m\left[\frac{\partial(\mu\hat{\Omega})}{\partial\mu}V + (1-\mu^2)\frac{\partial\hat{\Omega}}{\partial\mu}\frac{\partial V}{\partial\mu}\right]$$
$$- 2\mu\hat{\Omega}(\hat{L}W) - 2(1-\mu^2)\frac{\partial(\mu\hat{\Omega})}{\partial\mu}\frac{\partial W}{\partial\mu} - 2m^2\frac{\partial\hat{\Omega}}{\partial\mu}W, \quad (2.5)$$

and the equation for entropy disturbances, $(\hat{\omega} - m\hat{\Omega})S = \hat{L}V$. In these equations, $\hat{\Omega} = \Omega/\Omega_0$ is the angular velocity normalized to its characteristic value Ω_0, $\hat{\omega} = \omega/\Omega_0$ is the normalized eigenvalue, $\mu = \cos\theta$ is cosine of the colatitude. Finite viscosity and thermal conductivity are neglected. They are not important for the hydrodynamical stability problem which is considered here.

The key parameter for the influence of stratification is the normalized length scale

$$\hat{\lambda} = \frac{N}{\Omega_0 k r}. \quad (2.6)$$

The ratio of N/Ω in radiation zones of solar-type stars is large. Figure 2.2 shows the ratio for the Sun and an A1 star with 2 $M_\odot$. It can be seen from (2.6) that the effect of stable stratification decreases with increasing radial wave number k of the disturbances. This is probably why the most rapidly growing modes of instability have small radial scales.

The linear stability equations allow two types of equatorial symmetry. We shall use the notation Sm and Am for modes having the potential W equatorially symmetric and antisymmetric, respectively; m indicates the azimuthal wavenumber in both cases. Note that the eigenmodes combine W of a given symmetry with S and V of *opposite* symmetry, that is, S and V are symmetric for A-modes and antisymmetric for S-modes. The velocity field for S-modes has symmetric u_θ and antisymmetric u_r and u_ϕ, and the converse for S-modes. The symmetry convention is the same as in Charbonneau et al. (1999).

Without rotation ($\Omega \to 0$), the above equations give the spectrum of eigenvalues,

$$\omega^2 = \frac{l(l+1)N^2}{r^2 k^2}, \quad l = 1, 2, \ldots, \tag{2.7}$$

which corresponds to g-modes of stable oscillations for positive N^2, or unstable convective modes for negative N^2 (in the short-wave approximation in radius). More relevant to the stability problem is another case of very large $\hat{\lambda}$. The case of $\hat{\lambda} \gg 1$ leads to the 2D approximation originally used by Watson (1981) in the stability problem of latitudinal differential rotation.

2.1.2
2D Approximation

The stability of differential rotation has been extensively studied in the 2D approximation of purely toroidal disturbances. Symmetry types and growth rates of the 2D unstable modes are known (Dziembowski and Kosovichev, 1987; Gilman and Fox, 1997; Charbonneau et al., 1999; Gilman and Dikpati, 2000; Cally, 2003; Cally, Dikpati, and Gilman, 2003; Dikpati, Cally, and Gilman, 2004), and the weakly nonlinear evolution of the instability has also been described (Garaud, 2001; Cally, 2001). The 2D approximation is justified for stable oscillations with not too short radial wavelengths so that the $\hat{\lambda}$ parameter (2.6) is large. Its validity in the case of instability is more problematic because the value of kr for growing modes is not *a priori* known, and it is usually found to be so large that $\hat{\lambda} \lesssim 1$. Nevertheless, the 2D approximation provides important insights which help to understand the results of the more complicated 3D formulation.

The ratio of N^2/Ω^2 is stellar radiation zones can be so large (see Figure 2.2) that $\hat{\lambda}^2$ (2.6) can be large as well in spite of the short-wave approximation $kr \gg 1$. In the limit of large $\hat{\lambda}$, the linear stability problem reduces to its 2D version. For the leading order in $\hat{\lambda}^2$, (2.5) gives $S = 0$. It then follows that also $V = 0$, and (2.4) reduces to the standard equation of 2D theory, that is

$$(\hat{\omega} - m\hat{\Omega})(\hat{L}W) = -m \frac{\partial^2 \left[(1-\mu^2)\hat{\Omega}\right]}{\partial \mu^2} W. \tag{2.8}$$

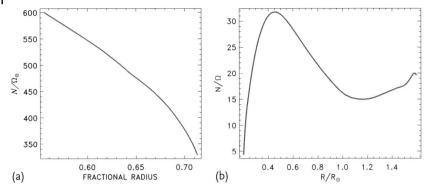

Figure 2.2 The ratio N/Ω in the upper part of the solar radiation zone (a), according to the model of Stix and Skaley (1990). The model convection zone includes the overshoot layer so that $N/\Omega_\odot$ is large immediately beneath the convection zone. The same for a star of 2 $M_\odot$, metallicity $Z = 0.02$ and rotation period of 10 days (b). The A1 star has radius 1.67 $R_\odot$ at the age of 100 Myr. From Kitchatinov and Rüdiger (2008a). Reproduced with permission © ESO.

For rigid rotation, this equation provides the spectrum of the r-modes,

$$\omega = m\Omega \left[1 - \frac{2}{l(l+1)}\right] \tag{2.9}$$

with $m = 1, 2, \ldots$ and $l = m, m+1, m+2, \ldots$ (Papaloizou and Pringle, 1978), the global vortices drifting westward in longitude. Instability of latitudinal shear can thus be understood as a loss of stability to excitation of the r-modes (Watts et al., 2003).

Equation (2.8) multiplied by the complex-conjugate potential W^* and integrated over latitude gives the relation

$$\int_{-1}^{1}(1-\mu^2)\left|\frac{\partial W}{\partial \mu}\right|^2 d\mu = m \int_{-1}^{1} \frac{|W|^2}{\hat\omega - m\hat\Omega} \frac{\partial^2\left[(1-\mu^2)\hat\Omega\right]}{\partial \mu^2} d\mu \, . \tag{2.10}$$

Its imaginary part is

$$\Im(\hat\omega) \int_{-1}^{1} \frac{|W|^2}{\left|\hat\omega - m\hat\Omega\right|^2} \frac{\partial^2\left[(1-\mu^2)\hat\Omega\right]}{\partial \mu^2} d\mu = 0 \, . \tag{2.11}$$

The *necessary* condition for instability, therefore, is that the second derivative $d^2[(1-\mu^2)\hat\Omega]/d\mu^2$ must change sign (Watson, 1981). For the angular velocity profile $\hat\Omega = 1-k\mu^{2n}$, this condition requires[1] $k > 1/(4n+1)$. The instability, therefore, needs a finite and not very small differential rotation. Already at this point it becomes clear that the instability cannot explain the solar tachocline. The latitudinal differential

---

1) We do not consider 'antisolar' rotation with negative k.

rotation can be reduced to the threshold amplitude for the instability but not to zero.

The instability is rather sensitive to details of the differential rotation profile: it should be $k > 0.2$ for instability of the $\cos^2 \theta$ profile ($n = 1$) of angular velocity, but the instability of $\cos^4 \theta$ profile may start with smaller value of $k > 1/9$. Watson (1981) considered the rotation law with $n = 1$ to find the smallest marginal value $k = 0.29$ for the A1 mode[2]. Dziembowski and Kosovichev (1987) and Charbonneau et al. (1999) found about two times smaller instability thresholds for S1 mode for the solar rotation law that combines $\cos^2 \theta$ and $\cos^4 \theta$ terms. The following 3D analysis provides even smaller instability thresholds.

2.1.3
Stability Maps

The latitudinal dependence of the angular velocity on the Sun may be approximated by the expression

$$\Omega = \Omega_0 \left\{ 1 - k\left[(1-f)\cos^2\theta + f\cos^4\theta\right] \right\}, \qquad (2.12)$$

where k is the normalized equator–pole difference of the rotation rate, and f is the fraction of the $\cos^4 \theta$ term contributing to the differential rotation.

In a 3D theory the radial displacements in the eigenvalue equations (2.4) and (2.5) are finite and the eigensolutions must be computed numerically. Figure 2.3 shows – as in 2D theory – that the results are sensitive to the details of the rotation law. For the $\cos^2 \theta$ profile (f = 0), the most unstable mode is A1. Apart from this mode, only S2 can be unstable at larger rotational shear. When the $\cos^4 \theta$-term is included (f = 0.5), four modes with $m = 1$ and $m = 2$ become unstable, and the S1 mode is the most easily excited. As it should be, the marginal amplitudes of differential rotation of 2D theory are reproduced for large $\hat{\lambda}$. It can be seen, however, that the most easily excited modes have $\hat{\lambda} < 1$, so allowance for radial displacements reduces the critical latitudinal shear for onset of the instability.

Even for large N/Ω, the small radial displacements are significant. The reason is that the most unstable modes have short radial scales. It can be seen from (2.2) that the assumption of zero radial velocity would exclude the entire class of poloidal disturbances. The ratio of horizontal ($u_\perp$) to radial velocities in a cell of poloidal flow of different radial (ℓ) and horizontal (H) scales can be estimated as $u_\perp/u_r \sim H/\ell$. Horizontal velocity of poloidal flow can thus remain important in spite of small u_r, if the radial scale is much shorter than the horizontal one. For the upper radiation zone of the Sun, the radial wave length can be estimated as $\lambda \simeq 10\hat{\lambda}$ in Mm. The most unstable disturbances in Figure 2.3, therefore, have the wavelengths $\lambda \simeq 6\,000$ km, which are shorter than the tachocline width $\simeq 30\,000$ km (Charbonneau et al., 1999) but not by much.

---

2) This is a bit larger than the true value of $k = 0.28$ found with higher numerical resolution.

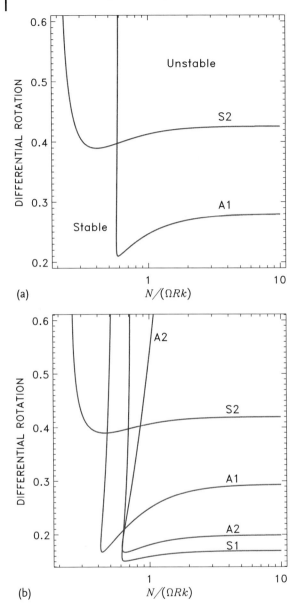

Figure 2.3 Threshold amplitude k of the latitudinal rotation law for the onset of the instability as function of $\hat{\lambda}$, (2.6). The neutral stability lines are marked by the symmetry type of the modes. The two panels show the results for $f = 0$ (a) and $f = 0.5$ (b) in rotation law (2.12). The instability region is above the lines. From Kitchatinov and Rüdiger (2009). Reproduced with permission © ESO.

The new feature of Figure 2.3, which the 2D theory cannot provide, is that stable stratification is necessary for the instability. The instability disappears when the stratification approaches the adiabatic one, $N \to 0$. It does not exist in convection zones therefore.

Figure 2.4 shows the stability map on the plane of the parameters k and f of the solar rotation law (2.12). The radial length scale $\hat{\lambda}$ was varied to find the maximum growth rates shown in this plot. The symmetry type of the eigenmodes, to which the maximum growth rates belong, varies over the stability map. The dashed line separates the regions of different symmetry types.

The latitudinal differential rotation of the outer Sun is well approximated by (2.12). Charbonneau et al. (1999) analyzed variations of the parameters of this rotation law with depth. The amplitude k of differential rotation and the fractional contribution f of the $\cos^4 \theta$-term vary slightly across the convection zone. The amplitude k varies smoothly inside the tachocline to decrease to practically zero at its base. The dimensionless f parameter falls sharply with depth near the base of the convection zone to vanish already in the upper tachocline (see Figure 10 in Charbonneau et al., 1999).

This behavior can be interpreted in terms of the instability. The differential rotation in the bulk of the convection zone is stable. In the region of penetrative convection near the base of the convection zone, the stratification changes to subadiabaticity and the instability can switch on. If it does, it reacts back on the differential rotation to change it towards the stable profile with $f = 0$. This interpretation is supported by the sense of latitudinal transport of angular momentum by the instability, illustrated in Figure 2.5. The linear theory cannot estimate the amplitude of the angular momentum flux, but its sign and latitudinal profile can be determined. The correlation $\langle u_\theta u_\phi \rangle$, after averaging in longitude, is negative in the northern

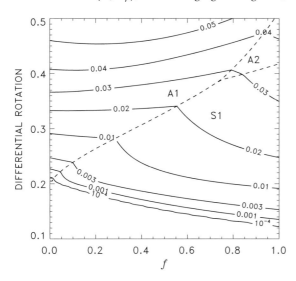

Figure 2.4 Isolines of the growth rates of the most rapidly growing modes on the plane of k and f in (2.12). Normalized growth rates $\Im(\omega)/\Omega_0$ are shown in the isoline gaps. The transition between different symmetry types of unstable modes is shown as a dashed line. From Kitchatinov and Rüdiger (2009). Reproduced with permission © ESO.

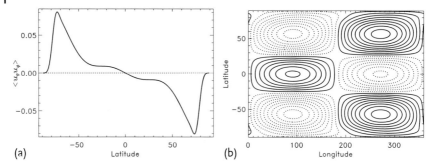

Figure 2.5 Meridional flux of angular momentum for slightly supercritical S1-mode for f = 0.2 (a). The averaging is over the longitude, the correlation is given in arbitrary units. Streamlines of toroidal flow in the same mode (b). The mode has a structure of transequatorial vortices in low latitudes. Reproduced with permission © ESO.

and positive in the southern hemisphere, indicating an angular momentum transport from the equator to the poles. The correlation takes its largest values at high latitudes, showing that rotational shear at those latitudes is primarily diminished.

The instability cannot, however, explain the existence of the tachocline. The critical amplitude of rotational shear remains finite, k > 0.1, whatever the rotation law is (Figure 2.4).

2.2
The Magnetic Tachocline

We proceed by discussing the possibility to explain the solar tachocline by an effect of a weak internal magnetic field subject to differential rotation in a stably stratified radiation zone. This concept, some alternative approaches and the relevant references are also described in the collected edition 'The Solar Tachocline' by Hughes, Rosner, and Weiss (2007).

2.2.1
A Planar Model

We start with rough estimations of the parameters expected for a magnetically originated tachocline. The estimations can be made with a simplified model in Cartesian geometry sketched in Figure 2.6a. The Cartesian coordinates x, y and z correspond to azimuth, latitude and depth beneath the convection zone, respectively. The plane $z = 0$ mimics the convection zone bottom where a shear flow $\bar{U} = (U(y, z), 0, 0)$ imitating the differential rotation is prescribed as $U = U_0 \sin ky$ at $z = 0$. Penetration of the shear flow into the radiative zone of $z > 0$ where a uniform (poloidal) field $\bm{B}_0$ along the y-axis is present imitates the magnetic tachocline. The shear flow produces a (toroidal) x-component $B(y, z)$ of the field so that $\bar{\bm{B}} = (B(y, z), B_0, 0)$. The parameters of the plane tachocline can be estimat-

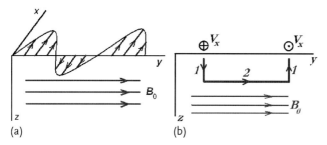

Figure 2.6 Simplified model of a plane magnetic tachocline. A shear flow is prescribed at $z = 0$ (a). Its penetration into the region $z > 0$ with magnetic field mimics the magnetic tachocline. Circulation of the 'angular' x-component of momentum in a steady magnetic tachocline (b). Viscous in and outflow of momentum (1) is linked by horizontal transport (2) by magnetic stress.

ed considering circulation of the 'angular' x-component of momentum illustrated by Figure 2.6b. The inflow of momentum along the z-axis in the region of positive U_x is balanced by the momentum outflow in the region of negative U_x. The in and outflows in the regions marked by '1' in Figure 2.6 are due to the viscous stress $\rho \nu U_0/D$, where D is the tachocline thickness. The momentum fluxes in z-direction per unit length in x and half-wave length $L = \pi/k$ in y can, therefore, be estimated as $\rho \nu L U_0/D$. The momentum circulation is closed on the part '2' of the chain by the magnetic stress, $B_x B_y/\mu_0$, that is, $B B_0 D/\mu_0 \sim \rho \nu L U_0/D$. The amplitude B of the 'toroidal' x-component of the field in this relation is still uncertain. It can be estimated from the induction equation by the product of the rate $B_0 U_0/L$ of the field production by the shear flow and the time D^2/η of the diffusive escape of the field from the tachocline: $B \sim B_0(U_0/L)(D^2/\eta)$. We finally get the estimations of the tachocline thickness and the amplitude of the resulting toroidal field in the tachocline as

$$D \sim \frac{L}{\sqrt{\text{Ha}}}, \quad B \sim \sqrt{\mu_0 \rho \text{Pm}}\, U_0, \qquad (2.13)$$

where $\text{Ha} = B_0 L/\sqrt{\mu_0 \rho \nu \eta}$ is the Hartmann number and $\text{Pm} = \nu/\eta$ is the magnetic Prandtl number.

The qualitative picture can be confirmed by the analytic solution of the plane tachocline equations

$$\nu \left(\frac{\partial^2 U}{\partial y^2} + \frac{\partial^2 U}{\partial z^2} \right) + \frac{B_0}{\mu_0 \rho} \frac{\partial B}{\partial y} = 0, \quad \eta \left(\frac{\partial^2 B}{\partial y^2} + \frac{\partial^2 B}{\partial z^2} \right) + B_0 \frac{\partial U}{\partial y} = 0, \quad (2.14)$$

which directly result from the general MHD equations

$$\rho \left[\frac{\partial U}{\partial t} + (U \cdot \nabla) U \right] = -\nabla P + \rho \nu \Delta U + \frac{1}{\mu_0} \text{curl}\, B \times B \qquad (2.15)$$

for the flow of the fluid and

$$\frac{\partial B}{\partial t} = \text{curl}(U \times B) + \eta \Delta B \qquad (2.16)$$

for the magnetic fields. The top boundary condition for the flow is given above. For the toroidal field we impose the vacuum condition $B = 0$ at $z = 0$ motivated by the very large turbulent magnetic diffusivity inside the convection zone compared with the microscopic diffusivity of the radiative interior. The remaining two conditions require U and B to vanish for $z \to \infty$. The solutions of (2.14) are

$$U(y, z) = U_0 \exp(-\lambda_1 k z) \cos \lambda_2 k z \sin k y ,$$
$$B(y, z) = \sqrt{\mu_0 \rho} \sqrt{\text{Pm}}\, U_0 \exp(-\lambda_1 k z) \sin \lambda_2 k z \cos k y , \qquad (2.17)$$

where

$$\lambda_1 = (1 + \text{Ha}^2)^{1/4} \cos\left(\frac{1}{2} \arctan \text{Ha}\right) ,$$
$$\lambda_2 = (1 + \text{Ha}^2)^{1/4} \sin\left(\frac{1}{2} \arctan \text{Ha}\right) \qquad (2.18)$$

depend on the Hartmann number defined in terms of the wave number k,

$$\text{Ha} = \frac{B_0}{k \sqrt{\mu_0 \rho \nu \eta}} . \qquad (2.19)$$

With $\text{Ha} = 0$, (2.18) gives $\lambda_1 = 1$, $\lambda_2 = 0$, and (2.17) converts into the nonmagnetic solution, $B = 0$, $U = U_0 e^{-kz} \sin k y$. The shear flow penetrates deep inside the $z > 0$ region in this case. The Hartmann number is, however, expected to be large. With solar parameters ($\rho \simeq 0.2\,\text{g/cm}^3$, $\nu \simeq 15\,\text{cm}^2/\text{s}$, $\eta \simeq 3000\,\text{cm}^2/\text{s}$) one finds $\text{Ha} \simeq 10^7 B_0$ with B_0 in gauss just below the convection zone. For large Ha, $\lambda_1 = \lambda_2 = \sqrt{\text{Ha}/2}$, and the solution (2.17) provides the two important consequences:

1. The tachocline thickness, D, is strongly reduced compared to its nonmagnetic value, D_0,

$$D = D_0 \sqrt{\frac{2}{\text{Ha}}} \simeq \sqrt{\frac{D_0}{B_0}} (\mu_0 \rho \nu \eta)^{1/4} . \qquad (2.20)$$

2. The amplitude of the toroidal magnetic field in the steady tachocline does not depend on the poloidal field strength, that is $B = U_0 \sqrt{\mu_0 \rho \text{Pm}}$. In terms of the Alfvén velocity, $V_A = B/\sqrt{\mu_0 \rho}$, one finds $V_A = \sqrt{\text{Pm}}\, U_0$ predicting $B \lesssim 1000\,\text{G}$ for the Sun.

The tachocline is so thin because its extension in the z-direction is only due to viscous stress while the smoothing in the y-direction is provided by the much stronger Maxwell stress. The ratio of magnetic to kinetic energy in the tachocline, therefore, equals Pm, which is a small number for microscopic diffusion. We shall see that the estimation (2.20) generally agrees with a numerical model in spherical geometry, though the model provides weaker toroidal field $\sim 200\,\text{G}$ in the solar tachocline (see Mestel and Weiss, 1987).

According to (2.20) even a weak poloidal field of only $B_0 \sim 10^{-3}\,\text{G}$ can reduce the tachocline thickness below 5% of the solar radius. A stronger field will tend to produce a still thinner tachocline. It is very probable, however, that an instability of radial rotational shear sets on when the tachocline is too thin, and the tachocline thickness cannot be reduced below the critical value for the instability.

Equation (2.20) also holds when B_0 is of some other origin than being a fossil field in the radiative core. Forgács-Dajka and Petrovay (2002) argued that the same effect can be produced by the poloidal field of the solar cycle which diffuses into the core due to turbulent mixing. In this case, however, the thickness of the tachocline would be cycle-dependent.

2.2.2
Magnetic Field Confinement by Meridional Flow

In order to produce the tachocline, the magnetic field should be almost horizontal in the tachocline region. In other words, the internal magnetic field should be confined in the radiation zone (MacGregor and Charbonneau, 1999). Finite magnetic diffusion can, however, change the field towards an open structure (Brun and Zahn, 2006). We shall see that the confined field geometry required for the tachocline formation can be produced by a small penetration of meridional flow from the convection zone into the radiative interior. If the electric conductivity is high enough, then the poloidal field lines in the penetration zone become parallel to the meridional flow (Mestel, 1999).

The global poleward flow observed on the solar surface persists to a depth of at least 140 000 km in the Sun (see Section 1.1.3). There must be a return flow towards the equator somewhere deeper which is not yet observed. The theory of differential rotation predicts an equatorward flow of some meters per second at the base of the convection zone (see Chapter 1). This flow can penetrate beneath the bottom of the convection zone into the radiative interior (Garaud and Brummell, 2008).

The penetration results from the viscous drag imposed by the meridional flow at the base of the convection zone on the fluid beneath, and this is opposed by the Coriolis force. Gilman and Miesch (2004) estimated the penetration depth to be at the Ekman scale $D_{\text{pen}} \sim \sqrt{\nu/\Omega}$. Figure 2.7 shows the penetration depth in dependence on the molecular viscosity ν beneath the convection zone. The model cannot work with the very small microscopic viscosity of the radiation zone, but the theoretical Ekman depth D_{pen} is closely reproduced. Figure 2.7a is well approximated by $D_{\text{pen}} \simeq 2.3\sqrt{\nu/\Omega}$. The penetration layer is, therefore, very thin for any reasonable value of turbulent viscosity, $D_{\text{pen}} < 1000$ km. To date, the shallow penetration cannot be resolved in 3D numerical simulations (Strugarek, Brun, and Zahn, 2011).

However, the shallow penetration can strongly influence the geometry of the internal poloidal field. The rate of meridional field production by the penetrating flow can be estimated as $B_r U^m / D_{\text{pen}}$, where B_r is the radial field and $U^m \sim 10$ m/s is the flow amplitude at the base of the convection zone. The meridional field B_θ in the penetration layer can be estimated by the product of this rate with the diffusion

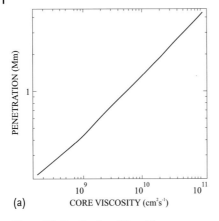
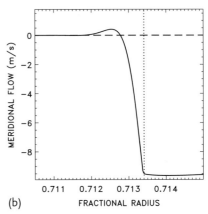

Figure 2.7 Depth of meridional flow penetration into the radiative core as a function of the viscosity below the convection zone (a). Depth profile of the meridional velocity in the penetration region at 45° latitude for $\nu = 1.3 \times 10^9$ cm²/s (b). The bottom of the convection zone is marked by the dotted vertical line. From Kitchatinov and Rüdiger (2006). Reproduced with permission © ESO.

time D_{pen}^2/η across the layer, that is $B_\theta \sim \text{Rm}\, B_r$, with

$$\text{Rm} = \frac{D_{\text{pen}} U^m}{\eta} \simeq 2.3 \frac{U^m \text{Pm}}{\sqrt{\nu \Omega}}. \tag{2.21}$$

For solar parameters, the magnetic Reynolds number can be estimated as $\text{Rm} \sim 10^6 \, \text{Pm}/\sqrt{\nu}$ with ν in cm²/s. The number is large, $\text{Rm} \sim 10^3$ for microscopic diffusion, and it remains above this value with eddy diffusivities ($\text{Pm} \sim 1$) up to 10^6 cm²/s. According to (2.21), the large Rm means that the latitudinal field inside the penetration layer is large compared to the radial field, so that the field has the confined geometry required for the tachocline formation. Only with large eddy diffusion $\nu \geq 10^{12}$ cm²/s does the influence of penetration on the internal field geometry become weak.

The effect of the shallow penetration on the internal poloidal field can be accounted for via the boundary condition imposed on the top of the radiation core. The condition can be formulated in terms of the potential A of the poloidal field

$$B_p = \text{curl}\left(\frac{A e_\phi}{r \sin \theta}\right) \tag{2.22}$$

to read

$$r \frac{\partial A}{\partial r} + \text{Rm}\, \hat{U}(\theta) \frac{\partial A}{\partial \theta} = r \left(\frac{\partial A}{\partial r}\right)_{\text{vac}} \quad \text{at} \quad r = R_{\text{in}}, \tag{2.23}$$

where R_{in} is the radius of the boundary, $\hat{U}(\theta)$ is the normalized profile of meridional flow at the base of the convection zone and Rm is defined by (2.21). The term

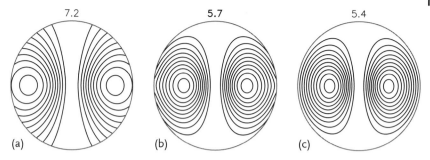

Figure 2.8 Field lines of the longest-living dipolar modes of the internal field for the Reynolds numbers varying as Rm = 0, 10, 1000 from left to right. The decay times in 7.2 (a), 5.7 (b) and 5.4 Gyr (c). The decay times exceed the age of the Sun. Reproduced with permission © ESO.

on the right-hand side of (2.23) is the radial derivative at the interface with vacuum,

$$\left(\frac{\partial A}{\partial r}\right)_{\text{vac}} = -\frac{\sin\theta}{r} \sum_{n=1}^{N} n A_n(r) P_n^1(\cos\theta),$$

$$A(r,\theta) = \sin\theta \sum_{n=1}^{N} A_n(r) P_n^1(\cos\theta). \tag{2.24}$$

The 'vacuum derivative' is formulated in terms of the amplitudes A_n of the expansion in the (associated) Legendre polynomials P_n^1. In the absence of the penetrating flow the condition (2.23) reduces to the vacuum boundary condition $r\partial A_n/\partial r + n A_n = 0$.

The penetrating meridional flow is accounted for via the boundary condition (2.23). Below the penetration layer, the meridional flow is absent and the weak internal poloidal field can be described by solving the eigenvalue problem for the diffusion equation

$$-\frac{A}{\tau_{\text{Ohm}}} = \eta \frac{\partial^2 A}{\partial r^2} + \eta \frac{\sin\theta}{r^2} \frac{\partial}{\partial \theta}\left(\frac{1}{\sin\theta}\frac{\partial A}{\partial \theta}\right). \tag{2.25}$$

The solution defines the internal field properties. The eigenfunctions define the internal field structure and the eigenvalue τ_{Ohm} is the Ohmic decay time of the eigenmodes. The linear theory cannot, however, define the amplitude of the field. The poloidal field amplitude, B_p, is a free parameter of the model. The eigenvalue problem can be solved for the microscopic magnetic diffusion of the solar radiation zone (see Figure 2.11). The solutions were obtained with the meridional flow profile $\hat{U} \propto P_2^1(\cos\theta)$.

Figure 2.8 shows how the structure of the internal field changes towards the confined geometry when the Reynolds number (2.21) increases. The internal field without a penetrating flow, Rm = 0, has an open structure. However, even a moderate flow with Rm = 10 changes the field considerably towards a confined geometry. The field for the solar value of Rm = 1000 is almost totally confined.

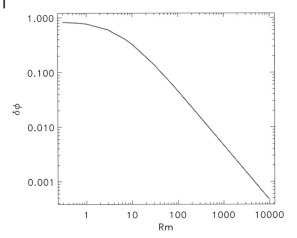

Figure 2.9 Confinement parameter (2.26) for the longest-living dipolar mode as a function of the magnetic Reynolds number (2.21). The dependence is close to $\delta\phi = 5/\text{Rm}$ for large Rm. Less than 1% of the poloidal field flux escapes the radiation zone for Rm > 1000. From Kitchatinov and Rüdiger (2006). Reproduced with permission © ESO.

The degree of confinement can be estimated with the parameter

$$\delta\phi = \frac{\max|A(r,\theta)|_{r=R_{\text{in}}}}{\max|A(r,\theta)|_{r\leq R_{\text{in}}}}. \tag{2.26}$$

Its meaning can be understood as follows. The value of A of (2.22) at a given point (r,θ) is proportional to magnetic flux Φ pervading the longitudinal circle going through this point, $\Phi = 2\pi A$. The value of $\max|A(r,\theta)|_{r=R_{\text{in}}}$, therefore, estimates the magnetic flux through the interface between the radiative core and the convection zone. Similarly, $\max|A(r,\theta)|_{r\leq R_{\text{in}}}$ is the characteristic value of the magnetic flux of the internal field. The ratio (2.26), therefore, estimates the fraction of magnetic flux leaving the core. Figure 2.9 shows the confinement parameter (2.26) for the longest-living dipolar mode in dependence on the Reynolds number; $\delta\phi$ steadily decreases for increasing Rm.

Figures 2.8 and 2.9 refer to the most slowly decaying dipolar modes. Other normal modes have smaller scales in radius or in latitude and also shorter decay times. The three modes following the dipole in order of decreasing lifetimes are shown in Figure 2.10. An internal field can be represented as a superposition of freely decaying normal modes. Relative weights of large-scale slowly decaying modes will increase with time. The decay times of Figures 2.8 and 2.10 are long enough to allow tachocline computations with steady poloidal fields.

The Ohmic decay may not be the only process that changes the internal field. The field may be subject to magnetic instabilities. Wright (1973) and Markey and Tayler (1973) found that poloidal fields in stellar radiation zones are unstable to almost horizontal adiabatic disturbances. The instability is, however, very slow for the fields of several milli-gauss required for the tachocline formation. Its growth time is estimated as $\Omega/\Omega_{\text{A}}^2$ for the case of weak ($\Omega_{\text{A}} < \Omega$) fields (Pitts and Tayler,

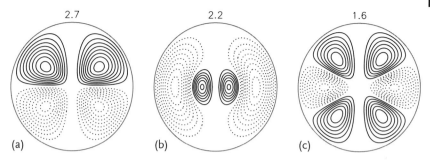

Figure 2.10 Normal modes of the internal field following the dipole in order of decreasing lifetimes. The decay times in 2.7 (a), 2.2 (b), and 1.6 Gyr (c) are shown. The confinement parameter (2.26) varies as 0.0051, 0.0073, 0.0026 from left to right. Rm = 1000. Reproduced with permission © ESO.

1985). The growth time is large compared to the characteristic time, r/V_A, of the tachocline formation and even larger than the age of the Sun.

The direction of the penetrating meridional flow does not play a role for sufficiently large Rm. Flows from equator to poles and from poles to equator yield equally efficient confinement of the internal field. The effect can be understood as a magnetic field expulsion from the region of circulating motion (Weiss, 1966). Another mechanism for the internal field confinement is provided by the diamagnetic pumping of mean field near the base of the convection zone (Garaud, 2007; Garaud and Garaud, 2008; Kitchatinov and Rüdiger, 2008b).

2.2.3
Tachocline Model in Spherical Geometry

A steady tachocline configured by an axisymmetric magnetic field is now considered. The poloidal field is not affected by the differential rotation and can be considered as given. Similar to (2.14) in plane geometry, we have two equations for the toroidal field B and angular velocity Ω,

$$\frac{\eta}{r}\frac{\partial}{\partial \theta}\left[\frac{1}{\sin\theta}\frac{\partial(B\sin\theta)}{\partial \theta}\right] + \frac{\partial}{\partial r}\left[\eta\frac{\partial(Br)}{\partial r}\right] = \frac{\partial \Omega}{\partial \theta}\frac{\partial A}{\partial r} - \frac{\partial \Omega}{\partial r}\frac{\partial A}{\partial \theta},$$

$$\frac{\rho v}{\sin^3\theta}\frac{\partial}{\partial \theta}\left(\sin^3\theta\frac{\partial \Omega}{\partial \theta}\right) + \frac{1}{r^2}\frac{\partial}{\partial r}\left(r^4\rho v\frac{\partial \Omega}{\partial r}\right)$$
$$= \frac{1}{\mu_0 r^2 \sin^3\theta}\left[r\frac{\partial A}{\partial r}\frac{\partial(B\sin\theta)}{\partial \theta} - \sin\theta\frac{\partial A}{\partial \theta}\frac{\partial(Br)}{\partial r}\right]. \qquad (2.27)$$

The differential rotation at the bottom of the convection zone imposes the angular velocity profile on the top boundary, $\Omega = 2.9(1 - 0.15\cos^2\theta)$ μrad/s, (Charbonneau et al., 1999). The remaining boundary conditions are the vacuum condition for the toroidal field on the top and regularity conditions for the field and the angular velocity at the center,

$$B|_{r=R_{in}} = B|_{r=0} = \left.\frac{\partial \Omega}{\partial \theta}\right|_{r=0} = 0. \qquad (2.28)$$

We do not expect turbulence in the radiative core. Therefore the microscopic magnetic diffusivity $\eta = 10^{13}\, T^{-3/2}$ cm²/s (Spitzer, 1962) is used and the viscosity $\nu = \nu_{\rm mol} + \nu_{\rm rad}$ with

$$\nu_{\rm mol} = 1.2 \times 10^{-16} \frac{T^{5/2}}{\rho}, \quad \nu_{\rm rad} = 2.5 \times 10^{-25} \frac{T^4}{\kappa \rho} \qquad (2.29)$$

(both in cm²/s, Kippenhahn and Weigert, 1994) including molecular ($\nu_{\rm mol}$) and radiative ($\nu_{\rm rad}$) parts, where κ is the opacity. The diffusivity profiles are shown in Figure 2.11. The tachocline equations can be solved numerically with these diffusivity profiles and poloidal field eigenmodes of Section 2.2.2. The poloidal field amplitude $B_{\rm p}$ remains a free parameter of the model.

Figure 2.12 shows the angular velocity distribution in the radiation zone for several $B_{\rm p}$ values. There is no tachocline without a magnetic field. Viscosity, irrespective of how large it is, cannot form a tachocline. However, even a weak poloidal field has strong effects. A field of 10^{-4} G considerably changes the angular velocity distribution. Poloidal fields of milligauss strength suffice to produce a rather thin

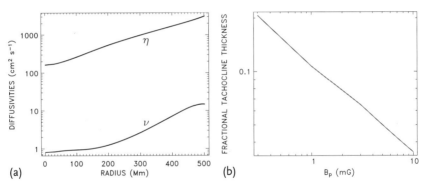

Figure 2.11 Microscopic magnetic diffusivity (Spitzer) and viscosity (2.29) in the solar radiation zone (a). Tachocline thickness in units of solar radius as function of the amplitude $B_{\rm p}$ of the dipolar internal poloidal field, Rm = 1000 (b).

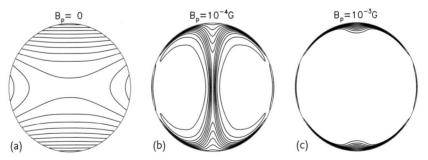

Figure 2.12 Angular velocity isolines in the radiation zone computed with dipolar internal field of Figure (2.9) for the magnetic Reynolds number (2.21) Rm = 1000. Corresponding amplitudes $B_{\rm p}$ of the poloidal field are marked (a–c). The outer circle is the bottom of the convection zone. From Rüdiger and Kitchatinov (1997).

tachocline. Note that B_p is the maximum strength of the poloidal field inside the core. The maximum is attained in the core center. The field in the tachocline region is weaker still.

The dependence of the tachocline thickness D_{tach} on the poloidal field amplitude is shown in Figure 2.11b. The thickness is defined as the depth of exponential decrease of the equator–pole difference of the angular velocity. D_{tach} falls below 5% of the solar radius for $B_p > 5$ mG. The plot is very close to the dependence $D_{tach} \sim B_p^{-1/2}$ envisaged in (2.20). The decrease of the tachocline thickness with increasing B_p may eventually be stopped by the onset of some nonaxisymmetric instability which cannot be seen in our axisymmetric tachocline model. It is not clear at the moment whether the solar tachocline thickness is marginal for some instability of radial rotational shear.

Another prediction of Section 2.2.1 is the near-constancy of the toroidal field amplitude, which is also confirmed by the numerical model. The toroidal field amplitude, $B_t \simeq 100-200$ G, hardly varies while the poloidal field amplitude changes by two orders of magnitude.

Figures 2.11b and 2.12 are valid for magnetic Reynolds number Rm = 1000. Rm = 100 already suffices for tachocline formation. For smaller Rm the poloidal field remains 'too open' to form the tachocline. In other words, a meridional flow not slower than 1 m/s at the base of convection zone is necessary for the magnetic tachocline theory to apply. Shallow penetration of the meridional flow from the convection zone into the radiative core influences the internal field geometry strongly enough and in such a way that the field becomes appropriate for the tachocline formation.

Not only the dipolar poloidal field but also higher-order modes of Figure 2.10 can produce a tachocline, provided that the fields have the confined geometry which is always the case for meridional flows with Rm $\gtrsim$ 100.

An open problem of the magnetic tachocline theory is that the times of relaxation to the steady tachocline solutions are too long. The regions near the O-points of the internal poloidal field are not connected by the field lines to the convection zone. The spin-down models with such magnetic fields show the vicinities of the O-points to rotate too fast at the solar age (Rüdiger and Kitchatinov, 1996). Only if the core viscosity is increased to the value of 10^5 cm^2/s or larger can a contradiction with helioseismology be avoided. A diffusivity of chemical species comparable with this turbulent viscosity, however, is in conflict with the observed lithium abundance. Denissenkov (2010) suggested that anisotropy of turbulent diffusion can resolve the problem. The radial diffusivity can be smaller than the upper bound of 10^3 cm^2/s imposed by the observed lithium content, but much larger horizontal viscosity can connect the O-points to slowly rotating regions across the field lines. Kitchatinov and Brandenburg (2012) found that if turbulence is present in the radiation zone, it should indeed be highly anisotropic,

$$\frac{\langle u_\phi^2 \rangle}{\langle u_r^2 \rangle} \approx \frac{\tau_{corr}^2 N^4}{\Omega^2}, \tag{2.30}$$

where τ_{corr} is the correlation time of the turbulence. The origin of the hypothetical turbulence in the radiation zone is, however, uncertain. What is clear though is that correlated magnetic fluctuations are able to transport angular momentum but without contributing to the mixing of chemicals (see Section 2.3.4).

Another possibility for resolving the O-point problem is a slight deviation of the internal field from axial symmetry. Ferraro's law (Ferraro, 1937) for such a field implies uniform rotation.

2.3
Stability of Toroidal Fields

Toroidal fields of the magnetic tachocline model are much stronger than the poloidal fields. We proceed by analyzing the toroidal field stability.

Which fields the stellar radiative cores can possess is mainly a matter of their stability. The resistive decay is so slow that primordial magnetic fields can easily be stored in the radiative cores (Cowling, 1945). Whether the fields of 10^5 G that can influence solar internal g-modes (Rashba, Semikoz, and Valle, 2006) or still stronger fields that can explain neutrino oscillations (Burgess *et al.*, 2003) can indeed survive inside the Sun mainly depends on their stability.

Toroidal fields can be expected to be the principal component of the internal field because the toroidal field is easily produced by differential rotation.

Among several instabilities to which the fields can be subjected (Acheson, 1978), the current-induced pinch-type instability of toroidal fields (Tayler, 1973) is probably the most relevant one for slowly rotating stars. This is because the instability proceeds via almost horizontal displacements, thus avoiding the stabilizing effect of buoyancy. Radial displacements, though necessary for the Tayler instability, are relatively small. Under the influence of differential rotation even current-free toroidal fields can become unstable (AMRI, see Section 5.6).

The presented stability analysis extends the treatment of hydrodynamical instability of differential rotation in Section 2.1 by including the toroidal magnetic field.

2.3.1
Equations

We start formulating the linear equations for small disturbances in a rotating radiation zone with toroidal magnetic field. The toroidal field can be expressed in terms of the Alfvén angular frequency Ω_A,

$$\bar{B} = e_\phi r \sin\theta \sqrt{\mu_0 \rho}\, \Omega_A(r, \theta) \qquad (2.31)$$

to ensure that the field vanishes at the rotation axis.

Similar to (2.2) for the velocity perturbations, the disturbances $\boldsymbol{b}$ of the magnetic field are expressed in terms of two scalar potentials P_m and T_m for the poloidal and toroidal magnetic disturbances

$$\boldsymbol{b} = \frac{1}{r}\left[\frac{1}{r}\hat{L} P_m, -\left(\frac{1}{\sin\theta}\frac{\partial T_m}{\partial \phi} + \frac{\partial^2 P_m}{\partial r \partial \theta}\right), \frac{\partial T_m}{\partial \theta} - \frac{1}{\sin\theta}\frac{\partial^2 P_m}{\partial r \partial \phi}\right] \quad (2.32)$$

($\hat{L}$ defined by (2.3)). The radial scale of the disturbances is assumed to be short compared to the local radius, to include the dependence on radius in the Fourier modes of the disturbances. Our stability analysis is, therefore, local in radius. It will be confirmed *a posteriori* that the most unstable modes do indeed have short radial scales. However, the modes occupy the largest available scale in the horizontal dimensions. Accordingly, the analysis remains global in horizontal dimensions.

The equations are written in terms of the normalized variables

$$A = \frac{k}{\Omega_0 r^2 \sqrt{\mu_0 \rho}} P_m, \quad B = \frac{1}{\Omega_0 r^2 \sqrt{\mu_0 \rho}} T_m, \quad \hat{\Omega}_A = \frac{\Omega_A}{\Omega_0}, \quad (2.33)$$

complementary to the hydrodynamical variables (2.1) of Section 2.1. The toroidal flow equation now becomes

$$(\hat{\omega} - m\hat{\Omega})(\hat{L} W) = -i\frac{\epsilon_\nu}{\hat{\lambda}^2}(\hat{L} W) - m\hat{\Omega}_A(\hat{L} B)$$

$$+ \frac{\partial^2 \left[(1-\mu^2)\hat{\Omega}_A\right]}{\partial \mu^2}\left[m B - (1-\mu^2)\frac{\partial A}{\partial \mu}\right] - \frac{\partial \left[(1-\mu^2)\hat{\Omega}_A\right]}{\partial \mu}(\hat{L} A)$$

$$- \frac{\partial^2 \left[(1-\mu^2)\hat{\Omega}\right]}{\partial \mu^2}\left[m W - (1-\mu^2)\frac{\partial V}{\partial \mu}\right] + \frac{\partial \left[(1-\mu^2)\hat{\Omega}\right]}{\partial \mu}(\hat{L} V), \quad (2.34)$$

with $\hat{\lambda}$ defined in (2.6). The new feature compared to Section 2.1 is that finite diffusivities are now included via the parameters

$$\epsilon_\nu = \frac{\nu N^2}{\Omega_0^3 r^2}, \quad \epsilon_\eta = \frac{\eta N^2}{\Omega_0^3 r^2}, \quad \epsilon_\chi = \frac{\chi N^2}{\Omega_0^3 r^2}. \quad (2.35)$$

Finite diffusion is important for the magnetic instability. We shall see that, in contrast to the Watson problem of hydrodynamic stability, 2D magnetic instability does not exist. Radial displacements are necessary for the instability. In this case, thermal diffusion produces a destabilizing effect by reducing buoyancy. In the ideal case of zero diffusion, the most unstable modes have indefinitely short radial scales. The radial scale is, therefore, controlled by finite diffusion.

Apart from (2.34) for the toroidal flow, the complete system includes four other equations of the eigenvalue problem. These are the equation for poloidal flow

$$(\hat{\omega} - m\hat{\Omega})(\hat{L}V) = -\hat{\lambda}^2(\hat{L}S) - i\frac{\epsilon_\nu}{\hat{\lambda}^2}(\hat{L}V) - m\hat{\Omega}_A(\hat{L}A)$$

$$- 2\mu\hat{\Omega}(\hat{L}W) - 2(1-\mu^2)\frac{\partial(\mu\hat{\Omega})}{\partial\mu}\frac{\partial W}{\partial\mu} - 2m^2\frac{\partial\hat{\Omega}}{\partial\mu}W$$

$$+ 2\mu\hat{\Omega}_A(\hat{L}B) + 2(1-\mu^2)\frac{\partial(\mu\hat{\Omega}_A)}{\partial\mu}\frac{\partial B}{\partial\mu} + 2m^2\frac{\partial\hat{\Omega}_A}{\partial\mu}B$$

$$+ 2m\left[\frac{\partial(\mu\hat{\Omega})}{\partial\mu}V + (1-\mu^2)\frac{\partial\hat{\Omega}}{\partial\mu}\frac{\partial V}{\partial\mu}\right]$$

$$- 2m\left[\frac{\partial(\mu\hat{\Omega}_A)}{\partial\mu}A + (1-\mu^2)\frac{\partial\hat{\Omega}_A}{\partial\mu}\frac{\partial A}{\partial\mu}\right], \quad (2.36)$$

the equation for the toroidal magnetic disturbances,

$$\hat{L}\left[(\hat{\omega} - m\hat{\Omega})B\right] = -i\frac{\epsilon_\eta}{\hat{\lambda}^2}(\hat{L}B) - m\hat{L}\left(\hat{\Omega}_A W\right) - m^2\frac{\partial\hat{\Omega}}{\partial\mu}A + m^2\frac{\partial\hat{\Omega}_A}{\partial\mu}V$$

$$- \frac{\partial}{\partial\mu}\left[(1-\mu^2)^2\frac{\partial\hat{\Omega}}{\partial\mu}\frac{\partial A}{\partial\mu}\right] + \frac{\partial}{\partial\mu}\left[(1-\mu^2)^2\frac{\partial\hat{\Omega}_A}{\partial\mu}\frac{\partial V}{\partial\mu}\right], \quad (2.37)$$

the equation for the poloidal magnetic disturbances,

$$(\hat{\omega} - m\hat{\Omega})(\hat{L}A) = -i\frac{\epsilon_\eta}{\hat{\lambda}^2}(\hat{L}A) - m\hat{\Omega}_A(\hat{L}V), \quad (2.38)$$

and the equation for entropy disturbances,

$$(\hat{\omega} - m\hat{\Omega})S = -i\frac{\epsilon_\chi}{\hat{\lambda}^2}S + \hat{L}V. \quad (2.39)$$

The system (2.34), (2.36)–(2.39) can be solved numerically for given profiles of both angular velocity and toroidal field.

The eigenmodes possess certain equatorial symmetries provided that the background toroidal field does so as well. As in Section 2.1, we shall use the notations Sm and Am for the modes with equatorially symmetric and antisymmetric toroidal flow potential W; m is the azimuthal wave number. The symmetry properties of the eigenmodes depend on the equatorial symmetry of the toroidal field. For the case of the toroidal field antisymmetric about the equator we have

Sm modes: W, A symmetric ,   V, B, S antisymmetric ,
Am modes: W, A antisymmetric ,   V, B, S symmetric .

In this case, Sm modes have the magnetic field **b** mirror-symmetric about the equatorial plane (symmetric b_ϕ, b_r and antisymmetric b_θ) and antisymmetric flow **u** (symmetric u_θ and antisymmetric u_ϕ, u_r). Am modes have mirror-symmetric flows u and symmetric fields b. This is the same symmetry convention as in Gilman, Dikpati, and Miesch (2007). For Ω_A symmetric with respect to the equator, the symmetry notations imply

Sm modes: W, B symmetric,   V, A, S antisymmetric,
Am modes: W, B antisymmetric,   V, A, S symmetric.

It can also be seen that the equation system possesses certain symmetry with respect to the sign reversal of the azimuthal wave number m. The equations are invariant under the simultaneous transformations $(m, \hat{\omega}, W, B, A, S) \to (-m, -\hat{\omega}^*, -W^*, V^*, -B^*, A^*, -S^*)$, where the asterisk indicates the complex conjugate. The symmetry means that there is no preferred handedness of the eigenmodes. If there is an eigenmode with certain m and a pattern of kinetic helicity $u \cdot \text{curl}\, u$ or current helicity $b \cdot \text{curl}\, b$, then a mode with the azimuthal wave number $-m$ exists which has the same growth rate $\omega_{\text{gr}} = \Im(\omega)$ and helicity patterns of opposite sign.

2.3.2
Nonexistence of 2D Magnetic Instabilities

Similar to the hydrodynamical analysis of Section 2.1, a 2D approximation can be reproduced in the case of very large $\hat{\lambda}$. For the leading order of $\hat{\lambda}^2$, (2.36) gives $S = 0$. It then follows successively from (2.39) and (2.38) that the poloidal flow vanishes, $V = 0$, and the poloidal magnetic disturbances also do so, that is $A = 0$. The equation system reduces to the two coupled equations

$$(\hat{\omega} - m\hat{\Omega})(\hat{L}W) = -m \frac{\partial^2 \left[(1-\mu^2)\hat{\Omega}\right]}{\partial \mu^2} W$$

$$+ m \frac{\partial^2 \left[(1-\mu^2)\hat{\Omega}_A\right]}{\partial \mu^2} B - m\hat{\Omega}_A(\hat{L}B),$$

$$(\hat{\omega} - m\hat{\Omega})B = -m\hat{\Omega}_A W, \tag{2.40}$$

describing purely toroidal disturbances in decoupled spherical shells (Gilman and Fox, 1997).

For the case of uniform rotation, the hydrodynamical 2D approximation provides the spectrum (2.9) of r-modes. If not only angular velocity Ω but also the Alfvén frequency Ω_A is uniform, the solution of (2.40) in terms of the Legendre polynomials $W, B \sim P_l^m(\mu)$ can be found as

$$\frac{\omega}{m} = \Omega \left[1 - \frac{1}{l(l+1)}\right] \pm \sqrt{\Omega_A^2 \left[1 - \frac{2}{l(l+1)}\right] + \frac{\Omega^2}{l^2(l+1)^2}} \tag{2.41}$$

describing the magnetically modified r-modes.

The case of uniform rotation is interesting because it excludes hydrodynamical instabilities[3]. Any instabilities, which we may find in this case, are magnetic by origin.

We shall see that a sufficiently strong field with uniform Ω_A is unstable in 3D. However, the spectrum (2.41) only shows stable oscillations. The 2D approximation, therefore, misses the instability. Though we find this for the particular case of constant Ω_A, the statement that the 2D equations (2.40) cannot show any magnetic instability is most probably valid in general. Disturbances b of the background field B can be expressed in terms of the displacement vector ξ defined by $b = \mathrm{curl}(\xi \times B)$. Two-dimensional uncompressive displacements can in turn be written as $\xi = \mathrm{curl}(r\psi)$ (ψ is a scalar function of position). It can be seen that an integral of the scalar product of $b \cdot B$ over a spherical surface of constant radius vanishes for a toroidal background field. This means that any 2D disturbance of the toroidal field increases the magnetic energy. No instability can, therefore, develop at the expense of the magnetic energy. This conclusion can be explained as follows. The toroidal field can be understood as consisting of closed flux tubes. Noncompressive disturbances conserve the volume and magnetic flux of the tubes, so that the magnetic energy of a tube is proportional to the square of its length (Zeldovich, Ruzmaikin, and Sokoloff, 1983). The 2D disturbances also conserve the area of a segment on a spherical surface encircled by a magnetic field line. The circular lines of background field have minimum length for any given encircled area. Any 2D disturbance increases the length, thereby increasing magnetic energy. There is no possibility of feeding a 2D instability by magnetic energy release.

In contrast to the hydrodynamical instability of differential rotation (Section 2.1), the Tayler instability of toroidal field has no 2D counterpart. The Tayler instability is of interchange type (Spruit, 1999). In addition to (dominating) horizontal displacements, radial velocities are necessary to interchange the toroidal field lines. If the instability operates in stellar radiation zones, a certain radial mixing of chemical species is unavoidable.

2.3.3
No Diffusion

The idealized case of vanishing diffusion serves well to illuminate some characteristic properties of the Tayler instability and to reconcile some controversial statements that can be found in the literature. We here still keep to the case of uniform rotation.

Two simple geometries of the toroidal field will be considered. The latitudinal profile consisting of two belts with equatorial antisymmetry, that is

$$\Omega_A = \Omega_{A0} \cos \theta , \qquad (2.42)$$

---

3) Uniform rotation has minimum kinetic energy for a given angular momentum. It is, therefore, not possible to tap energy from rigid rotation to feed any instability.

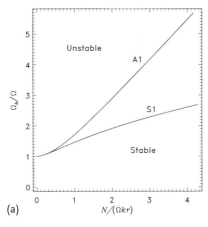 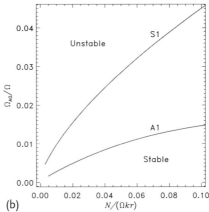

Figure 2.13 Neutral stability lines for constant Ω_A (a). The instability region is above the lines. The same for the equatorially antisymmetric background field profile (2.42) (b). Only nonaxisymmetric modes with $m = 1$ are unstable. From Kitchatinov and Rüdiger (2008a). Reproduced with permission © ESO.

is similar to the field structure found in the above magnetic tachocline models (Section 2.2). The other case of uniform Ω_A (one equator-symmetric belt) is used to consider the effect of the background field structure on the instability. Only nonaxisymmetric modes with $m = 1$ can be unstable for these two profiles (Goossens, Biront, and Tayler, 1981).

Figure 2.13 compares the stability maps for these two geometries of the toroidal field. Two features are remarkable: (i) the most unstable modes have indefinitely short radial scales, $\hat{\lambda} \to 0$, and (ii) the threshold amplitude of the field producing the instability strongly depends on the field geometry. Already Pitts and Tayler (1985) suggested that suppression of the instability by rotation may be exceptionally strong when the ratio of the Alfvén velocity to rotation velocity is uniform, but rotation is unlikely to suppress the instability for general configurations of the toroidal field. Figure 2.13a shows that for uniform Ω_A there is indeed no instability for subequipartition field, $\Omega_A < \Omega$. For the equatorially antisymmetric profile (2.42), however, instability for very short radial wavelengths persists even for small background fields.

Rotational quenching of the Tayler instability is a controversial issue. Spruit (1999) finds that rotation modifies the instability but does not switch it off. Cally (2003) concluded that the polar kink instability ($m = 1$) is totally suppressed if Ω_A is smaller than Ω near the pole. Simulations by Braithwaite (2006) also show that the instability does not develop when the toroidal field is below a critical value roughly equal to the equipartition level of $\Omega_A \simeq \Omega$. The controversy can be resolved by observing how sensitive the rotational quenching is to details of the toroidal field profile.

It has been claimed that the Tayler instability is of the polar type, that is, the unstable modes are concentrated near the poles. Figure 2.13 shows that this is at least not a general case. The marginal stability lines for different equatorial symmetries do not coincide. If the modes were sufficiently concentrated at the poles, their equatorial symmetry should not play any role. Figure 2.13 also shows that the smallest field producing the instability corresponds to indefinitely small radial scales. Finite diffusion, therefore, must be included.

2.3.4
Growth Rates, Drift Rates and Radial Mixing

The strong difference between the cases of uniform Ω_A and the equatorially antisymmetric field (2.42) disappears when finite diffusion is allowed. Thermal diffusion has a destabilizing effect by reducing the stabilizing effect of buoyancy. As a result, the minimum field producing the instability for uniform Ω_A is strongly reduced with finite diffusion. The instability becomes less sensitive to the toroidal field configuration. All further results of this section correspond to the 'dipolar' profile (2.42). Axisymmetric instability does not exist for this profile. Instability is again found for the kink mode of $m = 1$ only.

Using the diffusivities of Figure 2.11a and the profile of Figure 2.2, the diffusion parameters (2.35) for the upper radiation zone of the Sun can be estimated as

$$\epsilon_\nu = 2 \times 10^{-10}, \quad \epsilon_\eta = 4 \times 10^{-8}, \quad \epsilon_\chi = 10^{-4}. \tag{2.43}$$

All the following results are obtained with these values.

The stability map is shown in Figure 2.14. Its new feature is that the instability on the short wavelength side of the map is now stopped by finite diffusion. However,

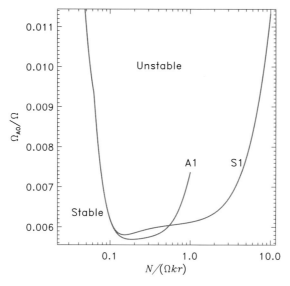

Figure 2.14 Stability map for finite diffusion of (2.43). The difference with Figure 2.13 for vanishing diffusion is that disturbances with very short radial scales are now stable. From Kitchatinov and Rüdiger (2008a). Reproduced with permission © ESO.

the radial scale of unstable modes remains short. The minimum field producing the instability corresponds to disturbances with $\hat{\lambda} = 0.1$–0.2. The radial wavelength in the upper solar radiation zone can be estimated by $\lambda = 10\hat{\lambda}$ [Mm]. The wavelength is thus of the order of 1000 km only. As the field strength grows above the critical value, the radial scale of the most rapidly growing modes becomes even shorter but remains close to $\hat{\lambda} \simeq 0.1$ (Kitchatinov and Rüdiger, 2008a).

Figure 2.15 shows the growth rates $\omega_{\rm gr} = \Im(\omega)$ and the rates of longitudinal drift as functions of the toroidal field amplitude. For strong fields ($\Omega_{\rm A} > \Omega$) the basic rotation is not important and the only characteristic frequency to scale the growth rates is $\Omega_{\rm A}$. The dependence of Figure 2.15 does indeed approach the relation $\omega_{\rm gr} \sim \Omega_{\rm A}$ in the strong-field limit; $\omega_{\rm gr}$ is the physical growth rate. The growth rate drops by almost two orders of magnitude when $\hat{\Omega}_{\rm A}$ reduces below unity. The parabolic law $\omega_{\rm gr} \sim \Omega_{\rm A}^2/\Omega$ applies in the weak-field regime (Spruit, 1999). The growth rates for weak fields, $\Omega_{\rm A} < \Omega$, are rather small. Such weak fields are destabilized by finite thermal conductivities. The growth rates of this double-diffusive instability are so small that neither time-stepping codes nor laboratory experiments can confirm its existence so that its physical relevance has been questioned (Cally, 2003). Nevertheless, the growth rate of, for example, 10^{-5} means an e-folding time of about 1000 yr for the Sun, which is short, however, compared to evolutionary timescales. We shall see that the instability of weak fields is actually too vigorous to be compatible with the observed lithium abundance.

Drift rates in Figure 2.15b also differ significantly between the cases of weak and strong fields. Small negative rates for $\Omega_{\rm A} < \Omega$ mean the slow longitudinal drift of unstable modes in counterrotation direction in the frame corotating with the

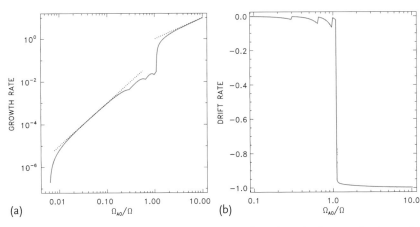

Figure 2.15 Normalized growth rate $\omega_{\rm gr}/\Omega$ as a function of normalized amplitude $\Omega_{\rm A0}/\Omega$ of the toroidal field (a). The dotted lines show approximations by power laws $0.1(\Omega_{\rm A0}/\Omega)^2$ for weak fields, $\Omega_{\rm A0} < \Omega$, and $\Omega_{\rm A0}/\Omega$ for strong fields with $\Omega_{\rm A0} > \Omega$. The plot is for the fastest growing S1 modes with $\hat{\lambda} = 0.1$. Same dependence for the normalized rate of azimuthal drift $\Re(\omega)/\Omega$ (b). The jumps in the drift rates are because the results for the most rapidly growing modes are shown and the maximum growth rates belong to different modes for different $\Omega_{\rm A0}$.

fluid. The normalized drift rates are close to −1 for strong fields. This means that the instability pattern does not follow the basic rotation. The unstable modes are resting in the inertial frame of reference. Judging from the growth and drift rates of Figure 2.15, Tayler instability in the strong field regime 'ignores' the background rotation. If the instability is operating in this regime near surfaces of some non-convective stars, the observed rotation of the stellar magnetic spots would be very slow and not reflecting a faster rotation of the star itself.

The amplitude of the toroidal field for the upper solar radiation zone can be estimated with $\rho \simeq 0.2\,\mathrm{g/cm^3}$, that is

$$B_\phi \simeq 10^5 \frac{\Omega_{A0}}{\Omega} \quad [\mathrm{G}]. \tag{2.44}$$

It follows from Figure 2.14 that the threshold field strength for the onset of the instability is slightly below 600 G. We conclude that the axisymmetric models for the magnetic tachocline (Section 2.2) are stable against the nonaxisymmetric Tayler instability.

The flow field of the instability also mixes chemical species in the radial direction. Such an instability can thus be relevant to the radial transport of light elements. The effective diffusivity, $D_T \simeq u_{\mathrm{rms}} \ell_{\mathrm{corr}}$ (u_{rms} and ℓ_{corr} are rms velocity and correlation length in the radial direction) can be roughly estimated from our linear computations, assuming that the instability saturates when the mixing frequency $u_{\mathrm{rms}}/\ell_{\mathrm{corr}}$ approaches the growth rate, and $\ell_{\mathrm{corr}} \simeq \lambda/2$. It follows that $D_T \simeq 7 \times 10^9\, (\omega_{\mathrm{gr}}/\Omega)\,\mathrm{cm^2/s}$ with the normalized growth rate of Figure 2.15. The resulting diffusivity is shown on the right axis of Figure 2.16 as a function of the toroidal field amplitude. For the range of $0.01\Omega < \Omega_{A0} < 0.2\Omega$ where the normal-

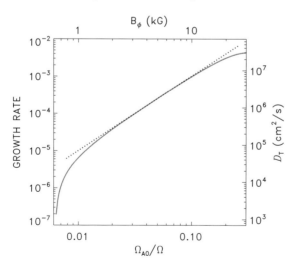

Figure 2.16 Same growth rates as in Figure 2.15. The right scale shows the corresponding values of the turbulent magnetic diffusion (2.35). The scale on the top shows the toroidal field amplitude of (2.44). From Kitchatinov and Rüdiger (2008a). Reproduced with permission © ESO.

ized growth rate is closely approximated by the parabolic law $\omega_{\rm gr} \simeq 0.1 \Omega_{\rm A0}^2/\Omega$, the eddy diffusivity can be rewritten in terms of B_ϕ as

$$D_{\rm T} \simeq 7 \times 10^4 \left(\frac{B_\phi}{1\,{\rm kG}}\right)^2 \quad [{\rm cm}^2/{\rm s}] \tag{2.45}$$

(see (2.44)). Diffusivities in excess of 10^3 cm^2/s in the upper radiative core are known to be incompatible with the observed solar lithium abundance. Hence, the toroidal field amplitude can only slightly exceed the marginal value of about 600 G. The observed solar lithium abundance seems to be not compatible with the concept of a hydromagnetic dynamo driven by Tayler instability in the upper radiation zone of the Sun.

2.4 Stability of Thin Toroidal Field Belts

The transition zones ('tachoclines') between convection zones with differential rotation and stellar cores with rigid rotations might be rather thin. The toroidal fields which are induced by the nonuniform rotation in such layers exceed the poloidal fields by many orders of magnitudes. We ask how their stability behavior (including differential rotation) differs from the stability of the radially smooth toroidal field profiles which have been considered above. However, the full system of equations proved so far as too complicated so that Arlt, Sule, and Rüdiger (2007) only considered the incompressible approximation using the spherical spectral MHD code by Hollerbach (2000a). Instead of the three diffusivities (2.35) the following theory only possesses two diffusivities which lead to the single Prandtl number Pm (here Pm = 0.1–1). The Navier–Stokes equation contains a Lorentz force but neglects the effect of buoyancy.

The integration domain is a spherical shell between $x = 0.6$ and 0.7. The boundary conditions for the magnetic field b are vacuum conditions for both radial boundaries. The boundaries for the flow are stress-free implying that there will be no strong shear near the inner or outer surfaces.

The toroidal field belt which is located in this layer (Figure 2.17) is perturbed with disturbances with $m > 0$. The magnetic Reynolds numbers of (equatorial) rotation Rm and the Lundquist number

$$S = \sqrt{\rm Pm}\,{\rm Ha} \tag{2.46}$$

of the toroidal field are fixed and the resulting stability map and growth rates are calculated. The linear, normalized incompressible MHD equations are

$$\frac{\partial \boldsymbol{u}}{\partial t} = {\rm Rm}\left[\boldsymbol{u} \times {\rm curl}\,\boldsymbol{U} + \boldsymbol{U} \times {\rm curl}\,\boldsymbol{u} - \nabla(\boldsymbol{u} \cdot \boldsymbol{U})\right]$$
$$+ S\left[{\rm curl}\,\boldsymbol{b} \times \boldsymbol{B} + {\rm curl}\,\boldsymbol{B} \times \boldsymbol{b}\right] - \nabla P + {\rm Pm}\Delta \boldsymbol{u},$$

$$\frac{\partial \boldsymbol{b}}{\partial t} = {\rm curl}({\rm Rm}\,\boldsymbol{U} \times \boldsymbol{b} + S\boldsymbol{u} \times \boldsymbol{B}) - \Delta \boldsymbol{b} \tag{2.47}$$

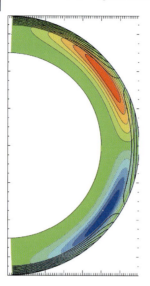

Figure 2.17 The two toroidal magnetic field belts (red, blue) as induced by a latitudinal rotation law in a thin layer from a large-scale fossil poloidal dipolar field. Solid: the isorotation line $\Omega = $ const. The stability of these belts against nonaxisymmetric perturbations is the basic problem of the tachocline theory. Courtesy of R. Arlt.

and $\mathrm{div}\,\boldsymbol{u} = \mathrm{div}\,\boldsymbol{b} = 0$. Here the time is measured in diffusion times, the velocity fluctuations are measured in units η/R and the magnetic fluctuations are measured in units $\sqrt{\mu_0 \rho}\,\eta/R$. The amplitudes of the angular velocity and the toroidal field are included in the definitions of Rm and S.

2.4.1
Rigid Rotation

First let the spherical shell rotate rigidly. The stability of the modes with $m = 1$ is checked. Figure 2.18a demonstrates the differences between the two possible

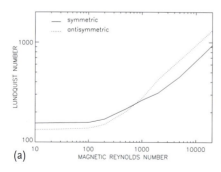
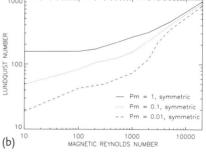
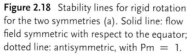

Figure 2.18 Stability lines for rigid rotation for the two symmetries (a). Solid line: flow field symmetric with respect to the equator; dotted line: antisymmetric, with Pm = 1. The symmetric modes for various magnetic Prandtl numbers (b), with $m = 1$. From Arlt, Sule, and Rüdiger (2007). Reproduced with permission © ESO.

symmetries with respect to the equator as small. The plot also shows the rotational suppression of the Tayler instability (TI): the faster the rotation the stronger the magnetic field must be in order to become unstable. Figure 2.18b demonstrates that for fast rotation the differences for Pm $\lesssim$ 1 also become small.

If the slopes in Figure 2.18 remain constant for faster rotation then $S_{\text{crit}} = 1.6\text{Rm}^{0.63}$ for the fast-rotation part of the Pm = 0.01 line (Arlt, Sule, and Rüdiger, 2007). The relation of the critical S and Rm is thus weaker than linear. The immediate consequence is that Ω_A/Ω goes to zero rather than to a finite value for Rm $\to \infty$. Even a very fast rotation cannot completely stabilize the TI.

2.4.2
Differential Rotation

Next a latitudinal differential rotation is assumed to exist in the spherical shell. The rotation law is given by $\Omega = \Omega_0(1 - k\cos^2\theta)$ with the shear variable k. Such a rotation law becomes unstable for k > 0.28 (see Section 2.1). There is thus the situation that one and the same curve of marginal instability connects the two points: (i) Watson instability without magnetic field and (ii) Tayler instability with rigid rotation. The two points can easily be recognized by the Figure 2.19. Watson's value k = 0.28 is reproduced for high Rm.

One finds the Watson instability (slightly) suppressed by the toroidal field. On the other hand, the stronger the shear of the rotation law, the stronger the magnetic field must be to become unstable. Both lines of marginal instability starting at the mentioned points at the vertical and the horizontal axis, therefore have positive slopes. The result is a stability 'balloon' which means that the mode m = 1 be-

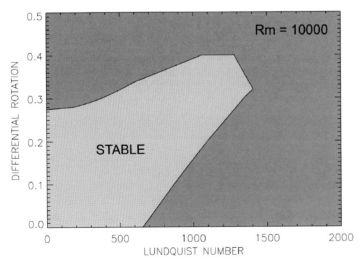

Figure 2.19 The lines of marginal stability for the interaction of toroidal magnetic field belts and differential rotation. The vertical axis gives the rotational shear k, with Rm = 10 000, Pm = 1.

comes stable for higher field amplitudes (higher than known from TI) and stronger shear than known (higher than known from the Watson instability).

2.4.3
High Fourier Modes

The differential rotation acts, stabilizes and enlarges the domain of stability. This is true for perturbations with $m = 1$. If the fields are concentrated in narrow belts one must check, however, whether it is also true for perturbations with higher m. Arlt, Sule, and Filter (2007) obtained the result that combinations of Rm and S which are stable against perturbations with $m = 1$ can be unstable against perturbations with $m > 1$. Figure 2.20 shows the main result for their maximal values of the rotation rate. The growth rates are normalized with Alfvén frequency. The resulting growth rates in its dependence on the azimuthal mode number m are marked with the normalized magnetic field amplitude $\Omega_A/\Omega \equiv S/Rm$. One finds the mode $m = 1$ to be stable but the modes with higher m to be unstable. The maximum growth rate may exist for $m \simeq 5$. The dependence of the growth rates on Ω_A/Ω is rather weak suggesting a linear relation $\omega_{gr}/\Omega_A = O(10^{-2})$. The data, however, are not sufficient enough to find out further details. It remains still unclear how the instability map (2.19) changes if the modes with $m > 1$ are included.

2.5
Helicity and Dynamo Action

Hydromagnetic dynamos can be understood as an instability of moving conducting fluids to weak seed magnetic fields. A finite initial field can be amplified by the fluid motion. There are, however, strong restrictions on the structure of the flows that

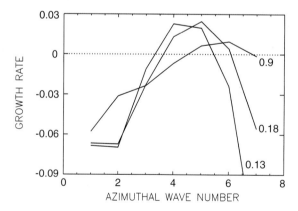

Figure 2.20 Growth rates in units of Alfvén frequency Ω_A for various azimuthal mode numbers m. The curves are marked with their values Ω_A/Ω. $S = 1800$, $k = 0.2$, $Pm = 1$. From Arlt, Sule, and Filter (2007).

can produce a dynamo (Elsasser, 1946) as well as on the structure of the generated fields (Cowling, 1933). As an example, differential rotation alone cannot lead to dynamo excitation. Generally, flows with finite kinetic helicity, $U \cdot \text{curl } U \neq 0$, are more probable to produce a dynamo (see Dudley and James, 1989).

An important question is whether a nonlinear dynamo-instability is possible, that is, whether a magnetic instability of a field of finite amplitude can produce a sufficiently complicated motion that together with a given not dynamo-effective background flow can support the magnetic field. Tout and Pringle (1992) suggested that nonuniformly rotating disks can produce a dynamo when magnetorotational (MRI) and magnetic buoyancy instabilities are active. Later, numerical simulations by Brandenburg *et al.* (1995) and Hawley, Gammie, and Balbus (1996) have shown that the MRI alone should be sufficient for the accretion disk dynamo. It remains uncertain, however, for the case of low magnetic Prandtl numbers whether the MRI dynamo has physical or numerical origin (Fromang and Papaloizou, 2007; Fromang *et al.*, 2007).

Another possibility was discussed by Spruit (2002) who suggested that differential rotation and TI can jointly drive a dynamo in stellar radiation zones. Such a dynamo would be important for the angular momentum transport in stars. Radial displacements converting toroidal magnetic field into poloidal field are necessary for any dynamo. The dynamo, therefore, always mixes chemical species in stellar interiors with consequences for the stellar evolution.

A dynamo effect is, however, not guaranteed by the joint action of differential rotation and a magnetic instability converting toroidal field into poloidal. The doubts especially concern the Tayler instability that, in contrast to MRI, develops at the expense of magnetic energy. Estimations of dynamo parameters are necessary to probe the dynamo effectivity of any magnetic instability. The ability of turbulence to produce a mean electromotive force (EMF) along the background magnetic field plays a basic role in turbulent dynamos, that is

$$\langle u \times b \rangle = \alpha \bar{B} - \ldots \tag{2.48}$$

which is called the α effect (here only written as a scalar). This ability is basically related to the kinetic helicity of the fluctuating flow. We proceed by evaluating the kinetic helicity and the α effect for the Tayler instability.

2.5.1
Helicity and Alpha Effect

Solving the linear stability problem of Section 2.3.1 may serve to estimate the sign and latitudinal profile of the kinetic helicity

$$\mathcal{H}^{\text{kin}} = \langle u \cdot \text{curl } u \rangle, \tag{2.49}$$

though the absolute value of (nonlinear) helicity cannot be determined by linear computations. The averaging in (2.48) and (2.49) is over the azimuth, and only the real part of the solution u of the eigenvalue problem should be used to evaluate the helicity.

Let the expressions

$$W = (w_R + iw_I)e^{im\phi}, \quad V = (v_R + iv_I)e^{im\phi} \tag{2.50}$$

represent the flow potentials W and V of (2.1). Then, the helicity (2.49) is

$$\mathcal{H}^{\rm kin} = \frac{1}{\sin^2\theta}\left[\left(mw_I + \sin\theta\frac{\partial v_I}{\partial\theta}\right)\left(mv_I + \sin\theta\frac{\partial w_I}{\partial\theta}\right)\right.$$
$$\left.+ \left(mw_R + \sin\theta\frac{\partial v_R}{\partial\theta}\right)\left(mv_R + \sin\theta\frac{\partial w_R}{\partial\theta}\right)\right]. \tag{2.51}$$

The following computations are performed for rigid rotation and the equatorially antisymmetric toroidal field of (2.42). The diffusivity values (2.43) for the upper solar radiation zone are adopted. If only the toroidal background field is present, we cannot expect any net helicity from the Tayler instability. The symmetry rules of Section 2.3.1 apply to this case. Applied to the helicity expression (2.51) it follows that the helicity has opposite signs for positive and negative azimuthal wave number m, that is

$$\mathcal{H}^{\rm kin}(m=1) = -\mathcal{H}^{\rm kin}(m=-1). \tag{2.52}$$

This means that for every unstable mode with finite helicity, there is another unstable mode with the same growth and drift rates but opposite helicity. If all modes are excited, the instability of purely toroidal field cannot produce finite kinetic helicity. The resulting net helicity vanishes. The same argument leads to the same conclusion for the current helicity $\mathcal{H}^{\rm curr} = \langle \mathbf{b} \cdot \mathrm{curl}\, \mathbf{b}\rangle$. Figure 2.21 shows the latitudinal profiles of kinetic helicities of unstable S ± 1 modes. As expected, the modes with $m = \pm 1$ have opposite helicities. The same plot shows the latitudinal profiles of the mean electromotive force (2.48). The EMF also reverses when the sign of m is

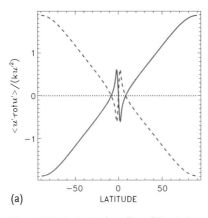
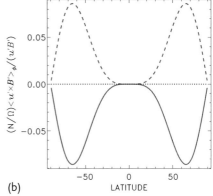
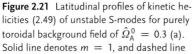

Figure 2.21 Latitudinal profiles of kinetic helicities (2.49) of unstable S-modes for purely toroidal background field of $\hat{\Omega}_A^0 = 0.3$ (a). Solid line denotes $m = 1$, and dashed line $m = -1$. EMF (2.48) for the same modes (b). From Rüdiger, Kitchatinov, and Elstner (2012). Copyright © 2012 RAS.

changed. The azimuthal EMF vanishes at the equator. This behavior is typical of the α effect but not for the $\Omega \times J$-term that could appear in the expression for the EMF as a consequence of a rotationally induced anisotropy of the diffusivity tensor. We, therefore, find that the $\Omega \times J$ effect due to the Tayler instability of toroidal fields does not exist. Its presence for more general field configurations cannot be excluded but it remains small in any case.

For purely toroidal fields no α effect exists. The contributions of the modes with opposite sign of m cancel each other not only with respect to the helicities but also with respect to the EMF. In the nonlinear regime, a spontaneous parity breaking may happen, as described by Chatterjee et al. (2011), Gellert, Rüdiger, and Hollerbach (2011) and Bonanno et al. (2012). In this case, however, it might be impossible to predict the sign and the amplitude of the α effect.

The 'equivalence' of positive and negative helicities is violated if one includes a poloidal background field. The background state has a certain handedness in this case and the symmetry rule (2.52) is violated.

In the short-wave approximation only the radial component of the poloidal field is significant. Similar to (2.31) we write $B_r = r\sqrt{\mu_0\rho}\,\Omega_{A,p}(\mu)$. The eigenequations of Section 2.3.1 change when the poloidal background field is allowed for. The toroidal flow equation now becomes

$$(\hat{\omega} - m\hat{\Omega})(\hat{L}W) = \cdots - \frac{1}{\hat{\lambda}}\left[\hat{\Omega}_{A,p}(\hat{L}B) + (1-\mu^2)\frac{\partial\hat{\Omega}_{A,p}}{\partial\mu}\frac{\partial B}{\partial\mu} - m\frac{\partial\hat{\Omega}_{A,p}}{\partial\mu}A\right] \quad (2.53)$$

(the dots mimic the RHS of (2.34)). The equation now includes the background poloidal field. The parameter measuring the effect of this field is

$$\hat{\Omega}_{A,p} = \frac{N}{\Omega_0}\frac{\Omega_{A,p}}{\Omega_0} = \frac{N\,B_r(\mu)}{\Omega_0^2 r\sqrt{\mu_0\rho}}\,. \quad (2.54)$$

This equation shows that the characteristic strength of poloidal field that can influence the magnetic instability ($\hat{\Omega}_{A,p} \sim 1$) is Ω_0/N times the toroidal field amplitude. This factor is of order 10^{-2}–10^{-3} for the Sun.

The computations will be performed only for the simplest profile

$$\hat{\Omega}_{A,p}(\mu) = \hat{\Omega}_{A,p}^0\mu\,. \quad (2.55)$$

Both components of the background field, B_r and B_ϕ, are antisymmetric about the equator in this case. The background current helicity $\mathbf{B}\cdot\text{curl}\,\mathbf{B}$ is, therefore, also equatorially antisymmetric. For positive Ω_A^0 and $\Omega_{A,p}^0$ it is positive in the northern hemisphere and negative in the southern hemisphere. We shall see that the background current helicity alone determines the behavior of the pseudoscalars $\mathcal{H}^{\text{kin}}$ and $\mathcal{H}^{\text{curr}}$. The basic rotation, which in stratified fluids can also form a pseudoscalar $\mathbf{g}\cdot\mathbf{\Omega}$ is of minor significance[4].

---

4) Another example of a possible pseudoscalar is the scalar product $\mathbf{\Omega}\cdot(\text{curl}\,\mathbf{B}\times\mathbf{B})$.

The α parameter of the electromotive force (2.48) will be characterized by the quantity

$$A = \frac{N}{\Omega} \frac{\langle \boldsymbol{u} \times \boldsymbol{b} \rangle_\phi}{u_{\rm rms} b_{\rm rms} \sin\theta \cos\theta} = \frac{\alpha B_0}{u_{\rm rms} b_{\rm rms}} \frac{N}{\Omega}. \tag{2.56}$$

The $\sin\theta\cos\theta$ in the denominator eliminates the latitudinal profile of the toroidal field and B_0 is the toroidal field amplitude. $u_{\rm rms}$ is the corresponding rms velocity fluctuation after horizontal averaging (longitude and latitude); $b_{\rm rms}$ is the rms magnetic fluctuation. The factor N/Ω is introduced because the horizontal velocity and magnetic fluctuations are larger than the radial fluctuations by just this factor. With this normalization, the characteristic value of the parameter (2.56) does not depend on the actual value of N/Ω.

The equation for the poloidal flow (2.36) is modified with allowance for the background poloidal field as

$$(\hat{\omega} - m\hat{\Omega})(\hat{L}V) = \cdots - \frac{1}{\hat{\lambda}}\left[\hat{\Omega}_{\rm A,p}(\hat{L}A) + (1-\mu^2)\frac{\partial \hat{\Omega}_{\rm A,p}}{\partial \mu}\frac{\partial A}{\partial \mu} - m\frac{\partial \hat{\Omega}_{\rm A,p}}{\partial \mu}B\right] \tag{2.57}$$

(the dots mimic the RHS of (2.36)). The remaining equations for the magnetic field disturbances are

$$\hat{L}\left[(\hat{\omega} - m\hat{\Omega})B\right] = \cdots - \frac{1}{\hat{\lambda}}\left[\hat{\Omega}_{\rm A,p}(\hat{L}W) + (1-\mu^2)\frac{\partial \hat{\Omega}_{\rm A,p}}{\partial \mu}\frac{\partial W}{\partial \mu} - m\frac{\partial \hat{\Omega}_{\rm A,p}}{\partial \mu}V\right] \tag{2.58}$$

and

$$(\hat{\omega} - m\hat{\Omega})(\hat{L}A) = \cdots - \frac{1}{\hat{\lambda}}\left[\hat{\Omega}_{\rm A,p}(\hat{L}V) + (1-\mu^2)\frac{\partial \hat{\Omega}_{\rm A,p}}{\partial \mu}\frac{\partial V}{\partial \mu} - m\frac{\partial \hat{\Omega}_{\rm A,p}}{\partial \mu}W\right] \tag{2.59}$$

(the dots mimic the RHS of (2.37) and (2.38)). The entropy equation (2.39) is not changed by the poloidal background field. Solution of the linear equations provides the growth rates and the structure of the eigenmodes for simultaneously present axisymmetric azimuthal and poloidal background fields.

Figure 2.22 shows the growth rates as function of the (normalized) radial scale $\hat{\lambda}$ (2.6) of unstable modes for different amplitudes of the background poloidal field. As might be expected, the poloidal field stabilizes (decreases the growth rates) and it increases the radial scales of unstable modes. The latter is due to radial coupling by the poloidal field. More significant (also for dynamo theory) is that the growth rates for opposite signs of m are no longer identical. The linear stability equations (2.57)–(2.59) do not obey the transformation rule. The modes with negative m typically grow faster. Their maximum growth rates decrease by three orders of magnitude as the poloidal field increases by a factor of 6, demonstrating a strong stabilization of the magnetic instability for helical background fields. We are thus

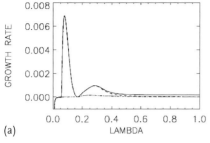
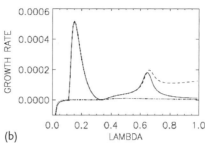

Figure 2.22 Fractional growth rates $\omega_{\rm gr}/\Omega$ for the modes A ± 1 and S ± 1 as functions of the normalized radial wavelength $\hat{\lambda}$; $\hat{\Omega}^0_{A,p} = 0.05$, 0.1 (a,b). The modes with $m = -1$ (solid line: S-mode, dashed line: A-mode) grow faster than the modes with $m = 1$ (dashed-dotted: S-mode, dotted: A-mode). In all cases $\hat{\Omega}^0_A = 0.3$. From Rüdiger, Kitchatinov, and Elstner (2012). Copyright © 2012 RAS.

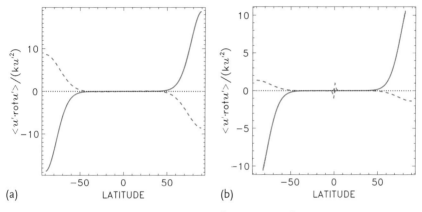

Figure 2.23 The case of helical background field ($\hat{\Omega}^0_A = 0.3$ and $\hat{\Omega}^0_{A,p} = 0.1$). Latitudinal profiles of the helicity by the unstable modes S ± 1 (a), A ± 1 (b). Solid lines: $m = 1$, dashed lines: $m = -1$. The net helicity no longer vanishes. Copyright © 2012 RAS.

confronted with the dilemma that only background fields with finite current helicity generate fluctuations with finite α effect, but the corresponding poloidal field suppresses the Tayler instability.

The helicity profiles are shown in Figure 2.23. The helicity of $m = 1$ and $m = -1$ modes are still opposite in sign, but they no longer coincide in absolute values when the poloidal field is finite. Comparison of Figure 2.23a,b shows that the rotation has a minor effect. Obviously, the pseudoscalar $\mathbf{B} \cdot$ curl $\mathbf{B}$ governs the unstable modes' helicity rather than the pseudoscalar $\mathbf{g} \cdot \mathbf{\Omega}$ formed by the global rotation and the stratification vector $\mathbf{g}$. This is due to the very short radial scales of the unstable modes.

Figure 2.24 shows latitudinal profiles of the A parameter (2.56) of the α effect for several amplitudes of the poloidal field. All the profiles are concentrated at high latitudes. The concentration of the α effect towards the poles reflects the basic

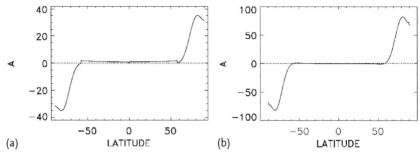

Figure 2.24 Latitudinal profiles of the normalized α effect parameter (2.56) for the fastest growing modes ($m = -1$, identical for A- and S-modes); $\hat{\Omega}^0_{A,p} = 0.05, 0.1$ (a,b). The α effect is strongly concentrated toward the poles. The amplitude of A is proportional to $\Omega^0_{A,p}$. For the given magnetic configuration (positive $\Omega^0_{A,p}$ and Ω^0_A), the α effect is positive in the northern hemisphere. From Rüdiger, Kitchatinov, and Elstner (2012). Copyright © 2012 RAS.

property of the Tayler instability, that is, that the instability pattern is more present in polar rather than equatorial regions (Spruit, 1999; Cally, 2003).

The most rapidly growing $m = -1$ modes produce the α effect of the same sense as the current helicity of the background field, that is, the α effect is positive (negative) where $\bar{B} \cdot \text{curl}\, \bar{B}$ is positive (negative). The sign of this α effect is opposite to that of the current helicity of unstable modes. The same relations were found by Gellert, Rüdiger, and Hollerbach (2011) in nonlinear simulations of kink-type instability of an incompressible fluid in a cylinder setup (see Section 6.5.3).

The lower limit for the poloidal field amplitude producing the α effect seems to be $\hat{\Omega}^0_{A,p} = 0.01$. For smaller $\hat{\Omega}^0_{A,p}$, the α effect is also small and its sign fluctuates.

2.5.2
Dynamo Action

The derived dynamo parameters should be estimated to assess the possibility of dynamo action. We start by estimating the effective magnetic diffusion for the Tayler instability. Assuming that the instability saturates when the mixing frequency $u_{\rm rms}/\lambda$ approaches the growth rate $\omega_{\rm gr}$, the characteristic velocity can be estimated as $u_{\rm rms} \simeq \omega_{\rm gr}\lambda$. This leads to the following estimation for the turbulent diffusivity $\eta_T \simeq u_{\rm rms}\lambda$,

$$\eta_T \simeq \pi^2 \hat{\omega}_{\rm gr} \hat{\lambda}^2 \frac{\Omega^2}{N^2} \Omega r^2, \tag{2.60}$$

where $\hat{\omega}_{\rm gr}$ is the growth rate normalized to the angular velocity Ω_0. Figure 2.22 gives three examples with a common value of $\hat{\omega}_{\rm gr}\hat{\lambda}^2 \simeq 10^{-6}$. This leads to the estimate $\eta_T \simeq 3 \times 10^5$ cm^2/s for the upper solar radiation zone. For more massive stars η_T can be three orders of magnitude higher. Figure 2.2b gives an example for a 2 $M_\odot$ star with a 10 day rotation period.

The α effect is parameterized in dynamo models by the dimensionless number $C_\alpha = \alpha r/\eta_T$. Equation (2.56) together with the heuristic relation $u^2_{\rm rms}/\eta_T \simeq \omega_{\rm gr}$

provides

$$C_\alpha \simeq \frac{5}{\hat{\Omega}_A^2} \frac{\Omega}{N} \hat{\omega}_{gr} A .\qquad(2.61)$$

Hence, the product $\hat{\omega}_{gr}A$ defines the efficiency of the α effect. With $\hat{\omega}_{gr}A \simeq 0.03$, a value of $C_\alpha \simeq 10^{-2}$–10^{-3} results for the Sun. For the 2 $M_\odot$ star $C_\alpha \simeq 0.1$ is also small. A stationary α^2 dynamo is thus not possible in stellar radiation zones. An oscillatory $\alpha\Omega$-dynamo cannot, however, be excluded. The $\alpha\Omega$-dynamos can operate with small α only if parameter

$$C_\Omega = \frac{\Omega_0 r^2}{\eta_T}\qquad(2.62)$$

is sufficiently large. A consequence of very small α effect is that the ratio of toroidal to poloidal magnetic field components is very large. The background current helicity is small in this case, and the formation of the α effect becomes problematic. Another consequence of small α is very long growth time of the dynamo instability. The growth time of weakly supercritical $\alpha\Omega$-dynamo for $\eta_T \simeq 10^5$ cm^2/s is of the order of some Gyr for the Sun.

Another complicating characteristic of the α effect of Figure 2.24 is its high concentration at the poles. It is known from the theory of $\alpha\Omega$-dynamos that even a smooth $\cos\theta$ profile of the α effect produces magnetic fields that are 'too polar.' It may, therefore, be expected that reproducing the midlatitude belts of toroidal fields like that of (2.42) by an $\alpha\Omega$-dynamo with polar α effect is not possible.

Let us construct a nonlinear dynamo model with the latitudinal profile $\alpha = C_\alpha \cos^\ell \theta$ of the α effect. Polar concentration of the magnetically induced α effect can be reproduced with high (odd) values of ℓ. A rather weak solar-type differential rotation $\hat{\Omega} = 1 - 0.03\cos^2\theta$ is prescribed. The small rotational shear is hydrodynamically stable (cf. Section 2.1). Possible production of the α effect by hydrodynamic instability in the radiation zone such as that by Dikpati and Gilman (2001) was therefore excluded. The shear peaks at the same 45° latitude where the toroidal field (2.42) has its maximum value.

Figure 2.25a–d shows the field configurations produced by the $\alpha\Omega$-dynamo with $\ell = 5$ and 15. The dynamo model provides positive (negative) large-scale current helicity $\mathbf{B}\cdot\mathrm{curl}\,\mathbf{B}$ in the northern (southern) hemisphere. As the same constellation was assumed for the background field producing the α effect, the theory is consistent in this respect.

However, the modeled toroidal fields for high ℓ are strongly concentrated to the poles[5]. For growing ℓ, the belt's position drifts poleward. Such fields cannot close the loop of field amplification by reproducing the original axisymmetric toroidal field. The radiation zone dynamo by joint action of differential rotation and Tayler instability is therefore unlikely.

---

5) This behavior is not trivial as the toroidal field generation by the differential rotation does not depend on ℓ.

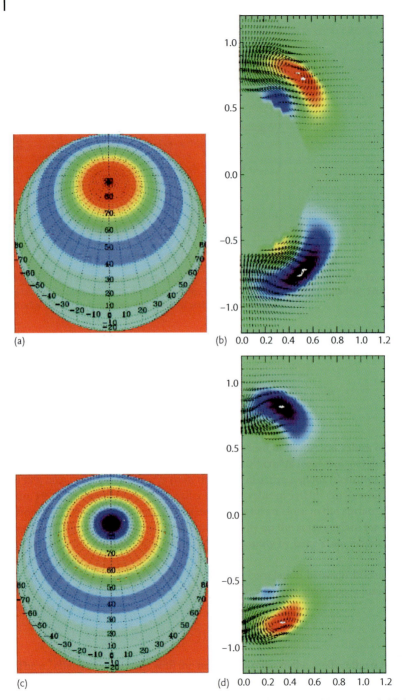

Figure 2.25 The field geometry in $\alpha\Omega$-dynamo models with $\ell = 5$ and $\ell = 15$. Radial field at the surface (a,c) and toroidal field pattern (b,d). The toroidal field belts drift radially outward during the cycles. From Rüdiger, Kitchatinov, and Elstner (2012). Copyright © 2012 RAS.

Zahn, Brun, and Mathis (2007) and Mathis, Zahn, and Brun (2008) arrived at a similar conclusion by performing 3D nonlinear numerical simulations of magnetic instabilities in differentially rotating radiation zones. They observed magnetic instabilities, which produced nonaxisymmetric magnetic configurations. The configurations did not, however, reproduce the background unstable field. The dynamo-loop was, therefore, not closed and a dynamo effect could not be found.

2.6
Ap Star Magnetism

The magnetic Tayler instability may also be relevant to the magnetism of Ap stars. Tayler (1973) noted that magnetic fields in stellar radiation zones have to be stable in order to survive over evolutionary timescales. Recently Wade et al. (2009) and Arlt and Rüdiger (2011a,b) invoked the instability to interpret the observed magnetic fields of Ap stars.

'Ap stars' is a common name of chemically peculiar stars of spectral types from late B to early F which typically have strong magnetic fields. The Ap stars constitute about 10% of all intermediate-mass stars. Auriére et al. (2007) found that all spectroscopically classified Ap stars posses magnetic fields with the lower bound on the field amplitude of about 300 G. The distribution of the field strengths ranges from this 0.3 to about 30 kG (Donati and Landstreet, 2009). Magnetic fields of Ap stars are topologically much simpler, that is of larger scale, compared to solar-type stars. An approximation of the fields by dipoles shows, however, the 'inclined dipoles' with their axes oblique to the axes of rotation. Magnetic fields of Ap stars are, therefore, both global and nonaxisymmetric.

Ap stars show statistically slower rotation than other A stars. The magnetic stars do not, however, represent the slow-rotation tail of a common distribution with other A stars, but form their own separate distribution (Abt and Morell, 1995).

There are different views on evolution of Ap stars. Hubrig, North, and Mathys (2000) documented that the Ap star phenomenon is much less frequent among relatively young A stars, which have not yet completed 30% of their life on the main-sequence. This result was, however, questioned by Landstreet et al. (2007), because of the difficulty to determine the ages of these stars.

There is no commonly accepted theory of Ap star magnetism. The currently leading concept is that their magnetic fields are relics of the early stellar evolution (Moss, 1987). The resistive decay in the radiation zones is slow enough for relic fields to survive over the main-sequence life of a star. Braithwaite and Spruit (2004) computed the evolution of relic fields from initial nonequilibrium states to stable equilibria. Another concept for the origin of Ap star magnetism is the field generation by hydromagnetic dynamos operating in convective cores (Schüssler and Pähler, 1978; Charbonneau and MacGregor, 2001). The dynamo hypothesis, however, has difficulties in explaining the observed inverse correlation with the rotation rate (Landstreet and Mathys, 2000; Bagnulo et al., 2002). Mathys (2008) argues that only the stars with rotation periods 100–1000 days show simple sinusoids with the rotation period of the star. Stars with both shorter and longer rotation periods have more complicated variations of the magnetic field modulus. In other words, the magnetic and rotation axes are mainly aligned for not too fast and not too slow rotation.

The structure of A stars differs significantly from that of the solar-type stars. The medium-mass stars have convective cores, extended radiative envelopes, and very thin surface 'convective skins.' The radiative envelopes are less subadiabatic compared with the cores of solar-type stars. Figure 2.2b shows the ratio N/Ω for a star of 2 $M_\odot$ at the age of 100 Myr rotating with a period of 10 days. The ratio N/Ω controls the radial scale of the unstable modes of the Tayler instability (Section 2.3). The ratio in the A1 star is much smaller compared with that of the solar radiation zone. If the instability operates in such a star, its radial scale is no longer small compared to the stellar radius.

The radial scale can be estimated as $\lambda \simeq 2\pi r \Omega/N$. This scale estimates the depth of the near-surface layer from which the instability can bring magnetic fields to the surface. This depth for the 2 $M_\odot$ star is about 30% of the stellar radius. The density and temperature at this depth are about 0.014 g/cm^3 and 1.6×10^6 K, respectively. Estimates of the microscopic magnetic diffusion and viscosity from (2.29) give $\nu \simeq 40 \text{ cm}^2/\text{s}$ and $\eta \simeq 5 \times 10^3 \text{ cm}^2/\text{s}$, not far above the values for the radiation zone of the Sun. The thermal diffusivity $\chi \simeq 2 \times 10^9 \text{ cm}^2/\text{s}$, however, is much higher than that in the Sun. We have seen in Section 2.3 that the thermal diffusion has a destabilizing effect on the magnetic instability. It is thus shown that conditions in the external radiation zones of medium-mass stars are indeed more favorable for current-driven instability.

Wade et al. (2009) suggested that observed magnetic fields of Ap stars have a low bound because of an instability. If strong fields are stable but weak fields are not, the fields below the instability threshold can be destroyed by the instability. However, this idea is in contrast with the common experience that the instability thresholds represent the *minimum* fields for magnetic instabilities. The suggestion is probed that the magnetic fields of Ap stars are remnants of Tayler instability of predominantly toroidal fields operating over early stages of stellar evolution. The instability brings the fields initially hidden in the stellar interior to the surface, thereby making them observable.

The scenario of the surface magnetic field formation is as follows. Rotation of the young medium-mass stars is not uniform. According to Stępień (2000), the medium-mass stars change angular momentum through disk accretion, magnetic star-disk coupling, and magnetized winds over their premain-sequence evolution. Ferrario et al. (2009) postulate a brief period of strong differential rotation if a small fraction of early-type binaries merge forming the later Ap stars.

The differential rotation winds strong toroidal fields from any available poloidal field so that the field becomes predominantly toroidal. If the toroidal field amplitude exceeds the threshold value for the Tayler instability, the nonaxisymmetric instability develops to display the internal field on the surface. The solid line in Figure 2.26 shows the amplification of a toroidal field under the influence of a

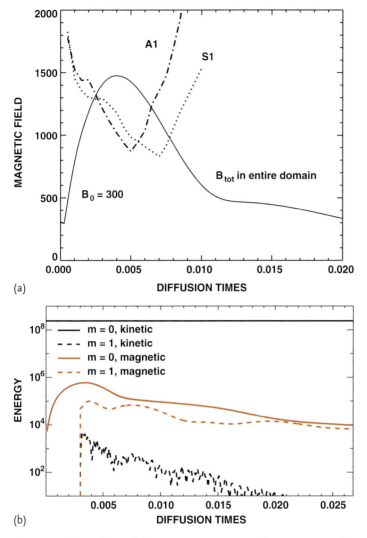

Figure 2.26 The evolution of the magnetic field strength resulting from the interaction of the rotation law (2.63) and a dipolar poloidal field with $S = 300$ (solid line). The instability window for such fields against nonaxisymmetric perturbations ($m = 1$) with equatorially symmetric flow pattern (dotted) and/or equatorially antisymmetric flow pattern (dash-dotted) (a). Evolution of the axisymmetric flow (black solid) and fields (red solid) and the nonaxisymmetric perturbations (dashed) (b). Pm = 1. From Arlt and Rüdiger (2011b). Copyright © 2011 RAS.

differential rotation with

$$\Omega = \frac{\Omega_{\text{pole}}}{\sqrt{1 + (R/R_{\text{out}})^q}}, \qquad (2.63)$$

(R the distance from the rotation axis) where the backreaction by the Lorentz force reduces the exponent q more and more. Ω_{pole} is the rotation rate of the stellar axis. At any instant the growing toroidal field is probed for instability against $m = 1$ perturbations. Figure 2.26a gives the lower limit of the toroidal field which becomes unstable. There is practically no difference of the threshold values for the two types of equatorial symmetry of the perturbation patterns. The initial poloidal field from which the toroidal field is generated has a Lundquist number of about 300. The MHD equations with which the stability behavior has been tested are fully incompressible. This physical deficit of the model may be overcompensated by the fact that the geometry of the toroidal field is consistently calculated rather than imposed.

For the lowest modes with $m = 0$ and $m = 1$ Figure 2.26b shows the behavior of their energy if a nonaxisymmetric perturbation is injected at a time which lies inside the instability window shown in Figure 2.26a. The energy of the magnetic $m = 1$ mode does not decay and even becomes nearly as large as that of the $m = 0$ mode of the background field. The energy of the kinetic $m = 1$ mode, however, seems to decay. The energy of the magnetic fluctuations always exceeds the energy of the kinetic fluctuations by more than one order of magnitude. The spectrum of the nonaxisymmetric modes is rather steep: the energy of the $m = 4$ mode is already two orders of magnitudes less than $m = 1$. Outside the mentioned instability window the nonaxisymmetric magnetic modes also decay.

The computations confirmed that the generated toroidal fields can be unstable over a limited range of time. The differential rotation eventually decays and the star is left with an already stable, nonaxisymmetric, slowly decaying magnetic field. Figure 2.27 shows the characteristic example of the surface field pattern produced with the model. The simulations are in qualitative agreement with observations. The dominant modes of the magnetic instability are nonaxisymmetric and global in the horizontal dimensions, similar to the observed fields. The relatively slow

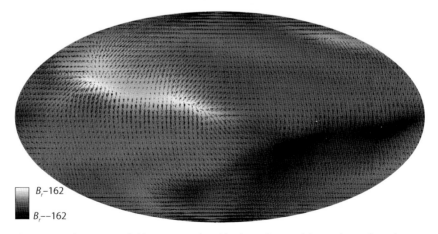

Figure 2.27 The magnetic field pattern produced by the Tayler instability on the surface of an Ap star. The contours represent the radial field and the arrows show the horizontal field.

rotation of Ap stars can be explained by the stabilizing effect of fast and rigid rotation on the Tayler instability. The estimated field strengths are also compatible with observations.

One can also ask for the generation of helicity in the domain where the interaction of differential rotation and poloidal field leads to a current helicity of the background field, which vanishes at the boundaries and shows both signs in the bulk of the radiative zone (Figure 2.28a). With the rotation law (2.63) with negative radial shear the negative (positive) sign appears to dominate in the northern (southern) hemisphere. In the bulk of the radiation zone a small-scale current helicity develops in both hemispheres which is *anticorrelated* with the large-scale current helicity. There is also indication that the kinetic helicity is correlated with the large-scale current helicity (i.e., anticorrelated with the small-scale current helicity) but the numerical values are strikingly small in the bulk of the radiation zone. We shall find a very similar constellation in Section 6.4, where a toroidal field is generated by an axial shear and an axial poloidal field in a cylindrical setup. In that model the resulting α_{zz} scales with the current helicity, which is also true for the present spherical model.

A special situation appears at the inner boundary where no current helicity of the background field exists. At this place negative (positive) values for both the small-scale helicities occur in the northern (southern) hemispheres which do not have any counterpart in the large-scale current helicity. In his hydrodynamic simulation of rotating convection Reshetnyak (2006) also finds negative kinetic helicity along

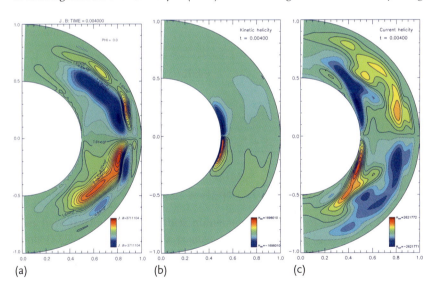

Figure 2.28 The current helicity of the background field (a, for negative radial shear), the kinetic helicity (b) and the current helicity (c) and averaged over the azimuthal direction for a model with the rotation law (2.63), $Rm = 20\,000$ and $Pm = 1$. Note that the small-scale helicities at the bottom of the domain are a boundary effect which has no counterpart in the plot of the current helicity of the background field. From Arlt and Rüdiger (2011b). Copyright © 2011 RAS.

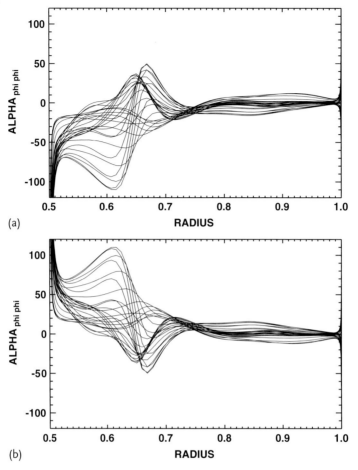

Figure 2.29 Snapshots for the $\alpha_{\phi\phi}$ of both hemispheres, for the north (a) and south (b). Note the clear antisymmetry with respect to the equator and the strong boundary layer effect at the bottom of the radiation zone with opposite signs. From Arlt and Rüdiger (2011b). Copyright © 2011 RAS.

the tangent cylinder in the northern hemisphere. In our case, however, the current helicity (not existing in hydrodynamical calculations) dominates the kinetic helicity which also proves to be negative (see Figure 2.28c). The corresponding $\alpha_{\phi\phi}$ at the tangent cylinder again has the same negative sign as the small-scale current helicity (Figure 2.29).

All in all, the calculations of the quadratic correlations $\mathcal{H}^{\text{kin}}$, $\mathcal{H}^{\text{curr}}$ and $\langle \boldsymbol{u} \times \boldsymbol{b} \rangle$ lead to the results that

- in the middle of the radiation zone $\mathcal{H}^{\text{curr}}$ is anticorrelated with the (very small) $\mathcal{H}^{\text{kin}}$ and the large-scale current helicity,

- $\alpha_{\phi\phi}$ has the same sign as $\mathcal{H}^{\text{curr}}$,
- at the inner tangent cylinder $\mathcal{H}^{\text{kin}}$, $\mathcal{H}^{\text{curr}}$ and $\alpha_{\phi\phi}$ are all negative (positive) at the northern (southern) hemisphere, which maybe interpreted as an unavoidable boundary layer effect.

2.7 The Shear–Hall Instability (SHI)

It is known that the general induction equation (2.16) possesses unstable solutions for the magnetic field for certain flow fields U without any turbulence. An early example of such a possibility is the well-known Herzenberg dynamo which works on the basis of the magnetic field induction within and between two separated spheres (Herzenberg, 1958; Brandenburg, Moss, and Soward, 1998). Dudley and James (1989) found many types of combinations of toroidal and poloidal flows in a single sphere which lead to dynamo-excited large-scale magnetic fields which, according to the Cowling theorem, must be nonaxisymmetric. Even purely meridional circulations have been demonstrated as dynamo-active (Gailitis, 1970, 1994; Moss, 1990, 2008). No solution, however, exists if the flow is purely toroidal in a finite domain. Its shear may induce toroidal fields from poloidal ones, but there is no transformation back from toroidal fields to poloidal ones. There is, however, a surprising possibility to close the dynamo-circle for media with Hall effect. Formally, the latter can also be described by a conductor with (anisotropic) conductivity tensor with off-diagonal elements. Under these conditions, indeed, a feedback of the toroidal field component to the poloidal field component can happen so that the magnetic field becomes unstable. The phenomenon is called the Shear–Hall instability (SHI) because of the basic effect that a stable rotational shear in a medium with Hall effect can destabilize a magnetic field (Wardle, 1999; Balbus and Terquem, 2001).

The complete induction equation reads as

$$\frac{\partial \mathbf{B}}{\partial t} = \text{curl}(\mathbf{U} \times \mathbf{B}) + \eta \Delta \mathbf{B} - \beta_{\text{H}} \, \text{curl}(\text{curl } \mathbf{B} \times \mathbf{B}) \,. \tag{2.64}$$

The last term describes the Hall effect; nonuniformities of the conductivities are neglected. The energy equation

$$\frac{1}{2} \frac{d}{dt} \int B^2 dV = -\eta \int (\text{curl } \mathbf{B})^2 dV + \int \mathbf{B} \cdot \text{curl}(\mathbf{U} \times \mathbf{B}) dV \,, \tag{2.65}$$

derived from (2.64) and integrated over the entire volume, shows that the Hall term exactly conserves the magnetic energy. The only positive source of energy in (2.65) is formed by the shear flow. The Hall term is basically a catalyst which does not feed its own instability (Wareing and Hollerbach, 2009). This is analogous to the MRI where the differential rotation is the energy source rather than the mean magnetic field which does not even occur in the energy of the magnetic fluctuations (see Section 5.3).

According to Urpin and Rüdiger (2005), for a rotation law depending on both R and z (in cylindrical coordinates), it is necessary for unstable modes that

$$(\boldsymbol{k} \cdot \bar{\boldsymbol{B}}) \left(k_z \frac{\partial \Omega}{\partial R} - k_R \frac{\partial \Omega}{\partial z} \right) < 0 , \tag{2.66}$$

with $\boldsymbol{k}$ as the wave number. For $\Omega = \Omega(R)$ this becomes $B_z \mathrm{d}\Omega/\mathrm{d}R < 0$, so that for a given sign of the shear an instability only occurs for B_z of the other sign. There is no instability if both signs of the field amplitude are equal.

The open question is the growth time of this instability. The reason is that the SHI only exists if the timescale of the Hall effect is shorter than the diffusion time and longer than the shear time. Otherwise the diffusion and/or the Hall effect would dominate and no instability occurs. For neutron stars the ratio

$$R_{\rm B} = \frac{\tau_{\rm diff}}{\tau_{\rm Hall}} = \frac{\beta_{\rm H} B_0}{\eta} \tag{2.67}$$

lies between 1 and 100 (Figure 2.30) which leads to Hall times of (say) 10^8 yr. On the other hand, the shortest rotation time of newborn neutron stars is of order 0.1 s so that an enormous gap exists between the two characteristic timescales. It can be shown, however, that the growth time of SHI is determined by the rotation time so that the instability is very fast.

This result leads to one of the most interesting applications of the SHI. In the first few minutes of a newborn neutron star its crust rotates differentially as shown by many numerical simulations of a SN explosion (Mönchmeyer and Müller, 1989; Janka and Mönchmeyer, 1989; Dimmelmeier, Font, and Müller, 2002; Kotake et al., 2004; Ardeljan, Bisnovatyi-Kogan, and Moiseenko, 2005; Burrows et al., 2007). Ott et al. (2005) have demonstrated how the differential rotation of the collapsing core of a SN leads to differential rotation in the newly formed neutron star. Due

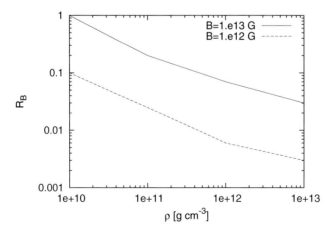

Figure 2.30 The Hall parameter $R_{\rm B}$ for 10^{10} K (after A. Y. Potekhin, www.ioffe.ru/astro/conduct/) reaches $R_{\rm B} = 1$. At this temperature the crust of a newborn neutron star should still be fluid. From Kondić, Rüdiger, and Arlt (2012).

2.7 The Shear–Hall Instability (SHI)

to its rapid cooling by neutrino emission a crust develops in the neutron star with an initial temperature of 10^9 K (Dall'Osso, Shore, and Stella, 2009). The extremely fast cooling happens with a timescale of only a few minutes (Page, Geppert, and Weber, 2006). Due to the short rotation period of the neutron star of order 50 ms the number of rotations within the lifetime of the fluid crust is high enough for instabilities which develop on the rotational timescale.

A spherical model is considered solving the linearized equation (2.64) in a sphere with a cylindrical rotation law under the presence of a homogeneous external axial field B_0. The magnetic Reynolds number is $\text{Rm} = R_*^2 \Omega_0/\eta$, with R_* as the stellar radius and Ω_0 the maximal angular velocity. Scaling lengths by R_*, times by R_*^2/η, U by $R_* \Omega_0$ and B by B_0 the linearized induction equation becomes

$$\frac{\partial \boldsymbol{b}}{\partial t} = -R_B \operatorname{curl}(\operatorname{curl} \boldsymbol{b} \times \boldsymbol{e}_z) + \text{Rm } \operatorname{curl}(\boldsymbol{U} \times \boldsymbol{b}) + \Delta \boldsymbol{b}. \tag{2.68}$$

The rotation law is prescribed as $\Omega = \Omega_0/\sqrt{1+(R/R_\Omega)^2}$ with $R_\Omega = R_*/2$. Such a flow does not generate an Ω effect interaction with the imposed axial field B_0. Insulating boundary conditions are imposed at both $x_{\text{in}} = 0.7$ and $x_{\text{out}} = 1$. Figure 2.31a presents the resulting instability curves for the modes S0, A0, S1 and A1 for weak and strong Hall effect (Kondić, Rüdiger, and Hollerbach, 2011). For not too large R_B the axisymmetric modes are slightly preferred over the nonaxisymmetric ones, but the differences are small. The minimum value of Rm for instability occurs for $R_B \simeq 2$. The shear frequency there is around 400 times the Hall frequency Ω_{Hall}, consistent with the above statement that the instability cannot exist for dominant Hall effect. For both $R_B \ll 1$ and $R_B \gg 1$ the differential rotation must thus become very strong to produce instability.

One concludes from Figure 2.31 that in the nonlinear regime several modes are simultaneously excited, which can easily be considered as a solution in the sense of an oblique rotator so that the magnetic configuration at the surface strongly differs for both hemispheres. Consequently, the Hall effect must lead to different temperatures and X-ray emissions (Schaaf, 1990; Heyl and Hernquist, 2001; Becker and Pavlov, 2001; Schwope et al., 2005) in the northern and southern hemispheres.

Figure 2.31b shows the growth rate scaled on the rotational timescale, $\hat{\omega}_{\text{gr}} = \tau_{\text{rot}} \omega_{\text{gr}}$. For $R_B = 1$ and $R_B = 5$ these growth rates prove to saturate at values of order unity. All the four modes behave comparably, though the axisymmetric modes do have larger growth rates (the S and A modes are very similar). The timescale of the SHI grows on the scale of the rotation time. It is independent of both τ_{diff} and τ_{Hall}. The reason is that its energy source is entirely drawn from the differential rotation. It differs strongly from the standard $\alpha\Omega$-dynamo where both the differential rotation and α effect contribute to the energy balance.

Now the momentum equation is added to the system, linearized around the background differential rotation and the axial background field. The critical Reynolds number depends on the two parameters S and R_B. First the magnetic field will be fixed and the R_B is varied. Only the single point (S = 70) is considered of the instability curves for axisymmetric modes in Figure 2.32. The parameter of the Hall effect is now written as $\beta_0 = R_B/S$ which is a small number for neutron stars. For

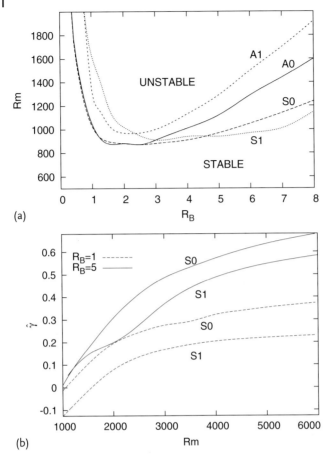

Figure 2.31 Stability curves for the axisymmetric and nonaxisymmetric modes (a). The normalized growth rates of the S0 and S1 modes, for $R_B = 1$ and 5 (b). From Kondić, Rüdiger, and Hollerbach (2011). Reproduced with permission © ESO.

cold protoplanetary disks it reaches the order of unity. For positive (negative) β_0 the field is parallel (antiparallel) to the rotation axis and for $\beta_0 = 0$ the standard MRI is reproduced (see the vertical middle axis of Figure 2.32). This result is in accordance to that of Hall-MRI in disks (Rüdiger and Kitchatinov, 2005, also Section 5.5) but the degree of destabilization for antiparallel fields remained open. Figure 2.32 shows the effect to be rather small. For negative β_0 a minimal Rm exists beyond which the Rm curve turns to the right. The result is that the unstable domain is drastically restricted for negative β_0: the Hall effect allows MRI to exist only in a domain between two limits of rotation (in opposition to the situation for $\beta_0 \geq 0$). A too strong Hall effect for antiparallel fields completely stops the instability.

Figures 2.33 and 2.34 present the curves of marginal instability for Hall-MRI for the model with $r_{\rm in} = 0.7$ and Pm = 1. For comparison in Figure 2.33a also the pure MRI solution without Hall effect is given. The critical magnetic Reynolds

2.7 The Shear–Hall Instability (SHI)

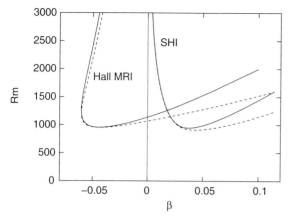

Figure 2.32 The stability map for axisymmetric modes with Hall effect for fixed $S = 70$. It is thus $R_B = 70\beta_0$. The SHI curves only exist for $\beta_0 > 0$. $r_{in} = 0.7$, $Pm = 1$. From Kondić, Rüdiger, and Arlt (2012).

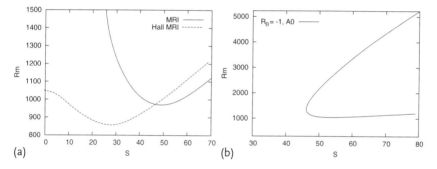

Figure 2.33 The stability map for Hall-MRI (only A0 modes) for $R_B = 0$ and $R_B = 1$ (a) and $R_B = -1$ (b). $x_{in} = 0.7$, $Pm = 1$.

number for A0 modes is $Rm_{min} \simeq 980$ (for S0 it is $Rm_{min} \simeq 950$). For $S \lesssim 20$ no MRI exists.

In contrast, for positive R_B the MRI exists for all S without any lower limit. The critical Rm is much smaller than that for $R_B = 0$. It approaches the same minimum value as it exists for SHI. With positive R_B the MRI is strongly destabilized by the support of SHI (which only exists for $R_B > 0$). Indeed, the SHI is easiest to excite as it does not generate extra modes of kinetic energy.

For negative R_B one finds a dramatic stabilization of the MRI. The minimum S is strongly increased but not the critical Rm (see also Section 5.5). More important, however, is the existence of upper limits of Rm above which no MRI exists. Here, too fast rotation avoids the MRI. In order to be unstable the considered rotation law $(1/R)$ must be subject to magnetic fields with $\Omega_A > 0.02\Omega$. With the characteristic density of neutron stars of 10^{13} g/cm^3 and the typical rotation rate $\Omega \simeq 100$ s^{-1} one finds $\Omega_A/\Omega \simeq 0.001 B_{12}$ with $B_{12} = B_0/(10^{12}$ G$)$ so that only fields exceeding

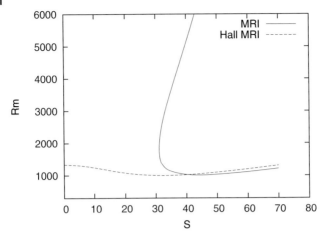

Figure 2.34 The excitation of nonaxisymmetric A modes for $m = 1$ for $R_B = 1$. For standard MRI the excitation fails for too weak fields and too fast rotation. This is not true under the presence of the Hall effect. For the S modes the results are very similar. Courtesy of T. Kondić.

10^{13} G can originate axisymmetric MRI for $R_B \simeq -1$. There is no such restriction for $R_B \geq 0$.

The plots of Figure 2.33 describe the excitation of axisymmetric modes with the Hall effect ($R_B = \pm 1$). The results for the S-modes (symmetric with respect to the equator) are very similar. For positive R_B the Hall-MRI is much easier to excite than the standard MRI. The coordinates for the excitation with the slowest rotation are smaller with the Hall effect than without. Due to the existence of SHI there is no minimal Lundquist number for the instability. There is always a solution for $S = 0$ whose Rm depends on the actual value of R_B. It can be taken for both equatorial symmetries from Figure 2.31a.

For the excitation of nonaxisymmetric modes there are even more puzzling results in the presence of the Hall effect. We shall demonstrate in Section 5.3 how delicate the excitation of nonaxisymmetric modes for MRI is. One needs stronger fields and higher rotation rates to excite modes with $m \neq 0$. It is obvious from Figure 2.34 that in the described MRI model without Hall effect no nonaxisymmetric modes exist for $S \lesssim 30$. The Hall parameter is fixed to $R_B = 1$ and for both symmetry types the Hall effect is destabilizing even for fields which are too weak for the standard MRI. The critical Rm hardly varies with the Lundquist number, and the upper rotation limit for the $m = 1$ modes disappears. The formal reason for this striking phenomenon is the existence of SHI also for nonaxisymmetric modes. Moreover, after Figure 2.31 there are only minor differences for SHI in the eigenvalues for the excitation of axisymmetric and nonaxisymmetric modes. Hence, the Hall effect enables the MRI to be much more nonaxisymmetric than it is without Hall effect. Because of the Cowling theorem this result should be of relevance for the excitation of turbulence and magnetic fields in cool protoplanetary disks.

3
Quasi-linear Theory of Driven Turbulence

The quasi-linear approximation has advantages and known deficits. There is the well-known difficulty of closure. The infinite system of equations, arising from the basic nonlinearity of the governing equations, must be made tractable by a process of linearization. Here we use the second-order correlation approximation (SOCA) in which the third and higher-order statistical moments are neglected, which is certainly allowed if

$$\mathrm{Min}(\mathrm{Rm}, \mathrm{St}) \ll 1 , \qquad (3.1)$$

where, as usual $\mathrm{Rm} = u_{\mathrm{rms}} \ell_{\mathrm{corr}}/\eta$ is the magnetic Reynolds number and $\mathrm{St} = \tau_{\mathrm{corr}} u_{\mathrm{rms}}/\ell_{\mathrm{corr}}$ is the Strouhal number. The SOCA does not trivially ignore the nonlinearity. Only terms such as $(\boldsymbol{u} \cdot \nabla)\boldsymbol{u} - \langle(\boldsymbol{u} \cdot \nabla)\boldsymbol{u}\rangle$, which describe the flux of momentum from one turbulent mode to another, are ignored. The interaction of the turbulence with the nonturbulent background fields remains untouched.

One must be able, however, to distinguish between a fluctuation and the mean-field component of the flow. This question concerns the applicability of the Reynolds rules. It might be argued that, within the framework of statistical physics, no such problem arises since averages are taken over a great number of identical examples. Yet this convenient theoretical concept is not a practical reality. What we really require is a further constraint, that of 'two-scaledness,' that is the fluctuations must be on much smaller scales than the mean fields. The Reynolds rules are then satisfied (see Krause and Rädler, 1980).

The zero-order terms represent the 'original' isotropic turbulence which we imagine in this chapter as driven by a system of homogeneous and isotropic fluctuating forces. Then the spectral tensor for the original turbulence is

$$\hat{Q}^{(0)}_{ij}(\boldsymbol{k}, \omega) = \frac{E(k, \omega)}{16\pi k^2} \left(\delta_{ij} - \frac{k_i k_j}{k^2} \right) - \mathrm{i}\epsilon_{ijk} k_k H(k, \omega) , \qquad (3.2)$$

where the positive-definite spectrum E gives the intensity of isotropic fluctuations, that is

$$\langle \boldsymbol{u}^{(0)2} \rangle = \int_0^\infty \int_0^\infty E(k, \omega) \mathrm{d}k \mathrm{d}\omega , \qquad (3.3)$$

Magnetic Processes in Astrophysics, First Edition. G. Rüdiger, L.L. Kitchatinov, and R. Hollerbach.
© 2013 WILEY-VCH Verlag GmbH & Co. KGaA. Published 2013 by WILEY-VCH Verlag GmbH & Co. KGaA.

and H gives the helical part of the homogenous and isotropic turbulence field. We apply the quasi-linear approximation to derive the higher-order terms in the equations. They are generally found by a perturbation method from the linearized equations. That is, the linearized momentum equation reads as

$$\frac{\partial u_i^{(n)}}{\partial t} - \nu \Delta u_i^{(n)} = -\frac{\partial}{\partial x_k}\left(\bar{U}_k u_i^{(n-1)} + \bar{U}_i u_k^{(n-1)}\right) - \frac{1}{\rho}\frac{\partial p^{(n)}}{\partial x_i}, \qquad (3.4)$$

where the upper index shows the order in the large-scale field or flow. With this equation taken for $n = 1$, the first-order correction, $u^{(1)}$, can be expressed in terms of the given original turbulence. As we shall demonstrate, this procedure proves to be successful in many applications. Both numerical simulations and observations often lead to a striking confirmation of the results of the quasi-linear approximation.

3.1
The Turbulence Pressure

For a driven turbulence under the influence of (say) a vertical background field $B_z = B_0$, the total pressure becomes

$$\frac{P^{\text{tot}}}{\rho} = \langle u_x^2\rangle + \frac{1}{2\mu_0\rho}B_0^2 + \frac{1}{2\mu_0\rho}\langle b_y^2 + b_z^2 - b_x^2\rangle. \qquad (3.5)$$

Note that $\langle b_y^2 + b_z^2 - b_x^2\rangle = \langle b_z^2\rangle$ when isotropy in the horizontal direction can be assumed. If the field is in the y-direction then one changes $\langle b_z^2\rangle \to \langle b_y^2\rangle$. Three positive quantities contribute to the total pressure. One can ask (Rogachevskii and Kleeorin, 2007; Brandenburg, Kleeorin, and Rogachevskii, 2010) whether the magnetic suppression of the kinetic pressure $\langle u_x^2\rangle$ for large B_0 is so strong that the total pressure under the influence of B_0 is smaller than the total pressure for $B_0 = 0$, that is

$$P^{\text{tot}}(B_0) < \rho u_0^2, \qquad (3.6)$$

where u_0^2 denotes the turbulence intensity only in the x-direction without magnetic field. To answer this question the spectral tensor of a driven turbulence under the presence of a large-scale background field can be analyzed for various magnetic Prandtl numbers and for high or low electrical conductivity. The expressions

$$\langle u_x^2\rangle = \frac{1}{16}\int_0^\pi\int_0^\infty\int_{-\infty}^\infty \frac{E(k,\omega)(1 + \cos^2\theta)\sin\theta\,d\theta}{1 + 2S^{*2}\Phi\cos^2\theta + S^{*4}\cos^4\theta}\,dk\,d\omega \qquad (3.7)$$

for the horizontal turbulence intensity and

$$\langle u_z^2\rangle = \frac{1}{8}\int_0^\pi\int_0^\infty\int_{-\infty}^\infty \frac{E(k,\omega)(1 - \cos^2\theta)\sin\theta\,d\theta}{1 + 2S^{*2}\Phi\cos^2\theta + S^{*4}\cos^4\theta}\,dk\,d\omega \qquad (3.8)$$

for the intensity parallel to the magnetic field with

$$S^* = \frac{B_0 k/\sqrt{\mu_0 \rho}}{\sqrt[4]{(\omega^2 + \nu^2 k^4)(\omega^2 + \eta^2 k^4)}}, \quad \Phi = \frac{\nu \eta k^4 - \omega^2}{\sqrt{(\omega^2 + \nu^2 k^4)(\omega^2 + \eta^2 k^4)}} \quad (3.9)$$

(Rüdiger, 1974) shall be analyzed numerically. The latter definitions can also be written as

$$S^* = \frac{S}{\sqrt[4]{(1 + \omega^{*2})(\text{Pm}^2 + \omega^{*2})}}, \quad \Phi = \frac{\text{Pm} - \omega^{*2}}{\sqrt{(1 + \omega^{*2})(\text{Pm}^2 + \omega^{*2})}} \quad (3.10)$$

with the Lundquist number

$$S = \frac{B_0}{\sqrt{\mu_0 \rho} k \eta} \quad (3.11)$$

and with the turbulence frequency $\omega^* = \omega/\eta k^2$, normalized with the diffusion time.

The integrals may be calculated by use of the simple spectral function

$$E = \frac{12}{\pi} u_0^2 \delta\left(k - \ell_{\text{corr}}^{-1}\right) \frac{w^3}{(\omega^2 + w^2)^2} \quad (3.12)$$

as a characteristic example of spectra monotonically decreasing for increasing frequency. Note that the latter factor forms a δ function for $w \to 0$. The characteristic frequency w is the inverse to the correlation time τ_{corr} of the turbulence, that is $w = 2/\tau_{\text{corr}}$. Transforming the last factor of (3.12) to ω^* as the frequency variable one finds

$$E = \frac{12}{\pi} \frac{u_0^2}{\eta k^2} \delta\left(k - \ell_{\text{corr}}^{-1}\right) \frac{w^{*3}}{(\omega^{*2} + w^{*2})^2} \quad (3.13)$$

with $w^* = w/\eta k^2$, which can also be written as $2\mu_0 \sigma \ell_{\text{corr}}^2 / \tau_{\text{corr}}$. Hence, large w^* represent the high-conductivity limit while small w^* represent the low-conductivity limit. In the low-conductivity limit the frequency spectrum can be approached by a δ function while the high-conductivity limit transforms the spectrum to a quasi-white noise behavior. The normalized frequency w^* then represents the magnetic Reynolds number of the fluctuations.

For 'large eddy' convection with (say) giant cells we have often used a mixing-length model of the spectrum with two basic characteristics. On the one hand, the life times of the eddies exceed by far the diffusion time, so that a Dirac function $\delta(\omega)$ for the frequency spectrum

$$E(k, \omega) = 2u_0^2 \delta\left(k - \ell_{\text{corr}}^{-1}\right) \delta(\omega) \quad (3.14)$$

should be applicable. On the other hand, the effective diffusivity in such a flow pattern can be described by $\eta_t \simeq \ell_{corr}^2/\tau_{corr}$, so that $w^* \simeq 1$ and St = Pm $\simeq 1$ should also be reasonable assumptions (Kitchatinov, 1991). We have used this 'mixing-length' model for the calculations of many diffusive and nondiffusive turbulence-induced quantities in convection zones.

The results of the integrations are given by Figure 3.1. The quantity $\langle u_x^2 \rangle / u_0^2$ is plotted vs. the Lundquist number S, for various values of the magnetic Prandtl number, and for the high-conductivity values $w^* = 5$ and $w^* = 10$. The parameter w^* can be understood as the magnetic Reynolds number $Rm' = u_0 \ell_{corr}/\eta$ divided by the Strouhal number St = $u_0/(\ell_{corr} w)$, that is

$$w^* = \frac{Rm'}{St} \simeq Rm', \qquad (3.15)$$

the latter taken for the standard case St $\simeq 1$.

One analytically finds the quantity $\langle u_x^2 \rangle / u_0^2$ for large Lundquist numbers S as magnetically quenched like 1/S, independently of the values of Pm and w^* (see Figure 3.1b). It can also be shown for all spectra which monotonically decrease for increasing frequency that for sufficiently small S there is always a magnetic-induced quenching of the turbulence (Figure 3.1a). For larger S there is an antiquenching phenomenon for high conductivity, that is for large w^*. The normalized value of $\langle u_x^2 \rangle$ increases rather than decreases. This is a consequence of the fact that $\Phi < 0$ for $\omega^* > 1$, so that the denominator in (3.7) becomes small. For low dissipation Alfvén waves are excited by the driven turbulence, and the turbulent pressure is formally increased by the magnetic field. For sufficiently strong fields the turbulence is always suppressed as 1/S (Figure 3.1b).

Equation (3.7) after a series expansion for small fields and for Pm = 1 becomes

$$\frac{\langle u_x^2 \rangle}{u_0^2} = 1 + \frac{2}{15} \frac{B_0^2}{\mu_0 \rho u_0^2} \int_0^\infty \int_{-\infty}^\infty \frac{\omega k^2}{\omega^2 + \eta^2 k^4} \frac{\partial E(k, \omega)}{\partial \omega} dk d\omega, \qquad (3.16)$$

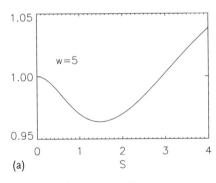
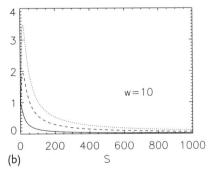

Figure 3.1 The magnetic influence on the Reynolds pressure $\langle u_x^2 \rangle / u_0^2$ for high conductivity. For weak fields the turbulence is slightly quenched for very weak fields and amplified for moderate fields. Pm = 1 (a). The turbulence is suppressed by strong magnetic fields like 1/S. The curves are for Pm = 0.1 (dotted), Pm = 1 (dashed) and Pm = 10 (solid). The amplification for moderate S only exists for Pm $\lesssim 1$ (b).

so that for all spectra with $\partial E/\partial \omega < 0$ the magnetic field reduces the turbulence intensity. After some algebra and use of the spectrum (3.12) one obtains

$$\frac{\langle u_x^2 \rangle}{u_0^2} = 1 - \kappa \beta^2 \tag{3.17}$$

with

$$\kappa \simeq \frac{4}{5} \frac{w^{*2}(1 + 3w^*)}{(1 + w^*)^3} \tag{3.18}$$

for St $\simeq 1$, which is small for low conductivity and reaches unity for high conductivity. In Figure 3.1a the quenching is maximum for S = 1.4, that is for $\beta = 0.3$. The magnetic quenching is thus (only) about 4%, which is indeed the reported value (Brandenburg, Kleeorin, and Rogachevskii, 2010) for the total pressure (3.5).

The Lundquist number S has a simple relation to the magnetic background field B_0 normalized with its equipartition value $B_{eq} = \sqrt{\mu_0 \rho} u_0$, that is

$$\beta = \frac{B_0}{B_{eq}} \simeq \frac{S}{w^* \mathrm{St}}. \tag{3.19}$$

For Strouhal number St = 1 the influence of β on the turbulent pressure is shown in Figure 3.2. Obviously, the equipartition value of the magnetic field does not play a particular role. For $\beta \simeq 1$ one finds an increase of the horizontal turbulence intensity $\langle u_x^2 \rangle$ in the high-conductivity regime and a slight reduction of the pressure for $w^* = 1$, that is for $\eta \simeq u_0 \ell_{\mathrm{corr}}$. As expected, there is no magnetic-induced change of the turbulence pressure in the low-conductivity regime $w^* < 1$. As can be seen in Figure 3.2 the relation (3.6) can be maximally fulfilled for a magnetic Reynolds number of $w^* \simeq 1$. In this case the magnetic quenching for Pm $\simeq 1$ is so strong already for $B_0 \lesssim B_{eq}$ that the Reynolds pressure under the influence of magnetic fields provides 'negative' contributions to the total pressure (Brandenburg, Kleeorin, and Rogachevskii, 2010; Brandenburg et al., 2011; Käpylä et al.,

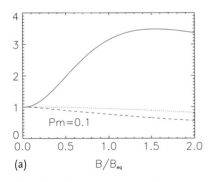
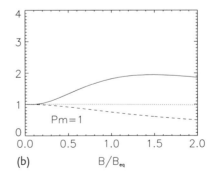

Figure 3.2 The normalized turbulent pressure $\langle u_x^2 \rangle / u_0^2$ under the influence of uniform magnetic fields for magnetic Prandtl numbers, 0.1 (a), 1.0 (b). For small Pm the turbulence is magnetically quenched only for medium conductivity while it is amplified for high conductivity and only weakly influenced in the low-conductivity regime. $w^* = 10$ (solid), $w^* = 1$ (dashed) and $w^* = 0.1$ (dotted).

2012; Kemel et al., 2012a,b). This phenomenon is quite typical for the magnetic suppression of turbulence, but only for $B_0 \lesssim B_{\mathrm{eq}}$. The question is whether this quenching is already overcompensated by the second term in (3.5), which can be written as $0.5\beta^2$ after normalization with u_0^2.

The curves in Figure 3.3 represent the sum of Reynolds stress and large-scale Maxwell stress, that is $\langle u_x^2 \rangle / u_0^2 + 0.5\beta^2$. The second term dominates the Reynolds pressure for not too small β so that almost everywhere

$$\frac{\langle u_x^2 \rangle}{u_0^2} + \frac{1}{2}\beta^2 \geq 1 . \tag{3.20}$$

An exception exists for Pm = 0.1 and $w^* = 1$. At $\beta \simeq 0.25$ the effective pressure is (slightly) smaller than the reference value u_0^2. In this case the magnetic quenching of the turbulence intensity from (3.17) is stronger than the Maxwell stress of the large-scale field. The effect, however, is small and is restricted to magnetic fields which are basically weaker than the equipartition value $\mu_0 \rho u_0^2$.

Next the positive-definite contribution of the Maxwell correlation, that is the last term in (3.5), is discussed. We are in particular interested in the ratio

$$q = \frac{\langle b_z^2 \rangle}{B_0^2} , \tag{3.21}$$

which according to Bräuer and Krause (1974) scales with Rm' in the high-conductivity limit and with Rm'^{2} in the low-conductivity limit. The q here only describes the magnetic fluctuations in the z-direction. The integral is

$$q = \frac{1}{8} \int_0^\pi \int_0^\infty \int_{-\infty}^\infty \frac{Ek^2 \cos^2\theta (1 - \cos^2\theta) \sin\theta \, d\theta}{(\omega^2 + \eta^2 k^4)(1 + 2S^{*2}\Phi \cos^2\theta + S^{*4} \cos^4\theta)} dk d\omega , \tag{3.22}$$

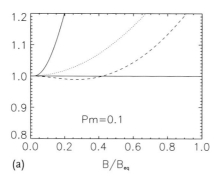

(a)

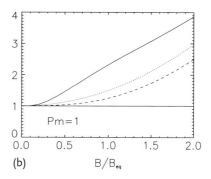

(b)

Figure 3.3 The two first parts of (3.5), that is Reynolds stress pressure plus large-scale Maxwell pressure. The curve for $w^* = 1$ crosses the indicated horizontal line of the original pressure at $\beta \simeq 0.40$ ('negative pressure' phenomenon). The third part of (3.5) is not included but is small for these parameters. $w^* = 10$ (solid), $w^* = 1$ (dashed) and $w^* = 0.1$ (dotted). Pm = 0.1 (a), 1.0 (b).

for which the relation $q \propto 2/5w^*$ for $\eta \to 0$ can immediately be derived. It is clear, therefore, that $q \gg 1$ is the rule rather than the exception in cosmical physics. For laboratory experiments with fluid conductors like sodium or gallium the magnetic Reynolds number typically does not exceed unity, so that the magnetic fluctuations do not play an important role. The same is true, according to our results, for fields that are too strong (such as in Ap stars and pulsars).

The numerical results are given in Figure 3.4. They show that (i) there are no remarkable values for q in the low-conductivity limit ($w^* < 1$) and (ii) for high conductivity ($w^* > 1$) the relation $q \propto 2w^*/5$ is fulfilled for weak fields. The magnetic suppression of q already starts for small β. The main result of the calculation is that the existence of $q > 1$ requires high conductivity ($w^* > 1$). MHD turbulence with dominating field fluctuations requires a high conductivity of the gas. The situation does not change for small Pm (Figure 3.4b), except that the magnetic suppression becomes weaker. Generally speaking, the Maxwell stress by small-scale magnetic fluctuations becomes very large only for large magnetic Reynolds numbers. For laboratory experiments the values of q are smaller than unity. For galaxies values of $q \lesssim 10$ have been derived from observations (Beck, 2007; Chyzy, 2008) and simulations (Gressel et al., 2008). Among the various galaxies the q increases with increasing star formation rate, that is with increasing magnetic Reynolds number. Beck (2007) finds the fields of stronger regularity in the interarm region of spiral galaxies also confirming the relation $q \propto Rm'$. By contrast, solar observations lead to $q \simeq 100-500$ (see Section 3.4.2).

Figure 3.5 shows the magnetic energy $\langle b_z^2 \rangle/(2\mu_0\rho)$ normalized with the equipartition value u_0^2 of the nonmagnetic turbulence, which indicates the contribution of the small-scale Maxwell tensor to the total pressure (3.5). This quantity also scales with the magnetic Reynolds number and thus becomes very large for very large w^*, so that this term will finally dominate the total pressure in astrophysical applications. For medium magnetic Reynolds numbers ($w^* \simeq 1$) this quantity is remarkably small. The normalized Maxwell pressure exceeds unity for $B = B_{eq}$ only for

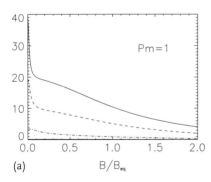
(a)

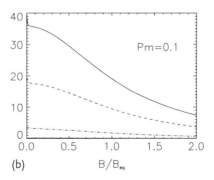
(b)

Figure 3.4 The ratio q of the energy of the fluctuating magnetic z-components to the energy of the background field vs. the normalized background field for various magnetic Reynolds numbers w^*. The lines are labeled as follows: dotted $w^* = 1$, dot-dashed $w^* = 10$, dashed $w^* = 50$, solid $w^* = 100$. Pm = 1 (a), Pm = 0.1 (b).

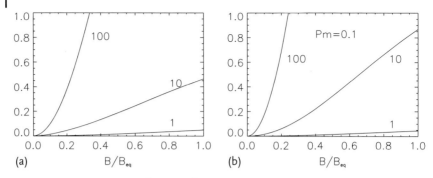

Figure 3.5 The energy ratio $\langle b_z^2 \rangle/(2\mu_0\rho u_0^2)$ vs. the normalized background field for various magnetic Reynolds numbers w^*. Note the smallness of the numerical values for $w^* \leq 1$ compared with the values of the first two parts of the total pressure. Pm = 1 (a), Pm = 0.1 (b).

$w^* \gtrsim 10$. For weaker fields, for example $\beta = 0.5$ one needs $w^* \gtrsim 50$ to exceed unity. For these magnetic Reynolds numbers we always have $P_{\text{tot}}/\rho > u_0^2$, so that no 'negative' turbulent pressure can exist by the too strong magnetic quenching of the turbulence.

Magnetoconvection has been simulated with the NIRVANA CODE in a box with overshoot region under the presence of a vertical field (Figure 3.6). The magnetic Prandtl number of the fluid is Pm = 0.1. The simulations have been described in more detail in Section 3.4.2 which, however, operate with much weaker external fields. Figure 3.6a describes the magnetic quenching of the horizontal turbulence intensity $\langle u_x^2 \rangle$. Its magnetic-induced reduction is proportional to the strength of the external field. The second term in (3.5) quadratically grows for growing external field so that for the large field in Figure 3.6b the magnetic reduction of $\langle u_x^2 \rangle$ is overcompensated. This is not the case, however, for the medium magnetic field (dotted line) as can be seen in Figure 3.6d. The pressure for the medium field (here $\bar{B}_z = 10$) is weaker than the pressure for $\bar{B}_z = 0$ (dashed) and/or for $\bar{B}_z = 20$ (solid). The reason is the smallness of the turbulent magnetic pressure which is (i) small and (ii) only slightly depends on the applied magnetic field (Figure 3.6c). Indeed, in Figure 3.6d the total pressure for modest magnetic fields can be smaller than the kinetic pressure without magnetic field and the total pressure with stronger magnetic fields (see Käpylä et al., 2012).

Brandenburg, Kleeorin, and Rogachevskii (2010) introduced the notation

$$P^{\text{tot}} = \rho u_0^2 \left[1 + (1 - \kappa_p) \frac{B_0^2}{2\mu_0 \rho u_0^2} \right], \qquad (3.23)$$

where κ_p describes the turbulence-induced reduction of the magnetic pressure by the background field. For strong fields both the kinetic and magnetic fluctuations are so strongly suppressed that $\kappa_p \to 0$. If somewhere $\kappa_p > 1$ then the magnetic part of the turbulence pressure would lead to a reduction of the pressure without magnetic field. Figure 3.7 gives the numerical value of κ_p for various amplitudes of the magnetic background field for the mentioned mixing-length model with very long correlation time (see the spectrum (3.14)) and $w^* = \text{St} = \text{Pm} = 1$. Just for

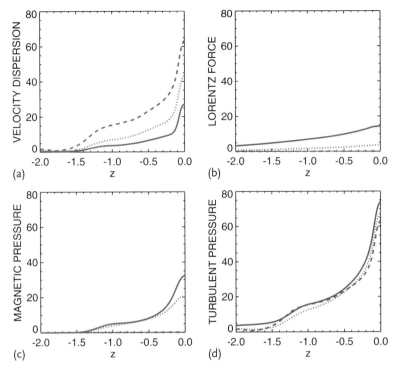

Figure 3.6 The terms in (3.5) taken from a box simulation of a convective box (z from -1 to 0) with overshoot zone ($z < -1$) for various external magnetic fields: $\bar{B}_z = 0$ (dashed), $\bar{B}_z = 10$ (dotted), $\bar{B}_z = 20$ (solid). The first (a) and the second term (b) of the RHS of (3.5). The small-scale Maxwell pressure (c) and the total pressure (d). Pm = 0.1. Courtesy of M. Küker.

this turbulence model the value of κ in (3.17) is 2/5, which is always overcompensated by the second term in (3.20), so that the sum (3.20) is positive-definite and no negative-pressure phenomenon can exist for this model.

The equations have been described in detail by Rüdiger, Kitchatinov, and Schultz (2012) where the total Lorentz force in magnetized turbulent media is written as

$$F = (1-\kappa)\bar{J} \times \bar{B} - \frac{1}{2\mu_0}(\kappa - \kappa_p)\nabla \bar{B}^2 \,. \tag{3.24}$$

The turbulence leads to a reduction of the 'laminar' Lorentz force and the appearance of an extra pressure term if the κ are unequal. It is obvious from Figure 3.7 that no magnetic field exists for which $\kappa \geq 1$ and always $\kappa_p > \kappa$. Kleeorin, Rogachevskii, and Ruzmaikin (1989) suggest κ_p even larger than unity so that the total magnetic pressure becomes negative. The resulting instability may produce structures of concentrated magnetic field and has been suggested to be important for the sunspot formation (Kleeorin, Rogachevskii, and Ruzmaikin, 1989; Brandenburg, Kleeorin, and Rogachevskii, 2010).

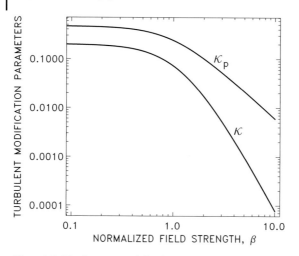

Figure 3.7 The functions κ defined by (3.23) and (3.24) for the mixing-length model (3.14) vs. the magnetic background field defined by (3.19). The turbulence model can be defined by $w^* = \mathrm{St} = \mathrm{Pm} = 1$.

3.2
The η-Tensor

The magnetic-diffusivity tensor η_{ijp} relates the mean electromotive force (EMF) $\mathcal{E} = \langle \boldsymbol{u} \times \boldsymbol{b} \rangle$ to the gradients of the mean magnetic field via the relation

$$\mathcal{E}_i = \eta_{ijp} \bar{B}_{j,p} . \tag{3.25}$$

It is clear from its definition that the η-tensor is a pseudotensor. In the simplest case it is formed by the ϵ-tensor, $\eta_{ijp} = \eta_T \epsilon_{ijp}$, with η_T as the turbulent diffusivity. We shall show in the following how this simple result is modified if (i) the basic rotation of the stellar turbulence and (ii) higher orders of the mean magnetic field are included. In case (i) we speak about the rotational quenching of the turbulent diffusivity, whereas in case (ii) the expressions describe the SOCA theory of the magnetic η-quenching.

3.2.1
Rotating Turbulence

Another simple example is $\eta_{ijp} = \cdots + \eta_{\|}\epsilon_{ijm}\Omega_p\Omega_m/\Omega^2$ which leads to the total diffusivity $\eta_T + \eta_{\|}$ in front of the z-derivatives in the Δ-operator, that is

$$\mathrm{curl}\,\mathcal{E} = \eta_T \left(\frac{\partial^2 \bar{B}}{\partial x^2} + \frac{\partial^2 \bar{B}}{\partial y^2} \right) + (\eta_T + \eta_{\|}) \frac{\partial^2 \bar{B}}{\partial z^2} , \tag{3.26}$$

if z is the coordinate parallel to the rotation axis. It is natural to define the total diffusivity in the z-direction, that is parallel to the rotation axis, as $\eta_{zz} = \eta_T + \eta_{\|}$.

On the other hand, terms containing odd powers of Ω can only exist without ϵ-tensor. Following Kitchatinov, Pipin, and Rüdiger (1994) we write

$$\eta_{ijp} = \eta_T \epsilon_{ijp} + (a-b)\delta_{ip}\Omega_j - b\delta_{ij}\Omega_p + \eta_{\|}\epsilon_{ijm}\frac{\Omega_p \Omega_m}{\Omega^2} + c\frac{\Omega_i \Omega_j \Omega_p}{\Omega^2}. \tag{3.27}$$

While the term with $a - b$ for rigid rotation forms a gradient and vanishes after taking the curl, we have $\mathcal{E}_i = \cdots + b(\boldsymbol{\Omega} \times \bar{\boldsymbol{J}})_i$ so that the numerical value of b represents the so-called $\boldsymbol{\Omega} \times \boldsymbol{J}$ effect (Rädler, 1969), which is known to vanish for delta-like frequency spectra (the mixing-length model).

To find the influence of field gradients on the mean electromotive force $\langle \boldsymbol{u} \times \boldsymbol{b} \rangle$ at linear order it is enough to solve the induction equation

$$\frac{\partial b_i}{\partial t} - \eta \Delta b_i = x_p B_{jp} u_{i,j} - u_j B_{ij}, \tag{3.28}$$

where the inhomogeneous mean magnetic field has been introduced by $\bar{B}_j = B_{jp}x_p$, hence $\bar{B}_{j,p} \equiv B_{jp}$. Note that the global rotation only appears in the Navier–Stokes equation for the fluctuation $\boldsymbol{u}$, which remains homogeneous if only expressions linear in B_{jp} are envisaged. One can thus work with

$$u_i(\boldsymbol{x}, t) = \iint \hat{u}_i(\boldsymbol{k}, \omega) e^{i(\boldsymbol{k}\boldsymbol{x}-\omega t)} d\boldsymbol{k} d\omega,$$

$$b_i(\boldsymbol{x}, t) = \iint \left[\hat{b}_i(\boldsymbol{k}, \omega) + x_l \hat{b}_{il}(\boldsymbol{k}, \omega)\right] e^{i(\boldsymbol{k}\boldsymbol{x}-\omega t)} d\boldsymbol{k} d\omega. \tag{3.29}$$

The result is

$$\hat{b}_i = -\frac{B_{ij} + (2\eta k_l k_m B_{lm}\delta_{ij})/(-i\omega + \eta k^2)}{-i\omega + \eta k^2}\hat{u}_j, \quad \hat{b}_{il} = \frac{ik_j B_{jl}}{-i\omega + \eta k^2}\hat{u}_i \tag{3.30}$$

with $ik_j \hat{b}_j + \hat{b}_{jj} = 0$. Hence, for homogeneous turbulence

$$\eta_{ijp} = \iint \left(\epsilon_{ijq}\hat{Q}_{qp} + \epsilon_{ikq}\frac{2\eta k_j k_p}{i\omega + \eta k^2}\hat{Q}_{qk}\right)\frac{d\boldsymbol{k}d\omega}{i\omega + \eta k^2}. \tag{3.31}$$

The spectral tensor of rotation under the influence of rigid rotation is well known, that is $\hat{Q}_{ij} = D_{ip}D^*_{jq}\hat{Q}^{(0)}_{pq}$ with $\hat{Q}^{(0)}_{pq}$ as the spectral tensor (3.2) of the undisturbed 'original' turbulence[1]. The rotation operator D is

$$D_{ij} = \frac{\delta_{ij} + \frac{2(\boldsymbol{k}\boldsymbol{\Omega}/k)}{-i\omega + \nu k^2}\epsilon_{ijp}\frac{k_p}{k}}{N} \tag{3.32}$$

---

1) D^* and N^* are conjugate complex to D and N.

with $N = 1 + (2k\boldsymbol{\Omega}/k)^2/(-i\omega + \nu k^2)^2$. One finds[2]

$$\eta_T = \iint \frac{\eta k^2}{\omega^2 + \eta^2 k^4}\left[1 + \frac{4\Omega^2 \cos^2\theta}{\omega^2 + \nu^2 k^4}\right] \frac{\hat{Q}^{(0)}_{11}}{NN^*} dk d\omega,$$

$$\eta_{zz} = \iint \frac{\eta k^2}{\omega + \eta^2 k^4}\left[1 + \frac{4\Omega^2 \cos^2\theta}{\omega^2 + \nu^2 k^4}\right] \frac{\hat{Q}^{(0)}_{33}}{NN^*} dk d\omega,$$

$$b = \iint \frac{16\omega^2 \eta^2 k_1^2 k_3^2}{(\omega^2 + \eta^2 k^4)^2(\omega^2 + \nu^2 k^4)} \frac{\hat{Q}^{(0)}_{jj}}{NN^*} dk d\omega. \tag{3.33}$$

Basic quantities in the mean-field MHD of turbulent media like the turbulent eddy viscosity, the eddy diffusivity and the α effect remain finite for ideal media with their infinite Reynolds numbers, for example

$$\eta_T = \frac{1}{6}\int_0^\infty \int_{-\infty}^\infty \frac{\eta k^2 E(k,\omega)}{\omega^2 + \eta^2 k^4} dk d\omega \to \tau_{\text{corr}} u_0^2 \quad \text{for} \quad \eta \to 0. \tag{3.34}$$

This is not true for the modifications of these quantities under the influence of magnetic fields and/or basic rotation (see below).

The integrals are evaluated with the spectrum (3.12) for various rotation rates $\Omega^* = 2\tau_{\text{corr}}\Omega$ and for high conductivity ($w^* = 10$), medium conductivity ($w^* = 1$) and low conductivity ($w^* = 0.1$). Figure 3.8 shows that for fast rotation the relation $\eta_{zz} \simeq 2\eta_T$ is a suitable approximation. For slow rotation it is simply $\eta_{zz} \simeq \eta_T$, that is the diffusion

$$\eta_T \simeq \frac{2w + \eta k^2}{(w + \eta k^2)^2} u_0^2 \tag{3.35}$$

is the same in all directions and approaches in the high-conductivity limit $\tau_{\text{corr}} u_0^2$ as in (3.34). In the low-conductivity limit the eddy diffusivity scales linearly with the magnetic Reynolds number w^*, and is thus very small. Calculated with the same quasi-linear method the eddy viscosity follows from the relation $\nu_T = \text{Pm}_T \cdot \eta_T$ with $\text{Pm}_T = 0.4$ (Nakano et al. (1979), (for details see Yousef, Brandenburg, and Rüdiger, 2003)).

For rapid rotation the diffusion coefficients are quenched (only) as $1/\Omega^*$, and the influence of the microscopic conductivity value disappears (see Figures 3.8). The anisotropy of the turbulent diffusion in cylindrical geometry plays an important role for the excitation of nonaxisymmetric magnetic modes in dynamo models. In direct simulations of rotating cylinders with a Roberts flow Tilgner (2004) found the occurrence of axisymmetric solutions under the influence of an increased diffusion in the direction parallel to the rotation axis. Chabrier and Küker (2006) also included the effects of the anisotropic diffusion in their simulations of the dynamo problem for fully convective stars. Donati et al. (2006) analyzed a rapidly rotating fully convective dwarf star by surface imaging from time series data and found the

[2] $\eta_T \equiv \eta_{312}, \eta_{zz} \equiv \eta_{123}, b\Omega \equiv -\eta_{223}$.

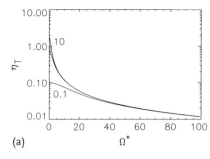
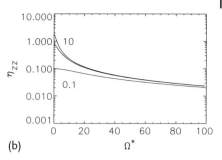

Figure 3.8 The eddy diffusivity for the horizontal coordinates (a) and for the coordinate parallel to the rotation axis (b), both normalized with $\eta_0 = u_0^2 \tau_{\text{corr}}$. The curves are marked with their characteristic magnetic Reynolds number w^*. $\eta_{zz} \simeq 2\eta_T$ proves to be a suitable approximation, but only for rapid rotation. Pm = 1.

magnetic field axis aligned with the rotation axis. This result suggests that fully convective stars are able to produce axisymmetric fields even without differential rotation which should be rather exceptional for thick convection zones (Rüdiger, Elstner, and Ossendrijver, 2003). The problem has thus been reformulated on the basis of an anisotropic η-tensor. Table 3.1 gives the results. For isotropic η-components ($\epsilon = 0$) the magnetic mode with the lowest eigenvalue is a nonaxisymmetric one. For anisotropic η-components ($\epsilon = 2$) one finds axisymmetric dipoles as the favored solutions. The transition for the simple dynamo model of fully convective stars happens for ϵ-values somewhere between 1 and 2. It is thus shown that the rotationally induced increase of the diffusion along the z-axis indeed favors the excitation of axisymmetric dynamo solutions which are normally suppressed by the anisotropic α effect.

The coefficient $\eta_b = b\Omega$ of the $\boldsymbol{\Omega} \times \boldsymbol{J}$ effect is positive-definite and very small for low conductivity (Figure 3.9, also Kitchatinov, Pipin, and Rüdiger (1994); Brandenburg et al. (2008)). Note the strong rotational quenching (as $1/\Omega^{*2}$). For the mixing-length model (3.14) it even completely vanishes due to the term ω^2 in the numerator. For high conductivity the b is larger but only for slow rotation as for high conductivity the rotational quenching is particularly strong.

The positivity of η_b for both hemispheres is confirmed by the numerical simulations of supernova-driven hydromagnetic interstellar turbulence in rotating galaxies (Gressel, Elstner, and Rüdiger, 2011; Gressel et al., 2008). The details of the

Table 3.1 Critical dynamo number for dynamo models for fully convective stars with anisotropic α effect ($\alpha_{zz} = 0.3\alpha_{\phi\phi}$) and with enhanced eddy diffusivity in the axial direction ($\eta_{zz} = (1 + \epsilon)\eta_T$). The minimum eigenvalues for dynamo excitation are marked in bold. From Elstner and Rüdiger (2007).

ϵ	A0	S0	A1	S1
0	10.88	11.19	9.48	**8.61**
1	14.08	15.04	14.80	**13.66**
2	**16.39**	17.83	17.96	18.00

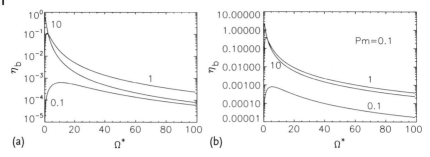

Figure 3.9 The coefficient $\eta_b = b\Omega$ normalized with η_0 of the $\Omega \times J$ effect vs. the Coriolis number Ω^* for Pm = 1 (a) and for Pm = 0.1 (b). The curves are marked with their characteristic magnetic Reynolds numbers w^*.

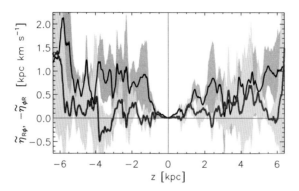

Figure 3.10 The η_b from the simulations of Gressel, Elstner, and Rüdiger (2011) is positive in both hemispheres. Shaded areas indicate 1σ-fluctuations.

simulations are described in Section 4.2. It has been demonstrated there that the η-components perpendicular to the rotation axis are both of order 4 kpc km/s for a rotation with a Coriolis number Ω^* of order unity. Figure 3.10 shows η_b of only 25% of η_T, in very good accordance with the results given in Figures 3.8 and 3.9. The differences of the nonlinear numerical results and the results of the SOCA approximation are here only small for applications with driven turbulence.

3.2.2
Nonrotating Turbulence but Helical Background Fields

The calculation of the EMF under the joint influence of a large-scale magnetic field *and* a large-scale electric current is more complicated than that for rotating turbulence. One also needs the spatial series expansion of the Fourier mode of the fluctuating inhomogeneous velocity. We write

$$u_i = \iint \left[\hat{u}_i^{(0)}(\mathbf{k}, \omega) + \hat{u}_i^{(1)}(\mathbf{k}, \omega) + x_p \hat{u}_{ip}^{(1)}(\mathbf{k}, \omega) + x_p x_q \hat{u}_{ipq}^{(1)}(\mathbf{k}, \omega) \right]$$
$$\times e^{i(\mathbf{k}\mathbf{x} - \omega t)} d\mathbf{k} d\omega ,$$

where $\hat{u}_i^{(0)}$ gives the spectral mode of the homogeneous 'original' turbulence. With (3.30) it follows in zeroth order in x that

$$\mathcal{E}_t = -\epsilon_{tjk} \iint \frac{B_{km}}{i\omega + \eta k^2} \hat{Q}_{jm} d\mathbf{k} d\omega . \tag{3.36}$$

This is the simplest expression which can be obtained for the turbulent EMF. One immediately finds $\langle \mathbf{u} \times \mathbf{b} \rangle = -\eta_T \text{curl}\, \bar{\mathbf{B}}$ with η_T according to the first line of (3.33) (for $\Omega = 0$), which even exists for ideal MHD ($w^* \to \infty$).

The terms of second-order in x are

$$\mathcal{E}_t = \epsilon_{tjk} x_p x_q \iint \left(L_{pq} B_{kn} \hat{Q}_{jn} - \frac{2k_j k_n}{k^2} L_{pm} B_{nq} \hat{Q}_{mk} \right) \frac{d\mathbf{k} d\omega}{i\omega + \eta k^2} \tag{3.37}$$

with $L_{pq} = k_r k_s B_{rp} B_{sq} / [(-i\omega + \nu k^2)(-i\omega + \eta k^2)]$. After some manipulations the final expression

$$\mathcal{E}_t = \frac{1}{5\mu_0 \rho} \int_0^\infty \int_{-\infty}^\infty \frac{\eta k^4 (\nu \eta k^4 - \omega^2) E d k d\omega}{(\omega^2 + \eta^2 k^4)^2 (\omega^2 + \nu^2 k^4)}$$

$$\times \left\{ \bar{\mathbf{B}}^2 \text{curl}\, \bar{\mathbf{B}} + \frac{1}{3} \left[\nabla \bar{\mathbf{B}}^2 + (\bar{\mathbf{B}} \cdot \nabla) \bar{\mathbf{B}} \right] \times \bar{\mathbf{B}} \right\} \tag{3.38}$$

is obtained (Kitchatinov, Pipin, and Rüdiger, 1994). There are only two basic terms. The first one represents the nonlinear quenching of the eddy diffusivity (3.35) while the second one forms an inverse diamagnetic pumping (Roberts and Soward, 1975). We do not find a term $(\bar{\mathbf{B}} \cdot \text{curl}\, \bar{\mathbf{B}}) \bar{\mathbf{B}}$ representing an α effect with the current helicity $\bar{\mathbf{B}} \cdot \text{curl}\, \bar{\mathbf{B}}$ of the helical background field as coefficient. For nonhelically driven nonrotating turbulence the current helicity has no impact with the α effect. Note that in this consideration the turbulence is forced only hydrodynamically rather than magnetically (see, however, Chapter 6).

The magnetic quenching of the eddy diffusivity may be written in the form

$$\eta_T = \tau_{\text{corr}} u_0^2 (1 - a_0 S^2) , \tag{3.39}$$

which can only be true for sufficiently small S. An extended calculation of the integral term in (3.38) for Pm $= 1$ and for the high-conductivity limit leads to $a_0 \simeq 0.3$. Obviously, the quadratic quenching expression (also heuristic expressions such as $1/(1+a_0\beta^2)$ are concerned) has a very restricted meaning: $S = 1$ leads to $\beta = 1/w^*$ which is very small. For $S = 1$ the magnetically quenched η_T is still 70%, which is also true for $\beta = 1$ for the mixing-length model (3.14).

The term $\nabla \bar{\mathbf{B}}^2$ represents an induction velocity which recalls the well-known diamagnetic effect γ in the α expression (3.83, below). The latter scales with $-\nabla \langle \mathbf{u}^2 \rangle$, and transports the magnetic flux towards domains of weaker turbulence. In opposition to that the term $\nabla \bar{\mathbf{B}}^2$ transports magnetic flux towards stronger magnetic fields, thus forming a phenomenon of antidiffusion. The magnetic η-quenching is thus increased and the factor a_0 in (3.39) is larger with this term than without.

On the other hand, for strong fields the magnetic suppression of this term is much weaker than the suppression of the standard expression (3.34) for η_T. Just this situation is demonstrated by Figure 3.11, where Figure 3.11a results from the decay of a helical magnetic field $(\cos z, \sin z, 0)$ (i.e., $\nabla \bar{B}^2 = 0$), while Figure 3.11b results from the decay of the nonhelical field $(\cos z, 0, 0)$ (i.e., $\nabla \bar{B}^2 \neq 0$). The influence of the antidiffusion term $\nabla \bar{B}^2$ is visible in Figure 3.11b by the two facts that (i) for small β the eddy diffusivity is smaller than that of Figure 3.11a and (ii) for large β the resulting diffusivity is larger than that of Figure 3.11a. Indeed, the behavior of the two curves (without and with $\nabla \bar{B}^2$) corresponds to the behavior of the two curves φ and φ_{tot} in Figure 3.12.

The magnetic influence on the turbulent magnetic Prandtl number can be taken from the same simulations (Yousef, Brandenburg, and Rüdiger, 2003). The

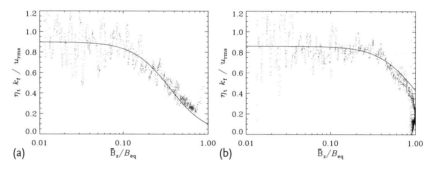

Figure 3.11 The magnetic quenching of the eddy diffusivity by driven turbulence under the influence of a helical field with $\nabla \bar{B}^2 = 0$ (a) and a nonhelical field with $\nabla \bar{B}^2 \neq 0$ (b). The vertical gradient of the magnetic energy reduces the effective eddy diffusivity so that $a^*_{eff} > a^*_0$. Rm = 20, Pm = 1. From Yousef, Brandenburg, and Rüdiger (2003). Reproduced with permission © ESO.

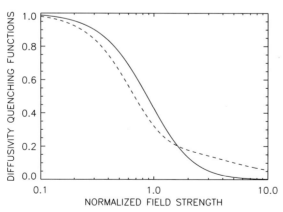

Figure 3.12 The two curves φ (solid) and φ_{tot} (dashed) which describe the eddy diffusivity without and with the second term on the RHS of (3.38) starting with $\nabla \bar{B}^2$. Note that with this (inverse) pumping term (dashed) the magnetic quenching compared to the quenching without this term (solid) is stronger for weak fields and weaker for strong fields.

result for weak magnetic fields ($\beta \ll 1$) is that both eddy viscosities, that is ν_T and η_T, show one and the same value. It is thus shown that for driven turbulence the turbulent magnetic Prandtl number (i) is larger than the molecular value but (ii) the resulting value does not exceed unity. It is thus not possible to explain the large values of the magnetic Prandtl number which are necessary to operate an advection-dominated solar dynamo (see Choudhuri, Schüssler, and Dikpati, 1995; Küker, Rüdiger, and Schultz, 2001; Dikpati and Gilman, 2001; Bonanno et al., 2002) simply by means of the convection-driven turbulence. We shall see in Section 6.3.3, however, that the flow and field patterns which are due to the action of unstable magnetic large-scale fields can lead to magnetic Prandtl numbers of order ten or even more.

3.3
Kinetic Helicity and DIV-CURL Correlation

Fortunately, the rotational influence on the turbulence – which is the heart of the cosmical electrodynamics – can directly be seen by the observed velocity pattern at the solar surface. Granulation and mesogranulation are the main candidates due to their nonmagnetic character. We shall show that the *horizontal* motions alone suffice to estimate the rotational influence. Just these observations are thus the true empirical basis of the mean-field theory of the cosmic dynamo.

As it is a pseudoscalar finite values of the kinetic helicity $\langle \boldsymbol{u} \cdot \text{curl } \boldsymbol{u} \rangle$ require the active influence of the basic rotation onto stratified turbulence. Here that part of the helicity is considered that results from the correlation of the vertical components of velocity and of the vorticity, that is

$$\mathcal{H}_{\text{kin}} = \left\langle w \left(\frac{\partial v}{\partial x} - \frac{\partial u}{\partial y} \right) \right\rangle, \tag{3.40}$$

where the turbulence pattern $\boldsymbol{u} = (u, v, w)$ is taken in a Cartesian coordinate system (x points east, y north, and z radially outward). In a stratified convection zone rising material expands and rotates because of the action of the Coriolis force. On the northern hemisphere the results are left-handed helical motions, that is $\mathcal{H}_{\text{kin}} < 0$ (see Figure 22 in Miesch et al., 2000). As simultaneous observations of the full velocity vector are difficult the correlation

$$C = \left\langle \left(\frac{\partial u}{\partial x} + \frac{\partial v}{\partial y} \right) \left(\frac{\partial v}{\partial x} - \frac{\partial u}{\partial y} \right) \right\rangle \tag{3.41}$$

is considered instead of (3.40) (see Wang et al., 1995).

The anelastic approximation, $\text{div}\rho\boldsymbol{u} = 0$, is adopted, so that

$$C \approx \frac{\mathcal{H}_{\text{kin}}}{H_{\text{m}}}, \tag{3.42}$$

where $H_{\text{m}} = -\partial z/\partial \log |\rho w|$ is the scale height for the vertical momentum fluctuations. We assume $H_{\text{m}} > 0$, that is the vertical momentum fluctuations decrease with height (see Simon and Weiss, 1997). Close to the bottom of the cells, however, H_m becomes negative. In the top of the cells a positive horizontal divergence corresponds to an updraft flow ($w > 0$) and a negative horizontal divergence corresponds to a downdraft flow ($w < 0$).

C is a pseudoscalar by definition because its sign depends on the coordinate system. The only pseudoscalar that can be constructed in anisotropic turbulence with the radial anisotropy direction g is the scalar product $g \cdot \Omega$. A nonvanishing helicity proxy C can thus only exist for rotating turbulence.

As the typical mesogranulation pattern at the solar surface only lives for a few hours, Ω^* is estimated to be of order of 0.1 so that C should be very small. Brandt et al. (1988) and Simon et al. (1994) present results of an overall inspection of horizontal flow patterns on mesoscales. The maximum velocities are ~ 750 m/s, and the maximum vertical vorticity is about 2×10^{-4} s^{-1} (see also Simon et al., 1988). There are indications for a negative correlation C (Wang et al., 1995). The correlations obtained by Duvall and Gizon (2000) are clearly negative at the northern hemisphere and they are antisymmetric with respect to the equator.

The expression (3.41) can be written as the radial component of the axial vector $C = \langle \text{div}\tilde{u}\,\text{curl}\,u\rangle$, with $\tilde{u} = u - (g \cdot u)g$ as the horizontal component of the random flow field. The turbulent flow is assumed as anelastic, so that div$m = 0$ with $m = \bar{\rho}u$. Following Roberts and Soward (1975) it is convenient to use the Fourier transform

$$m(x, t) = \iint \hat{m}(k, \omega) e^{i(kx - \omega t)} dk d\omega \tag{3.43}$$

of the momentum fluctuation m. One obtains

$$\rho^2 C_i = -\epsilon_{ijn} g_f g_m \iint k_j k_m \langle \hat{m}_f(k, \omega) \hat{m}_n^*(k, \omega)\rangle dk d\omega \tag{3.44}$$

for the components of C. To first order in Ω the vector C has only the radial component (3.41).

It remains to compute the tensor $\hat{M}_{ij} = \langle \hat{m}_i(k, \omega) \hat{m}_j^*(k, \omega)\rangle$ under the influence of rigid rotation. Its influence on the correlation tensor (3.2) has been computed. The resulting expressions must be introduced into (3.44), which after some manipulations for $\nu \to 0$ leads to

$$C = -\frac{4}{5} \frac{\alpha_{\text{MLT}}^2}{\gamma^2} \frac{\Omega}{\tau_{\text{corr}}} \cos\theta \tag{3.45}$$

for the radial component of the vector C – if only the density stratification is taken into account. The correlation C proves to be negative (positive) in the northern (southern) hemisphere, is proportional to $\Omega \cos\theta$, and is of order 10^{-10} s^{-2}.

The numerical relationship between C and the kinetic helicity $\mathcal{H}_{\text{kin}}$ is shown in Figure 3.13. Both the correlation C and the helicity $\mathcal{H}_{\text{kin}}$ are negative in the northern hemisphere. It is negative there at the top of the convection zone and positive but smaller at the bottom. The correlation C, therefore, forms a basic proxy for

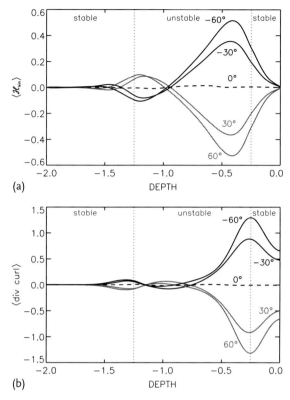

Figure 3.13 The helicity $\mathcal{H}_{kin}$ (a) and the DIV-CURL correlation $\mathcal{C}$ (b) after time-averaging vs. latitude and depth. The convection zone surface is at $z = -0.25$ and the bottom at -1.25. From Egorov, Rüdiger, and Ziegler (2004). Reproduced with permission © ESO.

the helicity. It is also negative at the top of the convection zone (in the northern hemisphere). The quasi-linear theory and the numerical simulations lead to the same result concerning the correlation (3.41). With his numerical modeling of the turbulence in the solar tachocline Miesch (2003) with a randomly driven turbulence in the rotating and (stably) stratified medium also obtained negative (positive) values for $\mathcal{C}$ in the northern (southern) hemisphere. Egorov, Rüdiger, and Ziegler (2004) correctly reproduced the DIV-CURL correlation observed by Duvall and Gizon (2000) for Ta $= 10^3$ which corresponds to the Coriolis number of $\lesssim 0.2$ (Figure 3.14).

Assuming a scale height of 10 000 km for the vertical momentum cells the above value of $\mathcal{C} = 10^{-10}$ s^{-2} leads to a prediction for the helicity of 0.1 cm/s^2. Indeed, the helicity derived by Komm et al. (2005) and Komm, Hill, and Howe (2008) leads to the same value. Multiplied with a correlation time of say 3 h (mesogranulation) one would find an α effect of about 10 m/s and positive at the northern hemisphere. Käpylä, Korpi, and Brandenburg (2009) find $\alpha \simeq 0.03 u_{rms}$ near the surface by their

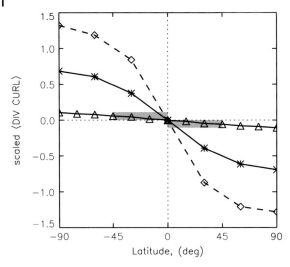

Figure 3.14 The computed correlation C for various Taylor numbers Ta $= 10^3$ to Ta $= 10^6$ (dashed) in comparison to the observations by Duvall and Gizon (2000) (shaded). From Egorov, Rüdiger, and Ziegler (2004). Reproduced with permission © ESO.

simulations, which with $u_{\rm rms} \simeq 300$ m/s perfectly agrees with the observational result. The maximal α-value in the box center is of order $0.3 u_{\rm rms}$.

3.4
Cross-Helicity

The decay of sunspots or larger active regions lead to numerical values of the turbulent magnetic diffusivity. One finds $\eta_{\rm T} \simeq 10^{11}$ cm^2/s from sunspot decay (Stix, 1989) or $\eta_{\rm T} \simeq 1.1 \times 10^{12}$ cm^2/s from the decay of active regions (Schrijver and Martin, 1990). These numbers are smaller than the value of 5×10^{12} cm^2/s which results from $\eta_{\rm T} \sim 0.3\, u_{\rm rms} \ell_{\rm corr}$ with parameter values taken close to the surface. There is no possibility to observe the turbulent diffusivity on the solar surface for the quiet Sun where the magnetic quenching of this quantity by large-scale magnetic fields is negligible.

The situation is quite different for another correlation between fluctuations of flow and field, that is the cross-helicity $\langle \boldsymbol{u} \cdot \boldsymbol{b} \rangle$, which is also a pseudoscalar. It is obvious that

$$\langle \boldsymbol{u} \cdot \boldsymbol{b} \rangle = \alpha_c \langle \boldsymbol{g} \cdot \bar{\boldsymbol{B}} \rangle - \beta_c \langle \boldsymbol{\Omega} \cdot \bar{\boldsymbol{J}} \rangle . \tag{3.46}$$

The α_c effect does not scale with the rotation. Similar to the α effect the α_c in (3.46) has dimensions of velocity. As the second term on the RHS of (3.46) only exists under the presence of rotation, it will be negligibly small at the solar surface.

3.4.1
Theory

If existing even for $\Omega = 0$ the pseudotensor $\langle u_i b_j \rangle$ can be finite only in the presence of a mean magnetic field $\bar{B}$ and for inhomogeneous fluids. The required inhomogeneity can be due to stratification of density or turbulent intensity as well as the inhomogeneity of the mean field itself. Only the symmetric part $\langle u_i b_j \rangle^s = (\langle u_i b_j \rangle + \langle u_j b_i \rangle)/2$ of the cross-correlation tensor $\langle u_i b_j \rangle$ is derived here.

The turbulent flow is taken as anelastic. One starts with the linear Fourier transform (3.43) for the flow pattern and uses

$$(-i\omega + \eta k^2)\hat{b}_i(\mathbf{k}, \omega) = ik_j \iint \left[\hat{m}_i(\mathbf{k} - \mathbf{k}', \omega - \omega') \left(\frac{\hat{B}_j}{\rho} \right)(\mathbf{k}', \omega') \right.$$
$$\left. - \hat{m}_j(\mathbf{k} - \mathbf{k}', \omega - \omega') \left(\frac{\hat{B}_i}{\rho} \right)(\mathbf{k}', \omega') \right] d\mathbf{k}' d\omega' \quad (3.47)$$

for the nonlocal relation between the fluctuating field components and the flow field, where $\hat{B}$ is the Fourier transform of the magnetic background field. For homogeneous background fields this relation simplifies to $\hat{b} = i(\mathbf{k} \cdot \bar{\mathbf{B}})\hat{u}/(-i\omega + \eta k^2)$, which easily can be multiplied with (3.43). Derivation of the cross-correlation yields

$$\langle u_i b_j \rangle^s = \frac{1}{2}\eta_T (G_i \bar{B}_j + G_j \bar{B}_i) + \left(\frac{1}{10}\eta_T + \frac{4}{15}\hat{\eta} \right) \delta_{ij} (\mathbf{G}' \cdot \bar{\mathbf{B}})$$
$$+ \left(\frac{1}{10}\eta_T - \frac{1}{15}\hat{\eta} \right) (G'_i \bar{B}_j + G'_j \bar{B}_i), \quad (3.48)$$

where $\mathbf{G} = \nabla \log \rho$ and $\mathbf{G}' = \nabla \log u_{\text{rms}}^2$ are relative gradients of density and turbulent intensity with η_T after (3.34) and

$$\hat{\eta} = \int_0^\infty \int_0^\infty \frac{\eta k^2 \omega^2 E(k, \omega)}{(\omega^2 + \eta^2 k^4)^2} dk d\omega \quad (3.49)$$

(Rüdiger, Kitchatinov, and Brandenburg, 2011). Both expressions are positive-definite. For high conductivity they remain finite while in the low-conductivity limit η_T scales linearly with the magnetic Reynolds number w^*, while $\hat{\eta}$ disappears in higher order as it vanishes for δ-like spectra. Hence,

$$\langle \mathbf{u} \cdot \mathbf{b} \rangle = \eta_T (\mathbf{G} \cdot \bar{\mathbf{B}}) + \left(\frac{\eta_T}{2} + \frac{2\hat{\eta}}{3} \right) (\mathbf{G}' \cdot \bar{\mathbf{B}}). \quad (3.50)$$

Observations, however, can only provide the particular correlation $\langle u_z b_z \rangle$. From (3.48) we find

$$\langle u_z b_z \rangle = \eta_T G \bar{B}_z + \left(\frac{3\eta_T}{10} + \frac{2\hat{\eta}}{15} \right) G' \bar{B}_z, \quad (3.51)$$

where G and G' are the only nonzero radial components of the stratification vectors. G is negative and close to the solar surface G' is positive. Further simplifications can be obtained by using the approximation for the turbulence spectrum,

$$E(k,\omega) = 2\rho^2 \langle u^2 \rangle \delta \left(k - \ell_{\text{corr}}^{-1}\right) \frac{w}{\omega^2 + w^2}, \tag{3.52}$$

where ℓ_{corr} is the mixing length. The ratio $\varepsilon = \hat{\eta}/\eta_{\text{T}}$ is then simply $\varepsilon = 3w^*/(2+2w^*)$, so that its value reaches from zero for low conductivity to 1.5 for high conductivity. One finds

$$\langle \boldsymbol{u} \cdot \boldsymbol{b} \rangle = \eta_{\text{T}} \left[G + \left(\frac{5}{10} + \frac{10\varepsilon}{15}\right) G' \right] \bar{B}_z ,$$

$$\langle u_z \cdot b_z \rangle = \eta_{\text{T}} \left[G + \left(\frac{3}{10} + \frac{2\varepsilon}{15}\right) G' \right] \bar{B}_z . \tag{3.53}$$

Hence, a finite correlation (3.53) indicates the presence of a large-scale vertical magnetic field. At the solar surface we have G negative and G' positive. As $\varepsilon > 0$ it is thus clear that the absolute value of $\langle u_z b_z \rangle$ always exceeds the absolute value of $\langle \boldsymbol{u} \cdot \boldsymbol{b} \rangle$, which is confirmed by the below simulations. The actual value of w^* plays only a minor role. The result is

$$\frac{15 + 20\varepsilon}{6 + 16\varepsilon} \langle u_z b_z \rangle - \frac{9 + 4\varepsilon}{6 + 16\varepsilon} \langle \boldsymbol{u} \cdot \boldsymbol{b} \rangle = -\frac{\eta_{\text{T}}}{H_\rho} \bar{B}_z . \tag{3.54}$$

The magnetic eddy diffusivity can thus be determined if the left side of (3.54) can be observed and the density scale height H_ρ is known from numerical models of the solar atmosphere. It is clear, however, that only $\langle u_z b_z \rangle$ can be observed so that the relation between $\langle u_z b_z \rangle$ and $\langle \boldsymbol{u} \cdot \boldsymbol{b} \rangle$ must also be known. The only possibility is the numerical simulation of the solar magnetoconvection in the top layers of the convection zone.

3.4.2
Simulations and Observations

Numerical simulations for magnetoconvection are possible with the NIRVANA CODE (Ziegler, 2002) for parameters including the strength of the imposed vertical field $\bar{B}_z$, the viscosity ν and the magnetic diffusivity η. The simulation domain is a Cartesian box. The stratification is along the z-coordinate and is piecewise polytropic, with the polytropic index chosen such that the hydrostatic equilibrium state is convectively stable in the lower and unstable in the upper half of the simulation box. An ideal, fully ionized gas is heated from below with a fixed temperature at the top of the simulation box. Periodic boundary conditions apply in the horizontal directions. The upper and lower boundaries are impenetrable and stress-free and a homogeneous vertical magnetic field is applied.

The rms value of the field fluctuations is strongly influenced by the strength of the large-scale magnetic field. Figure 3.15 contains the basic information about the

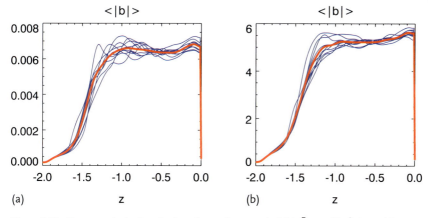

Figure 3.15 The numerical values for b_{rms} for weak magnetic field, $\bar{B}_z = 10^{-3}$ (a) and for strong magnetic field, $\bar{B}_z = 1$ (b). The ratio q of the magnetic energies sinks from $q \simeq 50$ (a) to $q \simeq 30$ (b). Reproduced with permission © ESO.

magnetic energy of the magnetoconvection, with the result that the ratio q hardly varies with the amplitude of the applied field. Normalized with the large-scale field one finds an averaged value of $b_{rms}/\bar{B}_z \simeq 7$ for weak fields and $b_{rms}/\bar{B}_z \simeq 5.5$ for stronger fields. The value for q (for weak fields) is about 50, which leads to an estimated magnetic Reynolds number of about 125. Indeed, the code works with $\eta \simeq 0.06$ in internal units and yields velocity fluctuations with $u_{rms} \simeq 10$ and about 1 cell per length unit so that the Rm $\simeq 160$ coincides well with the resulting q. The observed solar values for the ratio of the magnetic energies reach from $q \simeq 160$ (SST) to $q \simeq 500$ (HINODE).

Figure 3.16 holds for Ra $= 10^7$ and for weak and strong magnetic fields. The value of both Prandtl numbers is 0.1. Figure 3.16a,c shows the horizontal average of the cross-helicity as a function of the depth and Figure 3.16b,d shows the correlation $\langle u_z b_z \rangle$. One finds the vertical component actually being twice the cross-helicity. Equation (3.54) for $\varepsilon = 1.5$ (high-conductivity limit) can thus be written as

$$\frac{\langle u_z b_z \rangle}{\bar{B}_z} \simeq -\frac{4}{5} u_\eta \qquad (3.55)$$

with $u_\eta = \eta_T/H_\rho$. Velocities in the code are given in units of $c_{ac}/100$ with c_{ac} as the isothermal sound speed. Both simulations and observations provide a characteristic value of $u_\eta \simeq 0.75$ km/s. This value leads to $\eta_T \simeq 10^{12}$ cm^2/s (with density scale at the solar surface of about 100 km and sound speed of $c_{ac} \simeq 6.6$ km/s).

The data of the HINODE satellite lead to $\langle u_z b_z \rangle/\bar{B}_z \simeq -0.4$ km/s while the better resolution of the Swedish Solar Telescope (SST) yields $\langle u_z b_z \rangle/\bar{B}_z \simeq -0.72$ km/s. These data strongly coincide with the results of the numerical simulation. SOHO/MDI data from Doppler magnetograms also lead to cross-helicities of $O(1)$ G km/s (Zhao, Wang, and Zhang, 2011).

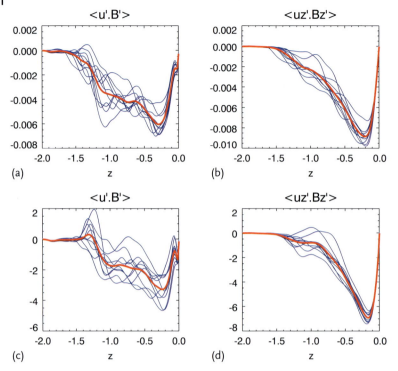

Figure 3.16 Snapshots of the cross-helicities $\langle \boldsymbol{u} \cdot \boldsymbol{b} \rangle$ (a,c) and $\langle u_z b_z \rangle$ (b,d) for weak magnetic field, $\bar{B}_z = 10^{-3}$ (a,b), and strong magnetic field, $\bar{B}_z = 1$ (c,d). Note that $\langle u_z b_z \rangle \simeq 2 \langle \boldsymbol{u} \cdot \boldsymbol{b} \rangle$ for the maximum values. The quenching of the helicities by the large-scale magnetic field is only weak. From Rüdiger, Küker, and Schnerr (2012). Reproduced with permission © ESO.

3.5
Shear Flow Electrodynamics

3.5.1
Hydrodynamic Stability of Shear Flow

Consider a shear flow

$$\bar{U}_y = Sx \tag{3.56}$$

with uniform vorticity in the vertical z-direction. It may exist in a turbulence field which does not possess any other anisotropy apart from that induced by the shear itself. The question is whether this flow is hydrodynamically stable under the presence of nonlinear shear terms in the correlation tensor Q_{ij} (Elperin et al., 2005). The one-point correlation tensor $Q_{ij} = \langle u_i(\boldsymbol{x},t) u_j(\boldsymbol{x},t) \rangle$ linear in the flow velocity reads as

$$Q_{ij} = P_\mathrm{T} \delta_{ij} - \nu_0 \left(\bar{U}_{i,j} + \bar{U}_{j,i} \right) . \tag{3.57}$$

Here v_0 is the isotropic eddy viscosity. The turbulent pressure, P_T, includes all coefficients of the Kronecker tensor δ_{ij}.

In a shear flow experiment by Champagne, Harris, and Corrsin (1970) the rms downstream velocities proved to be systematically larger than the rms velocities in the spanwise direction. The turbulence intensities for these two directions should, however, be equal according to (3.57) which by (3.56) can be read as $\langle u_y^2 \rangle = \langle u_x^2 \rangle = P_T$ and $\langle u_x u_y \rangle = -v_0 S$. The same remains true if higher-order derivatives such as $\bar{U}_{i,jll} + \bar{U}_{j,ill}$ are included. However, if the mean shear is indeed the only reason for anisotropy, one also has to involve second-order terms such as

$$Q_{ij} = P\delta_{ij} - v_0 (\bar{U}_{i,j} + \bar{U}_{j,i}) + v_1 \bar{U}_{i,k} \bar{U}_{j,k} + v_2 \bar{U}_{k,i} \bar{U}_{k,j} + v_3 (\bar{U}_{i,k} \bar{U}_{k,j} + \bar{U}_{j,k} \cdot \bar{U}_{k,i}) . \tag{3.58}$$

In this case the horizontal intensities can differ if the coefficients v_1 and v_2 of the nonlinear terms do not coincide, $\langle u_y^2 \rangle - \langle u_x^2 \rangle = (v_1 - v_2) S^2$. It should be $v_1 > v_2$ for agreement with the mentioned shear flow experiment.

The first-order term of (3.57) is the eddy viscosity

$$v_0 = \frac{2}{15} \int_0^\infty \int_{-\infty}^\infty \frac{v^3 k^6 E(k,\omega)}{(\omega^2 + v^2 k^4)^2} dk d\omega \tag{3.59}$$

(Krause and Rüdiger, 1974). The second-order corrections to the correlation tensor reproduce the nonlinear terms of (3.58). Only the coefficients v_2 and v_3 are relevant here, and end up as

$$v_i = \int_0^\infty \int_0^\infty K_i(k,\omega) E(k,\omega) dk d\omega , \tag{3.60}$$

with the kernels

$$K_2 = \frac{4(8v^6 k^{12} + 7v^4 k^8 \omega^2 - 46 v^2 k^4 \omega^4 + 3\omega^6)}{105(\omega^2 + v^2 k^4)^4},$$

$$K_3 = \frac{25 v^6 k^{12} + 49 v^4 k^8 \omega^2 - 149 v^2 k^4 \omega^4 + 19 \omega^6}{105(\omega^2 + v^2 k^4)^4} . \tag{3.61}$$

They have negative terms in their numerators but one can show that all coefficients (3.60) are 'almost always' positive. They are positive both for steep spectra (high viscosity) and also for the white-noise spectra for low viscosity.

For a simple slab model which is finite in the z-direction let the shear flow (3.56) be disturbed by the mode $\tilde{U}$ depending only on z. Then the cross-correlations defined by (3.58) become

$$Q_{xz} = -v_0 D \tilde{U}_x + v_2 S D \tilde{U}_y , \quad Q_{yz} = -v_0 D \tilde{U}_y + v_3 S D \tilde{U}_x . \tag{3.62}$$

The linear equation system for the disturbances is thus

$$\frac{\partial \tilde{U}_x}{\partial t} - \nu_0 D^2 \tilde{U}_x + \nu_2 S D^2 \tilde{U}_y = \frac{\partial \tilde{U}_y}{\partial t} + S \tilde{U}_x - \nu_0 D^2 \tilde{U}_y + \nu_3 S D^2 \tilde{U}_x = 0 \tag{3.63}$$

with $D = d/dz$. It reduces to $(\nu_2 \nu_3 S^2 - \nu_0^2) D^2 \tilde{U}_x + \nu_2 S^2 \tilde{U}_x = 0$ for marginal instability. This equation possesses an eigensolution only if the height H of the slab matches the condition $H = \pi \sqrt{(\nu_0^2 - \nu_2 \nu_3 S^2)/(-\nu_2 S^2)}$. The second term in the numerator must be small compared with the first one. Therefore, an instability only exists for $\nu_2 < 0$. The quantity ν_2, however, proved to be positive for almost all spectra.

3.5.2
The Magnetic-Diffusivity Tensor

The magnetic-diffusivity tensor η_{ijk} relates the mean electromotive force $\mathcal{E}$ to the gradients of the mean magnetic field via (3.25). If the influence of the shear flow is included to first order, the general structure of the diffusivity tensor is

$$\eta_{ijk} = \eta_0 \epsilon_{ijk} + \eta_1 \epsilon_{ijp} \bar{U}_{k,p} + \eta_2 \epsilon_{ijp} \bar{U}_{p,k} + \eta_3 \epsilon_{ikp} \bar{U}_{j,p} + \eta_4 \epsilon_{ikp} \bar{U}_{p,j}. \tag{3.64}$$

For the eddy diffusivities one finds the expression (3.34) and

$$\eta_i = \int_0^\infty \int_0^\infty C_i(k,\omega) E(k,\omega) dk d\omega, \quad i = 3,4, \tag{3.65}$$

with the nontrivial kernels

$$C_3 = -\frac{8\omega^2 \eta^2 k^4}{15(\omega^2 + \nu^2 k^4)(\omega^2 + \eta^2 k^4)^2},$$

$$C_4 = \frac{4\omega^2 \eta^2 k^4}{15(\omega^2 + \nu^2 k^4)(\omega^2 + \eta^2 k^4)^2} + \frac{5\omega^4 - 6\omega^2 \eta^2 k^4 - 3\eta^4 k^8}{15(\omega^2 + \eta^2 k^4)^3}. \tag{3.66}$$

The remaining expressions can be found in the original paper by Rüdiger and Kitchatinov (2006). C_3 is negative-definite. An important property is that the kernels C_3 and C_4 also exist in the limit $\nu \to 0$, that is

$$C_3 = -\frac{8}{15} \frac{\eta^2 k^4}{(\omega^2 + \eta^2 k^4)^2}, \quad C_4 = \frac{1}{15} \frac{5\omega^4 - 2\omega^2 \eta^2 k^4 + \eta^4 k^8}{(\omega^2 + \eta^2 k^4)^3}. \tag{3.67}$$

For sufficiently small magnetic Prandtl number the kernel C_4 is positive-definite. This is not true, however, for Pm of the order unity. For Pm $= 1$ one obtains from (3.66) that

$$C_4 = \frac{1}{15} \frac{(\omega^2 - \eta^2 k^4)(5\omega^2 + 3\eta^2 k^4)}{(\omega^2 + \eta^2 k^4)^3}, \tag{3.68}$$

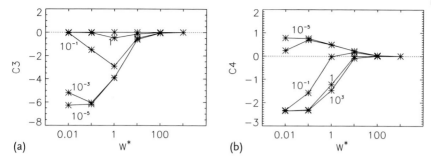

Figure 3.17 The integrals (3.65) (rescaled) for the frequency spectrum (3.12) for η_3 (a) and η_4 (b), marked with the magnetic Prandtl numbers. Note that for small Pm the factor η_4 is always positive so that from (3.73) no dynamo excitation is possible as $\eta_3 < 0$.

which has no definite sign. The high-frequency parts of the spectrum provide positive contributions to the integral and the low-frequency parts provide negative contributions. White noise (high conductivity) leads to positive-definite values of η_4, while the $\delta(\omega)$ profile for low conductivity leads to $\eta_3 \simeq 0$ and $\eta_4 < 0$. For small Pm the (negative) $|\eta_3|$ dominates (Figure 3.17).

Figure 3.22 shows the results of numerical simulations with the PENCIL CODE of the diffusivity of a turbulent shear flow (explained in more detail in Section 3.6.2). The values obtained for $\eta_{xy} = \eta_3 S$ and $\eta_{yx} = \eta_4 S$ are given in Figure 3.22d. They are both negative for positive shear S and $|\eta_{xy}| > |\eta_{yx}|$, both findings in accordance with the above results of the quasi-linear theory. Figure 3.22c demonstrates that the diagonal elements of the diffusivity tensor are almost identical. The numerical profiles in Figure 3.22c both represent the coefficient η_0 in (3.64).

Consider now the shear flow with the profile (3.56). For laboratory experiments it might be relevant that the relations

$$\mathcal{E}_x = \eta_0 D \bar{B}_y - \eta_4 S D \bar{B}_x , \quad \mathcal{E}_y = -\eta_0 D \bar{B}_x + \eta_3 S D \bar{B}_y \tag{3.69}$$

follow from (3.64) for a z-dependent field imposed in the horizontal direction. If the field is imposed in the x-direction, we have $\mathcal{E}_x = -\eta_4 S D \bar{B}_x$, and if the field is imposed in the y-direction then $\mathcal{E}_y = \eta_3 S D \bar{B}_y$. The sign of η_4 is thus opposite to the sign of the expression $\mathcal{E}_x S \bar{B}_{x,z}$, and the sign of η_3 is the same as the sign of $\mathcal{E}_y S \bar{B}_{y,z}$. Note that the EMF components are *perpendicular* to the mean electric current $\bar{J}$. Conversely, the standard diffusion-induced EMF (without shear) is always (anti)parallel to the mean current. The EMF due to the standard α effect is also *parallel* to the mean magnetic field.

3.5.3 Dynamos without Stratification

It has been argued that the general induction equation for large-scale fields,

$$\frac{\partial \bar{B}}{\partial t} = \mathrm{curl}(\bar{U} \times \bar{B} + \mathcal{E}) , \tag{3.70}$$

with the relations (3.69) as the components of the turbulent EMF possesses exponentially growing solutions, that is the turbulent shear flow – if fast enough – may operate as a kinematic dynamo (Rogachevskii and Kleeorin, 2003; Yousef et al., 2008). To probe this idea the dynamo equation (3.70) is applied to the standard 1D slab model. The dynamo region is finite in the z-direction but unlimited in x and y. The equations for the mean magnetic field with zero $\bar{B}_z$-component read as

$$\frac{\partial \bar{B}_x}{\partial t} - \eta_0 D^2 \bar{B}_x = -\eta_y S D^2 \bar{B}_y , \quad \frac{\partial \bar{B}_y}{\partial t} - \eta_0 D^2 \bar{B}_y = S \bar{B}_x - \eta_x S D^2 \bar{B}_x \tag{3.71}$$

with $\eta_x = \eta_4$, $\eta_y = \eta_3$. The equations are normalized in the sense that $\hat{S} = H^2 S/\eta_0$, $\hat{\eta}_x = \eta_x/H^2$ and $\hat{\eta}_y = \eta_y/H^2$ so that

$$\frac{\partial \bar{B}_x}{\partial t} - \frac{d^2 \bar{B}_x}{dz^2} = -\hat{\eta}_y \hat{S} \frac{d^2 \bar{B}_y}{dz^2}, \quad \frac{\partial \bar{B}_y}{\partial t} - \frac{d^2 \bar{B}_y}{dz^2} = \bar{B}_x \hat{S} - \hat{\eta}_x \hat{S} \frac{d^2 \bar{B}_x}{dz^2} \tag{3.72}$$

results. The vacuum boundary conditions $\bar{B}_x(0) = \bar{B}_y(0) = \bar{B}_x(1) = \bar{B}_y(1) = 0$ are applied. Figure 3.18 shows the numerically obtained stability map. Dynamo excitation only exists for

$$W = \hat{\eta}_y \left(1 + \pi^2 \hat{\eta}_x\right) > 0 . \tag{3.73}$$

Note, however, that according to (3.66) $\eta_y \equiv \eta_3$ is negative-definite and vanishes for low conductivity (see also Figure 3.17a). Planar shear flow dynamos are thus only possible with inclusion of η_x. Only a positivity of the product $\eta_x \eta_y$ could allow the existence of plane shear flow dynamos. However, the quantity $\hat{\eta}_x$ in (3.73)

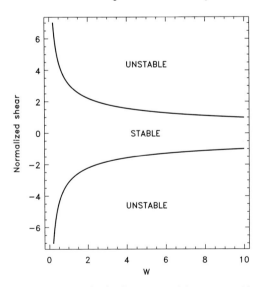

Figure 3.18 Map for the dynamo instability exists. Field excitation for sufficiently large positive values of (3.73). From Rüdiger and Kitchatinov (2006). Copyright © 2010 RAS.

is of order $(\ell_{\text{corr}}/H)^2$ which should be too small to influence the sign of the expression (3.73).

The quasi-linear theory provides a positive η_x for small Pm and a negative η_y in all cases. The negativity of η_y has been confirmed by numerical simulations (Figure 3.22). The dynamo condition (3.73) cannot be fulfilled (also Rädler and Stepanov, 2006; Sridhar and Subramanian, 2009). Mitra and Brandenburg (2012) and Brandenburg et al. (2013) explain a possible dynamo generation in turbulent shear flow by the concept of fluctuating α effect which in the average vanishes (Vishniac and Brandenburg, 1997).

Whether η_x is positive can easily be checked with laboratory experiments and/or with numerical simulations for various magnetic Prandtl numbers (Brandenburg, 2005b; Brandenburg et al., 2008; Rüdiger and Brandenburg, 2013). For magnetic Prandtl number of order unity which is used in most of the numerical MHD simulations η_x can only be negative for low conductivity. For Pm = 1 and $w^* \lesssim 10$ the factor C_4 is negative but too small to allow dynamo excitation according to (3.73).

3.6
The Alpha Effect

3.6.1
Helical-driven Turbulence

If the influence of a uniform magnetic background field $\bar{B}$ on a homogeneous and isotropic but not mirror-symmetric turbulence field is studied, then an EMF proportional to the field amplitude is the immediate consequence, that is $\mathcal{E}_i = \alpha_{ij}\bar{B}_j$. Let the spectral tensor of the homogeneous and isotropic turbulence be written as in (3.2) with H as the spectral function of the (negative) kinetic helicity and with $E > 0$ ('Bochner theorem,' see Krause and Rädler, 1980). By definition one has

$$\langle \boldsymbol{u} \cdot \text{curl}\, \boldsymbol{u}\rangle = -2 \iint k^2 H(k,\omega)\mathrm{d}k\mathrm{d}\omega \ . \tag{3.74}$$

It is easy to show with (3.47) that

$$\alpha_{im} = \mathrm{i}\epsilon_{ipq} \iint \frac{k_m}{-\mathrm{i}\omega + \eta k^2} \hat{Q}_{pq}(\boldsymbol{k},\omega)\mathrm{d}\boldsymbol{k}\mathrm{d}\omega \tag{3.75}$$

results, from which by (3.2)

$$\alpha = \frac{2}{3}\iint \frac{\eta k^4 H(k,\omega)}{\omega^2 + \eta^2 k^4}\mathrm{d}k\mathrm{d}\omega \tag{3.76}$$

directly follows, which remains finite for $\eta \to 0$ (high conductivity) and vanishes for $\eta \to \infty$ (low conductivity). Between these limits one finds with the spectrum (3.12) that

$$\alpha \simeq -\tau_{\text{corr}} \langle \boldsymbol{u} \cdot \text{curl}\, \boldsymbol{u} \rangle \begin{cases} 2 & w^* > 1 \\ w^* & w^* < 1. \end{cases} \quad (3.77)$$

The α effect is thus always anticorrelated to the kinetic helicity. Note that for low conductivity the α effect is very small, but for $w^* \simeq 1$ the second factor always has the order of unity. One must be careful, therefore, with experiments with fluid conductors, where the maximum magnetic Reynolds number w^* is typically of order unity[3]. The α effect is strongly reduced for experiments with $w^* < 1$. The use of gallium instead of sodium with its smaller conductivity reduces the expected α effect by a factor of $\simeq 3$.

For high conductivity the relation (3.75) can be rewritten as

$$\alpha_{im} = i\pi \epsilon_{ipq} \int k_m \hat{Q}_{pq}(\boldsymbol{k}, 0) d\boldsymbol{k} \quad (3.78)$$

because the kernel $\eta k^2/(\omega^2 + \eta^2 k^4)$ becomes a δ function $\pi \delta(\omega)$ for $\eta \to 0$. This means that

$$\alpha_{im} = i\pi \epsilon_{ipq} \iiint k_m Q_{pq}(\boldsymbol{\xi}, \tau) e^{-i\boldsymbol{k}\boldsymbol{\xi}} d\boldsymbol{k} d\boldsymbol{\xi} d\tau. \quad (3.79)$$

It follows the simple expression

$$\alpha_{im} = -\pi \epsilon_{piq} \int \langle u_p(\boldsymbol{x}, t) u_{q,m}(\boldsymbol{x}, t+\tau) \rangle d\tau \quad (3.80)$$

(Rädler and Rheinhardt, 2007) which for the trace of the α-tensor provides

$$\alpha_{ii} \simeq -\pi \langle \boldsymbol{u} \cdot \text{curl}\, \boldsymbol{u} \rangle \tau_{\text{corr}}. \quad (3.81)$$

Under the influence of strong magnetic fields the α effect of homogeneous turbulence becomes an anisotropic tensor, $\alpha_{ij} = \alpha_1 \delta_{ij} + \alpha_2 \bar{B}_i \bar{B}_j / \bar{B}^2$. Its structure, however, is so simple that it is enough to compute the sum $\alpha_\| = \alpha_1 + \alpha_2$, that is the induction parallel to the magnetic field. From Rüdiger (1974) one finds

$$\alpha_\| = 2 \iiint_0^\pi \frac{\eta k^4 H(k,\omega) \cos^2\theta \sin\theta \, d\theta}{(\omega^2 + \eta^2 k^4)(1 + 2S^{*2}\Phi \cos^2\theta + S^{*4}\cos^4\theta)} dk d\omega. \quad (3.82)$$

Adopting the spectrum (3.12) also for the helicity function $H(k, \omega)$, it is possible to evaluate the integrals. Figure 3.19 shows the results for w^* as the parameter. As it should for small w^* the α effect disappears, while the w^*-dependence disappears for large w^*. We find a mild magnetic quenching for small S and a strong magnetic quenching like $1/S^3$ for large S (Moffatt, 1972; Rüdiger, 1974; Gilbert and Sulem, 1990). This finding remains true also for Pm $\neq$ 1.

---

[3] Sodium experiments with velocities 1 m/s and characteristic length of 10 cm reach values of $w^* \simeq 1$.

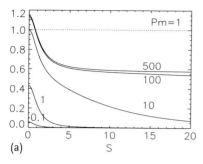
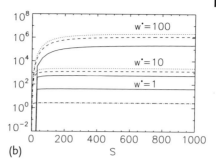

Figure 3.19 The magnetic quenching of the α effect (3.82). (a) Small Lundquist numbers. The curves are in increasing order for $w^* = 0.1$–500. $Pm = 1$. Note the plateau for medium S for high conductivity. (b) α multiplied with S^3 for various magnetic Reynolds number w^*. Dotted line denotes $Pm = 0.1$, dashed line $Pm = 1$, and solid line $Pm = 10$. The α effect is evidently suppressed by strong magnetic fields as $1/S^3$.

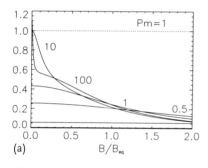
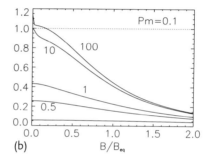

Figure 3.20 The α effect for various magnetic Reynolds numbers w^* (marked). $Pm = 1$ (a), and $Pm = 0.1$ (b).

Because of the connection (3.19) between the Lundquist number and the equipartition field β, the results are more dramatic if β is used as the independent variable. One finds that for $w^* > 1$ the quenching for weak fields is somewhat stronger than $1/\beta^2$, but the differences remain small. As demonstrated by the curve with the highest w^* there is a dramatic ('catastrophic') magnetic quenching for very weak fields (Cattaneo and Vainshtein, 1991), but this effect is stopped for $w^* \cdot \beta \gtrsim 1$. At $\beta \simeq 1$ the α-values hardly differ for various (large) values of w^*, so that the magnetic Reynolds number w^* for $\beta > 1$ does not dominate the quenching law of the α effect (Figure 3.20). This result strongly remembers the behavior of numerical simulations of helically driven MHD turbulence by Brandenburg (2001, Figure 15).

3.6.2
Shear Flow

If an α effect exists in a turbulent shear flow then it should exhibit a distinct tensorial structure. This tensor must be a pseudotensor, which is not the case for the shear tensor of the flow. Hence, an ϵ-tensor has to appear in the α-coefficients. The construction of $\mathcal{E}_i = \epsilon_{ijk}\langle u_j b_k \rangle$ is the only possibility for the ϵ-tensor to appear. Therefore, the subscript of $\mathcal{E}_i$ is always a subscript of the ϵ-tensor. As the ϵ-tensor is of third rank an inhomogeneity of turbulence with the stratification vector, $G' = \nabla \log u_{\rm rms}^2$, must also be present for the α effect to exist. If the shear flow is included to its first order, the general structure of the α-tensor is

$$\alpha_{ij} = \gamma \epsilon_{ijk} G'_k + \left(\alpha_1 \epsilon_{ikl} \bar{U}_{j,k} + \alpha_2 \epsilon_{ikl} \bar{U}_{k,j}\right) G'_l + \alpha_3 \epsilon_{ikl} G'_j \bar{U}_{l,k}$$
$$+ \alpha_4 \epsilon_{ikj} \bar{U}_{l,k} G'_l + \alpha_5 \epsilon_{ijk} \bar{U}_{k,l} G'_l . \tag{3.83}$$

If the stratification is along the vertical z-axis with $G' > 0$ and the shear is defined by (3.56), then it follows from (3.83) that

$$\alpha_{xx} = \alpha_2 G' S = \alpha_x S, \quad \alpha_{yy} = -\alpha_1 G' S = \alpha_y S,$$
$$\alpha_{xy} = -\alpha_{yx} = \gamma G' = \Gamma. \tag{3.84}$$

The anisotropy of the α-tensor is described by the difference between α_x and α_y. The turbulence-induced pumping is described by α_{xy}. The coefficients of (3.83) read as

$$\gamma = \frac{1}{6} \int_0^\infty \int_0^\infty \frac{\eta k^2 E(k,\omega)}{\omega^2 + \eta^2 k^4} dk d\omega \tag{3.85}$$

for the pumping term and

$$\alpha_i = \int_0^\infty \int_0^\infty A_i E(k,\omega) dk d\omega, \quad i = 1, 2, 3 \tag{3.86}$$

for the α effect with

$$A_1 = \frac{4\nu \eta^3 k^8 + 2\omega^2 \eta (\nu + \eta) k^4}{15(\omega^2 + \nu^2 k^4)(\omega^2 + \eta^2 k^4)^2} + \frac{\eta^2 k^4 (\eta^2 k^4 - 3\omega^2)}{15(\omega^2 + \eta^2 k^4)^3},$$
$$A_2 = -\frac{\eta^2 \nu^3 (4\eta - 5\nu) k^{12}}{60(\omega^2 + \nu^2 k^4)^2(\omega^2 + \eta^2 k^4)^2} - \frac{\omega^4 \eta (\eta + 36\nu) k^4 - 5\omega^6}{60(\omega^2 + \nu^2 k^4)^2(\omega^2 + \eta^2 k^4)^2}$$
$$- \frac{\omega^2 \nu (28\eta^3 - 4\eta^2 \nu + 12\eta \nu^2 + 5\nu^3) k^8}{60(\omega^2 + \nu^2 k^4)^2(\omega^2 + \eta^2 k^4)^2} \tag{3.87}$$

(Kitchatinov and Rüdiger, 2006) for the two most important kernels occurring in (3.84). For small Pm, the α_x and α_y are both of the same sign which is opposite to the sign of G'. The α_x appears to be smaller than the α_y.

For small Pm the expression for A_1 simplifies to

$$A_1 = \frac{1}{15} \frac{1}{(\omega^2 + \eta^2 k^4)^2} \left[\frac{4\nu \eta^3 k^8}{\omega^2 + \nu^2 k^4} + \frac{\eta^2 k^4 (3\eta^2 k^4 - \omega^2)}{\omega^2 + \eta^2 k^4} \right]. \tag{3.88}$$

The white-noise spectrum as the simplest approximation mimics the high-conductivity limit typical for cosmical applications. In this approach the spectrum does

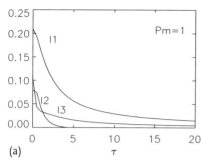
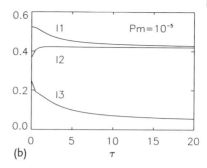

Figure 3.21 The numerical values of the coefficients I_1, I_2 and I_3 vs. the normalized correlation time $1/w^*$. Pm $= 1$ (a), Pm $= 10^{-5}$ (b). Note that in the high-conductivity limit ($\tau \to 0$) the coefficients I_1 and I_2 for Pm $\ll 1$ exceed the values for Pm $= 1$.

not depend on the frequency ω up to a maximum value $\omega_{\max}$ from which on the power spectrum vanishes. It is a turbulence model with the very short correlation time $\tau_{\text{corr}} \simeq 1/\omega_{\max}$. One finds from (3.88) after integration

$$\alpha_1 = \frac{\pi}{6\eta} \int_0^\infty \frac{E(k,0)}{k^2} dk \ . \tag{3.89}$$

The resulting integral can be written as $\tau_{\text{corr}} \ell_{\text{corr}}^2 u_{\text{rms}}^2$ so that $\alpha_1 = (\pi/6)\text{StRm}'\ell_{\text{corr}}^2$. The factor $\pi/6$ also appears in Figure 3.21 as the value of I_1 at the left vertical axis. For Pm $= 1$ the corresponding expression is $\alpha_1 = (\pi/15)\text{Rm}'\ell_{\text{corr}}^2$ so that the value for small Pm exceeds the value for Pm $= 1$. For given η, smaller ν lead to higher values of the EMF. It follows from (3.87)

$$\alpha_1 = \frac{1}{15}\left(1 + \frac{4}{\text{Pm}}\right) \text{Rm}'^2 \ell_{\text{corr}}^2 \tag{3.90}$$

for the low-conductivity limit so that Pm $= 1$ leads to $\alpha_1 = (1/3)\text{Rm}'^2\ell_{\text{corr}}^2$, and for small magnetic Prandtl number one finds $\alpha_1 = (4/15)\text{ReRm}'\ell_{\text{corr}}^2$ with Re $=$ Rm$'/$Pm as the Reynolds number.

The full expression might be written in the form $\alpha_1 \simeq I_1 \text{Rm}'\ell_{\text{corr}}^2$, where I_1 is numerically given in Figure 3.21. It is the result of a numerical integration under use of a spectral function of the form $\exp(-\tau_{\text{corr}}^2 \omega^2)$ with the dimensionless parameter $\tau = \eta k^2 \tau_{\text{corr}} \simeq 1/w^*$. While small τ represent the high-conductivity limit ($w^* \to \infty$), the larger τ represent the low-conductivity limit ($w^* \to 0$).

Similar calculations for α_2 lead to $\alpha_2 \simeq -I_2 \text{Rm}'\ell_{\text{corr}}^2$, with I_2 also plotted in Figure 3.21. For medium τ_{corr} and small Pm we find $I_2 \simeq I_1$, that is

$$\alpha_2 \simeq -0.5 \text{Rm}'\ell_{\text{corr}}^2 \ . \tag{3.91}$$

In the low-conductivity limit we have instead of (3.90)

$$\alpha_2 = -\frac{1}{60}\left(\frac{4}{\text{Pm}} - 5\right) \text{Rm}'^2 \ell_{\text{corr}}^2 \ . \tag{3.92}$$

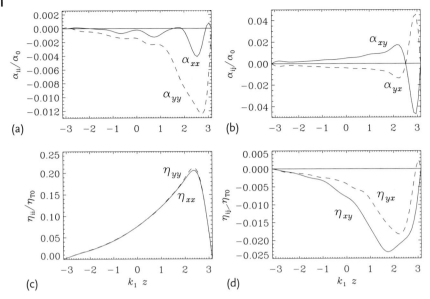

Figure 3.22 Simulations with positive shear ($s = 0.2$). The numerical values for α (a,c) and the two shear-current terms η_{xy} and η_{yx} (b,d). α_{ij} and η_{ij} are given in terms of $\alpha_0 = u_{\rm rms}/3$ and $\eta_{T0} = u_{\rm rms}/3k_{\rm f}$. It is $\eta_{xy} = \eta_3 S$ and $\eta_{yx} = \eta_4 S$. $H = L$, Rm = 0.2, Pm = 1. Courtesy of A. Brandenburg.

The plots in Figure 3.21 reveal an important influence of the magnetic Prandtl number. For small Pm in the low-conductivity limit the ratio of I_2/I_1 approaches unity, but it is very small for Pm = 1. Note that the influence of the magnetic Prandtl number is highly different for liquid metals (Pm $\ll$ 1) and for numerical simulations (Pm $\simeq$ 1).

Figure 3.22 shows in numerical simulations the α effect in the spanwise direction (α_{xx}) as smaller than in the streamwise direction (α_{yy}) but both components have the same sign. The simulations have been performed in a cubic domain of size L^3, so the minimal wavenumber is $k_1 = 2\pi/L$. The equations of compressible hydrodynamics have been solved with an isothermal equation of state and a constant sound speed $c_{\rm ac}$ (Rüdiger and Brandenburg, 2013). A numerical resolution of 64^3 mesh points was found to be sufficient for the PENCIL CODE.

The flow is driven by a random forcing function consisting of nonhelical waves with wave numbers whose modulus lies in a narrow band around an average $5k_1$. The amplitude of the forcing function increases the rms velocity with height while the maximum Mach number is fixed as small enough so that the effects of compressibility are negligible. Periodic boundary conditions in the y direction, shearing-periodic boundary conditions in the x direction, and stress-free perfect conductor boundary conditions in the z direction are adopted. The quantities S and G' are positive in the calculations, so that the basic velocity grows in the positive x direction while the turbulence intensity grows in the positive z direction. We focus here on low conductivity with Rm′ $\equiv u_{\rm rms}/\eta k_{\rm f} = 0.2$. The shear of the background flow is normalized with the speed of sound, that is $S = sc_{ac}k_1 = 2\pi s u_{\rm rms}/{\rm Ma}\, L$ with the Mach number ${\rm Ma} = u_{\rm rms}/c_{ac}$. The simulations worked with ${\rm Ma} = 0.05$. Hence, the SOCA theory leads to

$$\frac{\alpha_{yy}}{\alpha_0} = -12 I_1 s \frac{\ell_{\rm corr}^2}{L^2} \frac{{\rm Rm}'}{{\rm Ma}}, \qquad (3.93)$$

that is $\alpha_{yy}/\alpha_0 \simeq -0.01$. Both the streamwise and the spanwise α-tensor components, α_{yy} and α_{xx}, should be negative. Figure 3.22a shows these predictions to be fulfilled. About 10 cells exist in the average in the vertical direction. Both the α-tensor components are negative and $|\alpha_{yy}|$ dominates. Its amplitude is indeed 0.01. For ${\rm Pm} = 1$ the factor I_2 is strongly reduced relative to I_1, so that also the small amplitude of α_{xx} becomes understandable.

To be complete the vertical component of the α-tensor in shear flows, that is $\alpha_{zz} = \alpha_3 G' S$ must also be discussed. The integral kernel is

$$A_3 = -\frac{8\omega^2 k^4(\omega^2 + \nu\eta k^4)\eta(\nu+\eta)}{30(\omega^2 + \nu^2 k^4)^2(\omega^2 + \eta^2 k^4)^2} - \frac{3\eta^4 k^8 - 24\omega^2 \eta^2 k^4 + 5\omega^4}{30(\omega^2 + \eta^2 k^4)^3} \qquad (3.94)$$

(Rüdiger and Kitchatinov, 2006) which remains finite for $\nu \to 0$,

$$A_3 = -\frac{11\eta^4 k^8 - 16\omega^2 \eta^2 k^4 + 5\omega^4}{30(\omega^2 + \eta^2 k^4)^3}. \qquad (3.95)$$

By use of the definition $\alpha_3 = -I_3 {\rm Rm}' \ell_{\rm corr}^2$ one finds $I_3 > 0$ for the monotonically decreasing spectrum (3.52). Hence, for one and the same shear and turbulence intensity all diagonal components of the α-tensor have one and the same sign. The vertical component α_{zz} is the smallest among the three components (see Figure 3.21).

3.6.3
Shear-Dynamos with Turbulence-Stratification

Simple slab dynamos have already been considered in Section 3.5.3 without α effect. The α effect, however, exists for shear flows but only for stratified turbulence. It thus makes sense to consider dynamos with such α effect but without the nonlinear shear-induced dissipation terms in (3.64). The normalized equations ($\hat{\alpha} = \alpha/H$, $\hat{\Gamma} = H\Gamma/\eta_0$, $\hat{S} = H^2 S/\eta_0$) are

$$\frac{\partial \bar{B}_x}{\partial t} - \frac{d^2 \bar{B}_x}{dz^2} = \hat{\Gamma}\frac{d\bar{B}_x}{dz} - \hat{\alpha}_y \hat{S}\frac{d\bar{B}_y}{dz},$$

$$\frac{\partial \bar{B}_y}{\partial t} - \frac{d^2 \bar{B}_y}{dz^2} = \hat{\Gamma}\frac{d\bar{B}_y}{dz} + \left(\hat{\alpha}_x\frac{d\bar{B}_x}{dz} + \bar{B}_x\right)\hat{S}. \qquad (3.96)$$

For given shear, self-excitation of magnetic fields is possible for sufficiently large $\hat{\alpha}_y$. All the solutions are oscillatory. The cycle time is of the order of the diffusion time for the case of an isotropic α effect. In both considered cases a typical condition for dynamo excitation is $|\hat{\alpha}_y \hat{S}| > 10$ (Figure 3.23). A reformulation of this

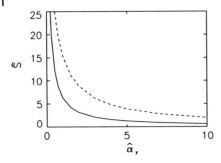

Figure 3.23 Stability map for a global dynamo solution of (3.96) for $\hat{\Gamma} = -0.5$. Solid line denotes $\hat{a}_x = \hat{a}_y$, dashed line $\hat{a}_x = 0.1\hat{a}_y$. From Kitchatinov and Rüdiger (2006).

relation with $S \simeq U/L$ and $G' \simeq 1/H$ yields $\mathrm{Rm} > 10 L^2/\alpha_1$, where $\mathrm{Rm} = LU/\eta_0$ is the magnetic Reynolds number of the mean flow defined with the eddy diffusivity (3.34). The estimate $\alpha_1 \simeq \ell_{\mathrm{corr}}^2 \mathrm{Rm}'$ provides $\mathrm{Rm} > 10/(\mathrm{Rm}' \varepsilon^2)$ with $\varepsilon = \ell_{\mathrm{corr}}/L$ as the number of cells across the channel. The channel height disappears from the excitation condition. As usual the magnetic Reynolds number of the velocity fluctuations is $\mathrm{Rm}' = u_{\mathrm{rms}} \ell_{\mathrm{corr}}/\eta$. For the most optimistic case ($\varepsilon \simeq 1, \mathrm{Rm}' \simeq 1$) the excitation condition reads as $\mathrm{Rm} > 10$. Hence, in a channel with a horizontal size of 1 m the horizontal velocity difference U of the channel must exceed 1 m/s as the microscopic magnetic diffusivity η of liquid metals is about 10^3 cm^2/s. This velocity is of the same order as in other dynamo experiments with a much more complicated geometry.

3.6.4
Alpha Effect by Density Stratification

We proceed with the consideration of the kinematic α effect of driven turbulence where the feedback of the magnetic field on the turbulence is neglected. The α effect is odd in both $\boldsymbol{\Omega}$ and $\boldsymbol{G} = \nabla \log \rho$, which means that the corresponding spectral tensor is also odd in $\boldsymbol{\Omega}$. The 'nonmagnetic' representation (3.32) for the tensor D is used as valid for arbitrary rotation rate $\boldsymbol{\Omega}$. The resulting α-tensor splits into two parts that separately involve the effects of the inhomogeneities of turbulence intensity and density, that is $\alpha = \alpha^\rho + \alpha^u$. Both tensors have the same structure given here only for the density stratification, that is

$$\alpha_{ij} = -(\boldsymbol{G} \cdot \boldsymbol{\Omega}) \left(a_1 \delta_{ij} + a_4 \frac{\Omega_i \Omega_j}{\Omega^2} \right)$$
$$- a_2 (G_i \Omega_j + G_j \Omega_i) + a_3 (G_i \Omega_j - G_j \Omega_i), \quad (3.97)$$

with

$$a_i = \iint \frac{\nu \eta k^4 \hat{Q}_{11}^{(0)}}{(\omega^2 + \nu^2 k^4)(\omega^2 + \eta^2 k^4)} A_n^\rho(\boldsymbol{\Omega}, k, \omega) \mathrm{d}k \mathrm{d}\omega, \quad i = 1, 2, 4 \quad (3.98)$$

and

$$\alpha_3^\rho = \iint \frac{\omega^2 \hat{Q}_{ll}^{(0)}}{(\omega^2 + \nu^2 k^4)(\omega^2 + \eta^2 k^4)} B_3^u(\Omega, k, \omega) d\mathbf{k} d\omega . \tag{3.99}$$

The full kernels A_n and B_n are given in the original paper by Rüdiger and Kitchatinov (1993). For slow rotation one finds

$$\alpha_1 = \frac{4}{15} \iint \frac{\nu \eta k^4 (3\nu^2 k^4 + 5\omega^2) \hat{Q}_{ll}^{(0)}}{(\omega^2 + \nu^2 k^4)^2 (\omega^2 + \eta^2 k^4)} d\mathbf{k} d\omega ,$$

$$\alpha_2 = -\frac{8}{15} \iint \frac{\nu^3 \eta k^8 \hat{Q}_{ll}^{(0)}}{(\omega^2 + \nu^2 k^4)^2 (\omega^2 + \eta^2 k^4)} d\mathbf{k} d\omega , \tag{3.100}$$

and $\alpha_4 = 0$. Note the opposite signs of α_1 and α_2. On the other hand, for rapid rotation $\alpha_1 = -\alpha_4$ and $\alpha_2 = \alpha_3 = 0$. Then one finds the striking relation

$$\alpha_{ij} = -\hat{\alpha} \frac{(G + G') \cdot \Omega}{\Omega} \left(\delta_{ij} - \frac{\Omega_i \Omega_j}{\Omega^2} \right) \tag{3.101}$$

with

$$\hat{\alpha} = \frac{\pi \eta}{4\nu^2} \iint \frac{(\omega^2 + \nu^2 k^4) \hat{Q}_{ll}^{(0)}}{k^2 (\omega^2 + \eta^2 k^4)} d\mathbf{k} d\omega . \tag{3.102}$$

The surprising result is that the α effect survives the limit $\Omega \to \infty$. This is in contrast to all other turbulence coefficients of the mean-field theory, which are subject to a distinct rotational quenching due to the suppression and/or deformation of the turbulence by rotation. Obviously, very fast rotation completely suppresses the zz-components of the α-tensor (but not the remaining diagonal components). It is known that such a two-dimensional α-tensor for rigid rotation leads to the preferred excitation of nonaxisymmetric modes (see Rüdiger and Elstner, 1994).

The limit (3.102) is the maximum value of the α effect. It scales linearly with the magnetic Reynolds number w^* for both Pm $= 1$ and for large w^* for Pm < 1. The integral (3.102) does not converge for too flat frequency spectra. It follows with (3.12) that

$$\frac{\hat{\alpha}}{u_0} = \frac{3\pi}{2} \frac{\text{St} w^*}{\text{Pm}^2} \left[1 + \frac{(\text{Pm}^2 - 1)(2w^* + 1)}{(1 + w^*)^2} \right] \tag{3.103}$$

with St $\simeq 1$ as the Strouhal number. Hence, for Pm $\ll 1$ and in the low-conductivity limit the expression (3.103) scales as w^{*3}/Pm^2, so that it is not too large. In the high-conductivity limit, however, we have

$$\frac{\hat{\alpha}}{u_0} \simeq \frac{3\pi}{2} \text{St} \frac{w^*}{\text{Pm}^2} \tag{3.104}$$

which yields the highest possible value. In the high-conductivity limit, therefore, the α effect for fast rotation can be very large; it is not limited by the relation

$\alpha \simeq u_0$. In the low-conductivity limit, (3.103) demonstrates the smallness of the α effect in laboratory experiments and in planetary cores (see Giesecke, Ziegler, and Rüdiger, 2005). Figures 3.24 and 3.25 demonstrate the behavior of the components of the α effect for slow and fast rotation. The component $\alpha_{\phi\phi}$ saturates for $\Omega \to 0$ while α_{zz} vanishes. It becomes clear also how strongly the α effect depends on the magnetic Reynolds number w^* of the turbulence, and that for high conductivity it can become much larger than the $u_{\rm rms}$ of the turbulence. The α_{zz} is always numerically smaller than $\alpha_{\phi\phi}$.

Both basic stratifications combine into a common gradient in (3.101). Since the G' dominates the G in the bottom layers of the convection zone – and in the overshoot region below the convection zone – one expects that the α effect becomes negative in these layers (see Ossendrijver, Stix, and Brandenburg, 2001; Käpylä, Korpi, and Brandenburg, 2009).

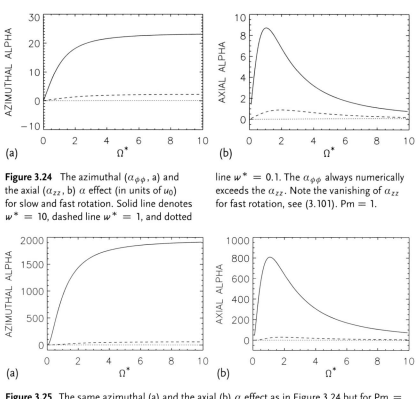

Figure 3.24 The azimuthal ($\alpha_{\phi\phi}$, a) and the axial (α_{zz}, b) α effect (in units of u_0) for slow and fast rotation. Solid line denotes $w^* = 10$, dashed line $w^* = 1$, and dotted line $w^* = 0.1$. The $\alpha_{\phi\phi}$ always numerically exceeds the α_{zz}. Note the vanishing of α_{zz} for fast rotation, see (3.101). Pm = 1.

Figure 3.25 The same azimuthal (a) and the axial (b) α effect as in Figure 3.24 but for Pm = 0.1. The α effect is much larger for small Pm.

3.7 The Current Helicity

The electric-current helicity $\mathcal{H}_{\text{curr}} = \langle \boldsymbol{b} \cdot \text{curl}\,\boldsymbol{b}\rangle$, which is obviously another interesting pseudoscalar, also plays an important role in studies of magnetoconvection. Observations show that it exists at the solar surface, and satisfies simple rules. It is antisymmetric with respect to the equator, and does not vary from cycle to cycle ('hemisphere rule.') The prevailing sign at the northern (southern) hemisphere is minus (plus) but at the beginning and/or end of a cycle there can be observations with the 'wrong' sign in each hemisphere (Zhang et al., 2010; Zhang, 2012, see Figure 3.26). It seems, therefore, that two independent sources of current helicity exist with opposite signs. Indeed, there are two independent pseudoscalars in the system, one due to the rotation $\boldsymbol{g} \cdot \boldsymbol{\Omega}$ and the other one due to the current helicity of the dynamo-generated large-scale magnetic field $\bar{\boldsymbol{B}} \cdot \text{curl}\,\bar{\boldsymbol{B}}$. The latter proves to be positive in the northern hemisphere beneath the solar surface (see Figure 3.27) if the toroidal field is due to the action of differential rotation with an accelerated equator. A negative $\bar{B}_z$ in the northern hemisphere produces a positive $\bar{B}_\phi$ so that with the boundary condition $\bar{B}_\phi = 0$ one finds that $\bar{B}_z \cdot \text{curl}_z\,\bar{B} > 0$. In the southern hemisphere this quantity is always negative. This part of the current helicity is due to the interaction of the nonhelical turbulence with the helical background field.

It is easy to show that the same sign rule holds for magnetic background fields with dipolar or quadrupolar symmetry. For a single magnetic mode with a clear type of symmetry the current helicity is always antisymmetric. Only for mixed modes, when symmetric toroidal fields meet antisymmetric poloidal fields or vice versa, does the quantity $\bar{\boldsymbol{B}} \cdot \text{curl}\,\bar{\boldsymbol{B}}$ possess parts which are symmetric with respect to the equator.

Let us start with the consequences of the helical part of the turbulence, which can be imagined as due to the influence of the basic rotation. For a homogeneous magnetic background field, (3.47) can be used so that the small-scale current helicity becomes

$$\mathcal{H}_{\text{curr}} = \langle \boldsymbol{b} \cdot \text{curl}\,\boldsymbol{b}\rangle = \epsilon_{pqj} \iint \frac{i(\boldsymbol{k}\cdot\bar{\boldsymbol{B}})^2 k_p}{\omega^2 + \eta^2 k^4} \hat{Q}_{jq} dk d\omega. \tag{3.105}$$

The current helicity is a pseudoscalar by definition, which is proportional to the helical part of the spectral tensor (3.2). One finds

$$\mathcal{H}_{\text{curr}} = -2\bar{B}^2 \int_0^\pi \iint \frac{k^4 H(k,\omega)\cos^2\theta \sin\theta\, d\theta}{(\omega^2 + \eta^2 k^4)(1 + 2S^{*2}\Phi\cos^2\theta + S^{*4}\cos^4\theta)} dk d\omega, \tag{3.106}$$

from which the simple relation

$$\mathcal{H}_{\text{curr}} = -\frac{\alpha_\| \bar{B}^2}{\eta} \tag{3.107}$$

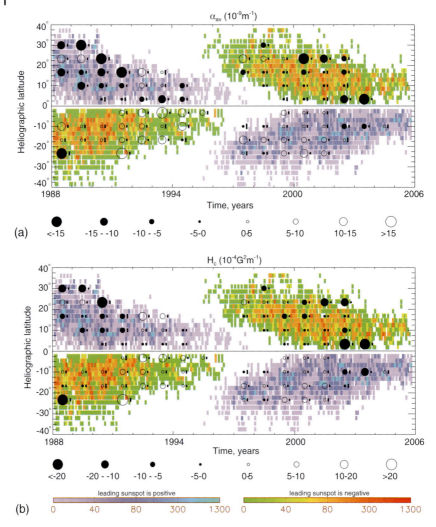

Figure 3.26 The distribution of α_{ff} (a) and the current helicity (b) of 88 groups plotted in the butterfly diagram during the recent solar cycles. The circle sizes represent the magnitudes of the effects. During the cycles the signs of the helicity fulfill the hemisphere rule, but between cycles a few 'wrong' signs appear. From Zhang et al. (2010). Copyright © 2010 RAS.

follows (see (3.82), Keinigs, 1983). The current helicity has thus a very close relationship to the α effect. It is (i) antisymmetric with respect to the equator and (ii) in each hemisphere it always has the same sign from cycle to cycle. It should have a maximum at the maximum of the solar activity, and it should be negative in the northern hemisphere as the α effect is expected to be positive there. All these consequences of (3.107) are indeed fulfilled by the observations (Figure 3.26). It also

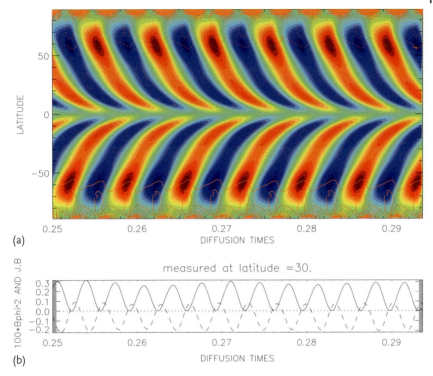

Figure 3.27 An advection-dominated $\alpha\Omega$-dynamo with a solar-type butterfly diagram (a) provides positive values of $\bar{\boldsymbol{B}} \cdot \mathrm{curl}\,\bar{\boldsymbol{B}}$ at mid-latitudes (30° north) and in the bulk of the convection zone only during the activity minima (b, dashed line). The dynamo operates with positive (negative) α effect in the northern (southern) hemisphere so that by (3.107) the main contribution to $\mathcal{H}_{\mathrm{curr}}$ is negative (positive) as observed (Figure 3.26). $C_\alpha = 20$, $C_\Omega = 230\,000$. Courtesy of R. Arlt.

directly follows from (3.82) and (3.106) that $\mathcal{H}_{\mathrm{curr}}$ and $\mathcal{H}_{\mathrm{kin}}$ possess the same sign (Brandenburg (2001), (see, however, Ossendrijver, Stix, and Brandenburg, 2001).

If we understand the magnetic field components in active regions as the fluctuations of the magnetic fields under the influence of stratification and rotation, and the fields are considered as force-free, that is $\mathrm{curl}\,\boldsymbol{b} = \alpha_{\mathrm{ff}}\boldsymbol{b}$, then $\langle \boldsymbol{b} \cdot \mathrm{curl}\,\boldsymbol{b}\rangle = \alpha_{\mathrm{ff}}\langle \boldsymbol{b}^2\rangle$ (for constant α_{ff}, Seehafer, 1990). From (3.107) follows

$$\alpha = -\eta\,\alpha_{\mathrm{ff}}\frac{\langle \boldsymbol{b}^2\rangle}{\bar{B}^2}\,, \tag{3.108}$$

or with (3.22), $\alpha \simeq (2/5)\alpha_{\mathrm{ff}}\eta_{\mathrm{T}}$, so that for the normalized α effect $C_\alpha = R_\odot \alpha/\eta_{\mathrm{T}}$ the simple result $C_\alpha \simeq 3$ follows if the typical value $\alpha_{\mathrm{ff}} \simeq 10^{-10}\,\mathrm{cm}^{-1}$ has been inserted (Rust and Kumar, 1994; Bao and Zhang, 1998). The characteristic value of the α effect becomes 40 cm/s, that is about 10% of the u_{rms} in the bulk of the convection zone. This is in excellent accordance with the simulations by Käpylä, Korpi, and Brandenburg (2009) who for the pole find $\alpha \simeq 0.25\,u_{\mathrm{rms}}$.

One can also assume on the other hand that the current helicity exists due to the influence of a helical background field with $\bar{B} \cdot \text{curl } \bar{B} \neq 0$. In this case the turbulence itself must *not* be helical. Hence, we consider a turbulence with a spectral tensor (3.2) with $H = 0$ under the presence of a field $\bar{B}_i = B_{ij} x_j$, which leads to $\bar{B} \cdot \text{curl } \bar{B} = \epsilon_{ilk} B_{ij} B_{kl} x_j$ as the only remaining pseudoscalar in the system. From (3.105) one finds

$$\mathcal{H}_{\text{curr}} = x_l \epsilon_{pqi} \iint \frac{k_j k_q}{\omega^2 + \eta^2 k^4} \left(B_{ps} B_{jl} \hat{Q}_{is}^{(0)} - B_{is} B_{jl} \hat{Q}_{sp}^{(0)} \right) dk d\omega . \quad (3.109)$$

It follows that

$$\mathcal{H}_{\text{curr}} = \frac{1}{6} \int_0^\infty \int_{-\infty}^\infty \frac{k^2 E(k, \omega)}{\omega^2 + \eta^2 k^4} dk d\omega \, \bar{B} \cdot \text{curl } \bar{B} \quad (3.110)$$

or $\mathcal{H}_{\text{curr}} \simeq 1/3 w^* \bar{B} \cdot \text{curl } \bar{B}$, so that $\mathcal{H}_{\text{curr}}$ can become much larger than $\bar{B} \cdot \text{curl } \bar{B}$ (similar as for the ratio q $= \langle b^2 \rangle / \bar{B}^2$). The integral in (3.110) does not exist for $\eta \to 0$. In these cases the turbulence parameter scales with the magnetic Reynolds number w^*. A characteristic value $\mathcal{H}_{\text{curr}} \simeq 2 \times 10^{-5}$ G^2/cm (Zhang et al., 2010) leads with $\bar{B} \cdot \text{curl } \bar{B} \simeq 10^{-7}$ G^2/cm to a magnetic Reynolds number of about 600. Here we have used $\bar{B}_r \bar{B}_\phi \simeq 10^3$ G^2 and a characteristic radial scale of 100 000 km. It is the same value as followed from the calculation of q (see (3.22)). For a single magnetic dynamo mode with dipolar symmetry one finds the current helicity due to the helicity of the field according to (3.110) as positive (negative) in the northern (southern) hemisphere. However, from (3.107) the action of the helicity of the turbulence has the opposite effect. The first effect scales with the amplitude of the product $\bar{B}_r \bar{B}_\phi$, while the second one scales with $\bar{B}_\phi^2$, which certainly dominates during the main portion of the cycle. The observed negative values of $\mathcal{H}_{\text{curr}}$ in the northern hemisphere over the cycle match this picture. Only at the beginning and/or end of a cycle the factor $\bar{B} \cdot \text{curl } \bar{B}$ becomes observable according to (3.110) – and seems indeed to be observed (Figure 3.26).

4
The Galactic Dynamo

The large-scale magnetic fields of spiral and elliptic galaxies (Figure 4.1) must be maintained by the inducing action of the partly ionized interstellar turbulence. Although the maximal field strength does not exceed (say) 10 µG, because of their huge dimensions galaxies are *the* supermagnets in the Universe. The magnetic flux $\pi R^2 B$ of a galaxy exceeds the total flux of all (say) 10^{11} stars by many orders of magnitude. Apart from the impressive numbers characterizing galactic magnetism, galaxies are also a very interesting realization of a dynamo machine as they are rather transparent, providing detailed information about the internal flow structures.

The observed magnetic field energy is of the order of the energy of the interstellar turbulence. Beck (2002) for the inner disk of NGC 6946 derived for both kinetic and magnetic energy the common value of 10^{13} erg/cm^3. The equipartition field strength

$$B_{\rm eq} = \sqrt{\mu_0 \rho} u_{\rm T} \tag{4.1}$$

with density of order 10^{-24} g/cm^3 and the turbulence velocity of about 10 km/s is 3.5 µG. As the observed values are indeed of this order a turbulent origin of the induced large-scale fields has been suggested. The coincidence between observed fields and their theoretical equipartition value strongly favors the canonical mean-field dynamo concept with quenching in a way as described above for various turbulence models.

4.1
Magnetic Fields of Galaxies

The interpretation of radio-polarization data of galaxies reveals the existence of large-scale magnetic fields with very special properties. Their explanation is of considerable interest because galaxies are astrophysical configurations with observable internal flow systems. In particular, there is a problem of understanding the relation between the considered flow field and the associated α effect. The observed butterfly diagram of the solar activity seems to indicate a very small α compared with the action of differential rotation in the solar convection zone. On the other

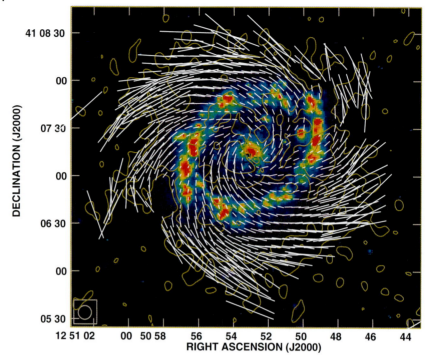

Figure 4.1 NGC 4736 is an early-type galaxy with a large-scale magnetic field, but without strong density waves. The ring marks the location of enhanced star formation which seems not to influence the magnetic field generation. Note the large pitch angles almost everywhere. Courtesy of K. Chyży.

hand, the observed large-scale structure of the known galactic magnetic fields requires only small differential rotation, or in other words, a rather large α effect. The key properties of the galactic large-scale magnetic field pattern are

- poloidal and toroidal field components are equally strong,
- field lines have pitch angles up to 35°,
- a bisymmetric spiral structure (M 81) or a distinct vertical orientation (NGC 4631) of the field geometry exists,
- the fields appear to be symmetric with respect to the midplane (see Figure 4.2),
- large-scale magnetic fields even exist in very young galaxies.

See also the review papers by Wielebinski and Krause (1993), Beck et al. (1996), Kulsrud (1999) and Fletcher (2010), particularly for the details of special galaxies. Krause and Beck (1998) discuss the empirical background to explore the symmetry of the magnetic field with respect to the galactic midplane. For many cases a dominant quadrupolar symmetry of the magnetic fields is deduced, together with symmetry with respect to the rotation axis. They mention a basic difference of dipolar and quadrupolar field symmetry. If the galactic magnetic fields have dipolar symmetry, the magnetic field amplitudes averaged over the whole sky must definitely

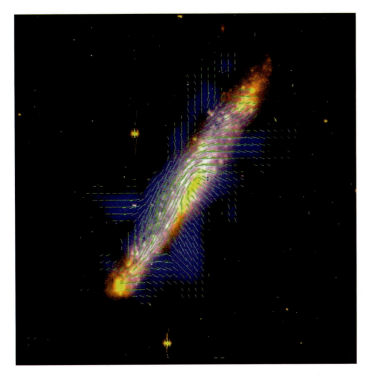

Figure 4.2 The magnetic geometry of the edge-on galaxy NGC 5775 indicates even (quadrupolar) symmetry with respect to the equator. Courtesy of M. Urbanik and M. Soida.

be zero. This would be not true for nonoscillating quadrupoles with a fixed magnetic orientation in the galactic plane.

On the other hand galaxies rotate with a characteristic velocity of 100–200 km/s. For field amplitudes of 10 µG one finds the small value

$$\frac{\Omega_A}{\Omega} \simeq 0.2 \qquad (4.2)$$

for the Alfvén frequency $\Omega_A = \bar{B}_\phi / \sqrt{\mu_0 \rho R^2}$ normalized with the characteristic rotation frequency $\Omega \simeq 10^{-15}$ s^{-1}. Galaxies are super-Alfvénic rotators.

The pitch angles $p = \arctan(\bar{B}_R / \bar{B}_\phi)$ reflect the ratio of the radial and the toroidal magnetic field strengths. Dynamos of $\alpha\Omega$-type are characterized by very small pitch angles, for example $\sim 0.05°$ for the Sun. For galaxies pitch angles of 10–40° are reported (Table 4.1). In NGC 4414 pitch angles up to 45° have been found. The pitch angles mostly decrease outward (Beck, 1993). Such observed values indicate that the differential rotation in galaxies does *not* play the dominant role in the induction equation. Frozen-in fields as a consequence of a small magnetic diffusivity are wound up by any differential rotation up to very small pitch angles.

Of special interest is the case of NGC 6946, possessing pitch angles between 20° and 30°. NGC 6946 is a standard example insofar as there is no companion,

Table 4.1 Comparison of the pitch angles of the large-scale magnetic field and the spiral arms. From Fletcher (2010).

Galaxy	Magnetic field	Spiral arms	Ref.
IC 342	20–16	19	Krause, Hummel, and Beck (1989)
M31	17– 8	7	Fletcher et al. (2004)
M33	48–42	65	Tabatabaei et al. (2008)
M51	20–18	20	Fletcher et al. (2011)
M81	14–22	11	Krause, Beck, and Hummel (1989)
NGC6949	38	36	Frick et al. (2000)

no strong density wave, and no active nucleus. The large-scale magnetic fields are *concentrated between* the optical spiral arms but the azimuthal field orientation is of the axisymmetric type (Beck and Hoernes, 1996). The turbulent component of the magnetic field in the spiral arm reaches 15 µG, while a regular field of 10 µG is located in the inter-arm region.

The large pitch angles are the most characteristic property of the galactic fields. Obviously, the galactic dynamo differs strongly from the solar/stellar $\alpha\Omega$-dynamo. They cannot be interpreted as *oscillating* stellar $\alpha\Omega$-dynamos that are made stationary by some extra effect like turbulent pumping (see Brandenburg, Moss, and Tuominen, 1992; Ferrière and Schmitt, 2000).

For one case (M 81) there is a clear azimuthal structure with polarity reversals ($m = 1$), so that in one magnetic arm the field spirals into the center and in the other it spirals outward. It is not easy to explain such an asymmetry with a mean-field dynamo theory. But a stability analysis against nonaxisymmetric perturbations for the toroidal fields including also the poloidal fields of the same order of magnitude leads to the appearance of modes with $m > 0$ (Section 4.4.3). In the majority of the examples the $m = 0$ mode dominates but only for a few of them this mode is the only one. Galaxies with the indication of a *spectrum* of modes seem now to be the rule rather than the exception (Fletcher, 2010, Table 2).

Almost no azimuthal structure exists for the flocculent galaxy NGC 4414. It is hard to imagine its large-scale magnetism as a result of the inducing action of the galactic differential rotation starting with an external uniform magnetic field. An external uniform magnetic field not parallel with the rotation axis subject to a differential rotation will mainly produce nonaxisymmetric field geometry. An axisymmetric field can only be produced from an initially even-m field with radial components in the equatorial plane, for example quadrupoles of type S0. The existence of such a rather artificial seed field is unlikely though, so that NGC 4414 seems to prove the existence of a large-scale galactic dynamo (Beck, 1996).

There is strong evidence for substantial amounts of highly turbulent gas at heights above 1 kpc (Münch and Zirin, 1961; Lockman, 1984; Reynolds, 1989). Inclusion of this halo gas has changed the traditional view of a very thin gas layer. Despite its low density this high-velocity component (Anantharamaiah, Radhakrishnan, and Shaver, 1984) is energetically important because it contains a significant fraction of the whole kinetic energy in the interstellar medium (Kulkarni and Fich, 1985). The gas as a whole has a nearly constant scale height of 230 pc. In the inner part of the galaxy, the interstellar medium consists mostly of molecular gas concentrated in a very thin layer. Its thickness decreases toward the center. More than 13% of HI emissions are from gas that lies mostly above 500 pc from the plane. The thickness of the HI layer is low there, below 100 pc.

4.2 Interstellar Turbulence

Consider the interstellar turbulence as driven by supernova (SN) explosions with their energy budget of $\lesssim 10^{51}$ erg. This energy feeds a total gas mass of about $10^{10}\,M_\odot$ over a time of (say) 50 yr. The resulting output power is $\varepsilon \simeq 0.03\,\mathrm{cm}^2/\mathrm{s}^3$. If the turbulent flow is incompressible and follows the Kolmogorov law $u^3 \simeq \varepsilon \ell$ with $1/\ell$ as the wave number then one obtains $u \simeq 10$ km/s for the characteristic scale $\ell_{\mathrm{corr}} \simeq 100$ pc. By use of the estimate $\eta_T \simeq u\ell_{\mathrm{corr}}$ one finds $\eta_T \simeq 10^{26}\,\mathrm{cm}^2/\mathrm{s}$ (Zeldovich, Ruzmaikin, and Sokoloff, 1983). This value is even exceeded by one order of magnitude in the numerical simulations presented below.

The basic assumption of this argumentation is the use of the Kolmogorov (1941) spectrum for incompressible velocity fluctuations of the interstellar turbulence. It can be generalized, however, including supersonic density fluctuations if $\rho u^3 \propto \ell$ is adopted. For driven isothermal turbulence a Kolmogorov exponent of 1.37 results instead of the standard value 5/3 (Kritsuk et al., 2007) which should not drastically change the above result. As the maximal length scale of correlated velocity fluctuations de Avillez and Breitschwerdt (2007) find $\ell_{\mathrm{corr}} \simeq 75$ pc which also fits well the values used by Zeldovich, Ruzmaikin, and Sokoloff (1983).

The other physical component which must be known to construct large-scale dynamos is the α-tensor. For its calculation it is necessary that the turbulence field (i) is stratified and (ii) rotates. Also here the calculations are based on the model of the SN-driven interstellar turbulence. We start to consider the dimensionless dynamo numbers for the galactic dynamo which operates with the observed shear $d\Omega/dR \simeq -\Omega/R$. The vertical half-thickness H of the disk is used to normalize the α effect and the shear, that is

$$C_\alpha = \frac{\alpha H}{\eta_T}, \quad C_\Omega = -\frac{\Omega H^2}{\eta_T}. \tag{4.3}$$

If the standard expression $\alpha \simeq \ell^2_{\mathrm{corr}}\Omega/H$ is used, together with the estimate $\eta_T \simeq \ell^2_{\mathrm{corr}}/\tau_{\mathrm{corr}}$, one finds

$$C_\alpha \simeq \tau_{\mathrm{corr}}\Omega, \quad |C_\Omega| \simeq \tau_{\mathrm{corr}}\Omega \left(\frac{H}{\ell_{\mathrm{corr}}}\right)^2, \tag{4.4}$$

so that

$$C_\alpha/C_\Omega| \simeq (\tau_{corr}\Omega)^2 \left(\frac{H}{\ell_{corr}}\right)^2, \quad \frac{C_\alpha}{|C_\Omega|} \simeq \left(\frac{\ell_{corr}}{H}\right)^2 \qquad (4.5)$$

results. The latter relation determines the pitch angle between the radial and the azimuthal component of the magnetic field. The observed large pitch angles can only be understood with $C_\alpha \simeq |C_\Omega|$, that is with the large correlation length $\ell_{corr} \lesssim H$. This finding, however, immediately leads to $C_\alpha/C_\Omega| \propto \Omega^{*2}$, which only exceeds any threshold value of order unity if $\tau_{corr} \gtrsim \tau_{rot}/2\pi \simeq 10$ Myr. Turbulent patterns with shorter correlation (turnover) times cannot work as a turbulent dynamo for galaxies.

The ratio of the two induction effects is given by (4.5)$_2$, that is

$$\frac{C_\alpha}{|C_\Omega|} \propto \Omega^{*2} \, \mathrm{Ma}_T^2 \,, \qquad (4.6)$$

with the turbulent Mach number $\mathrm{Ma}_T = u_T/c_{ac}$ and with $H\Omega \simeq c_{ac}$, which is true not only for accretion disks but also for self-gravitating galaxies. Obviously, for supersonic turbulence or longer correlation times the differential rotation no longer dominates. We expect larger pitch angles for longer correlation times. Indeed, only for rather long correlation times do the pitch angles reach values that are compatible with observations.

4.2.1
Hydrostatic Equilibrium and Interstellar Turbulence

Rotating turbulence does not produce a finite α effect if the turbulence is not stratified. A homogeneous rotating field of isotropic SN explosions produces eddy diffusivity but does not produce any α effect. Mean-field dynamo theory is thus the theory of density and/or intensity stratifications under the presence of a basic rotation. There is strong observational evidence for substantial amounts of highly turbulent gas at heights above 1 kpc. This suggests using the equation of vertical hydrostatic equilibrium to derive the anisotropies in the turbulence field (Fröhlich and Schultz, 1996). The main questions for this concept concern the sign of the α effect and the resulting amplitude.

For the density stratification the empirical H I distribution from Dickey and Lockman (1990) has been taken. The extended ionized gas, which is the source of the diffuse Hα emission at high latitudes, has been described by an exponential with a scale height 1500 pc and a midplane density of 0.025 cm^{-3} (Reynolds, 1989).

The turbulence pressure occurs in the equation for the vertical momentum, that is

$$\frac{dP_{tot}}{dz} = -\rho k_z \quad \text{with} \quad P_{tot} = \rho \langle u_z^2 \rangle + \frac{\bar{B}_x^2 + \bar{B}_y^2 - \bar{B}_z^2}{2\mu_0} + P_{CR} \,. \qquad (4.7)$$

Only the intensity of the vertical turbulence contributes. The pressure due to cosmic rays is P_{CR} (Parker, 1992; Hanasz et al., 2004; Kowal, Otmianowska-Mazur, and Hanasz, 2006; Hanasz et al., 2006). Equation (4.7) for prescribed turbulence and magnetic field gives the density profile $\rho = \rho(z)$. The k_z-force is essentially due to a self-gravitating isothermal sheet of stars with constant half-thickness $z_0 = 600$ pc, that is

$$k_z(R, z) = \frac{2u_*^2}{z_0} \tanh\left(\frac{z}{z_0}\right) \qquad (4.8)$$

with u_* being the vertical turbulence intensity of the old disk stars. For z-values exceeding the thickness of the stellar disk the contribution of the more spherical components of the galaxy becomes important. The separation of the disk potential into a vertical and a radial component is of course only reasonable in the thin-disk approximation. If the density profile is given, the vertical profile $\langle u_z^2 \rangle$ results from (4.7).

To calculate unquenched coefficients α and η_T one has to specify the correlation time τ_{corr} for the interstellar turbulence. The results for the characteristic value $\tau_{corr} = 10$ Myr are given in Figure 4.3. One finds the turbulence intensity increasing with the height z. This is due to the exponential tail of the vertical density profile. We shall see that interstellar turbulence models driven by SN explosions and/or MRI will lead to very similar results. The components of the α-tensor are computed after the relations (4.9) and (4.10). The α effect becomes positive (negative) in the northern (southern) disk plane. Obviously, the density gradient over-

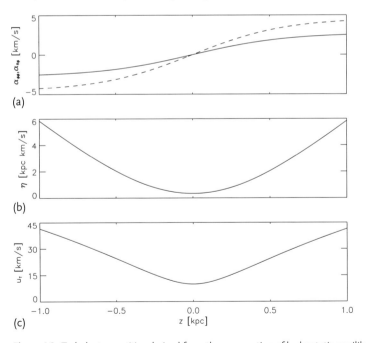

Figure 4.3 Turbulent quantities derived from the assumption of hydrostatic equilibrium, (4.7), and for $P \propto \rho^{1/3}$; $\alpha_{\phi\phi}$ (solid), $\alpha_{R\phi}$ (a), eddy diffusivity (b), turbulent intensity in the vertical direction (c). $\Omega^* = 0.6$. Courtesy of H.-E. Fröhlich.

compensates the influence of the turbulence, which shows the opposite stratification leading to the opposite signs of the α effect[1]. The turbulent pumping given in Figure 4.3a exceeds the given α effect numerically by a factor $\lesssim 2$ so that $\hat{\gamma} = \alpha_{R\phi}/\alpha_{\phi\phi} \lesssim 2$. Its positive sign indicates a large-scale advection of the magnetic field towards the midplane.

The numerical values of both the α effect and eddy diffusivity are rather high. Characteristic values are 1 km/s for $\alpha_{\phi\phi}$ and 1 kpc km/s for η_T, so that for a vertical scale of 1 kpc the dynamo number, $(4.3)_1$, takes values of $O(1)$. Such values are also characteristic for other completely different models of rotating stratified turbulence in the interstellar medium. Note that the α effect does not exceed the value of the turbulence intensity hence $C_\alpha \lesssim H/\ell_{\text{corr}}$, which should always (slightly) exceed unity. All the computations lead to *positive* values of $\alpha_{\phi\phi}$ in the upper part of the disk. The dynamo number of the differential rotation, see (4.3), with the mentioned value of the eddy diffusivity leads to $C_\Omega \simeq 30$. These numbers anticipate a massive problem of the dynamo theory to understand field geometries with large pitch angles.

We shall see that the majority of these results are confirmed by much more complicated simulations of the interstellar turbulence as driven by clustered SN explosions, also called a superbubble (Figure 4.4). Mac Low and McCray (1988) have shown how, due to the gravitational forces, the shape of the superbubble loses its symmetry between top and bottom (Figure 8 in Mac Low and McCray, 1988) so that

Figure 4.4 Superbubble N70 in the Large Magellanic Cloud, with a diameter of 100 pc, blown by stellar winds and SN explosions. Courtesy of FORS Team, ESO.

---

1) The density gradient overcompensates the turbulence stratification as long as the coefficient of the density gradient in (4.9) exceeds the value 0.22 (see the discussion in Brandenburg et al., 2013).

it is able to develop a net α effect. Explosions with symmetry between top and bottom would only provide an α effect if the intensity of the explosions systematically varied in the vertical direction. The signs of this α effect are opposite, however, if the turbulence intensity grows – as observed – with the galactic height.

4.2.2
Alpha Effect by Supernova Explosions

Ziegler (1996) numerically simulated SN explosions in the galactic disk and also calculated all components of the α-tensor for a random ensemble of explosions. Only isolated explosions under the influence of the galactic differential rotation have been simulated. From the time-dependent 3D MHD simulations the turbulent EMF is calculated and then averaged over an empirical SN distribution function to derive the α-tensor for a whole sample of explosions. An empirical SN rate per unit volume, $\Phi(x)$, is introduced (see Kaisig, Rüdiger, and Yorke, 1993), implying that the evolution of every remnant is independent of the others. The spatial distribution of SN shows a strong dependence on z, but varies only slowly with radius. Φ is thus assumed to be a function of z alone. The electromotive force $\mathcal{E}$ is simply the convolution of the EMF of one basic explosion and the distribution function of the SN. As the EMF of one explosion is known from the simulations the integrals can be evaluated. In all runs the interstellar medium is regarded as isothermal, with a uniform density of 10^{-24} g/cm^3 and a temperature of 5000 K, and an explosive energy release of 10^{51} erg. The uniform initial magnetic field is parallel to the coordinate axes. The resulting α effect is rather weak and highly anisotropic. The results of only a few m/s generally confirm the early analytical findings of Ferrière (1992a,b). The exact values of the amplitudes depend strongly on the scale height of the SN distribution. They vanish for very large-scale heights. In the upper disk layer $\alpha_{\phi\phi}$ is positive and α_{zz} is negative. The off-diagonal component $\alpha_{R\phi}$ exceeds the diagonal elements by two orders of magnitude. However, the pitch angle statistics require a strong α effect, and such a large escape velocity suppresses the dynamo action.

Ensembles of up to 1000 SN explosions in clusters of OB stars have been considered by Ferrière (1996) to increase the resulting α effect. Then Korpi et al. (1999a,b) provide a detailed discussion of the filling factors of the warm (10^4 K) and hot (10^6 K) component of the ISM reporting an increase of the correlation time to 30 Myr. The SN-created superstructures interact in the model and provide indeed a greatly enhanced α effect of several km/s. The nonlocal character of the SN-driven turbulence seems indeed to resolve the contradiction between the strong α effect found in Section 4.2.1 and the weak α effect that results from isolated SN explosions.

Clustered SN explosions are re-considered by Gressel et al. (2008) with a present-day MHD code. The injection of thermal energy of the explosions is combined with a radiation cooling function. The NIRVANA CODE (Ziegler, 2004) is used to simulate a shearing box of about $1 \times 1 \times 4$ kpc^3 initially in exact hydrostatic equilibrium. The rotation period with 60 Myr is short enough to allow dynamo activity. As it must, the rotation rate scales inversely to the distance from the rotation axis. The microscopic viscosity in the code is set to 5×10^{24} cm^2/s, and the magnetic Prandtl number is larger than unity (Pm = 2.5). The turbulence is mainly driven by the most frequent type-II SN explosions where the energy input of one explosion slightly exceeds 10^{51} erg. The frequency of the explosions is assumed as 15 Myr^{-1} kpc^{-2}, that is only one explosion per century for a galaxy with 30 kpc diameter.

The simulations provide a turbulence intensity u_T growing *upwards* despite the fact that the SN distribution peaks at the equator. The superbubbles, however, become highly deformed in the vertical direction due to the action of the vertical density gradient (Figure 4.5). The resulting net α effect (after averaging over z) overcompensates the opposite influence of the increasing halo turbulence. The eddy diffusivity also grows linearly with height, so that a correlation time can easily be derived as $\tau_{\text{corr}} \simeq 3$ Myr. Hence, the Coriolis number becomes $\Omega^* \simeq 0.5$. The α effect in the upper disk plane becomes positive due to the negative density gradient. Without the latter, the α effect would be negative such as the advection term which does not depend at all on the density gradient. Figures 4.6 and 4.7 show the results of the simulations derived with the test-field method (Schrinner *et al.*, 2005, 2007). One finds the eddy diffusivity also peaking at $z \gtrsim \pm 1$ kpc. The amplitude of $\alpha_{\phi\phi}$ is of order 5 km/s, that is 10% of the amplitude of the turbulence velocity. The numerical values also relate to the estimate $\alpha/u_T \simeq \Omega^*/2$.

MHD simulations of SN explosions also allow to study the magnetic suppression of the α-tensor and the eddy diffusivity. Ziegler (1996) derived the quenched coefficients $\alpha_{\phi\phi}$, α_{zz} and $\alpha_{\phi R}$ from the magnetically influenced flow pattern, and the magnetic fields of a single SN explosion and a vertically stratified ensemble of SN

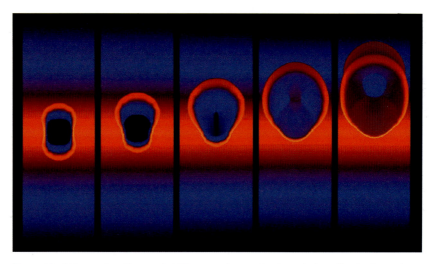

Figure 4.5 A time series of a superbubble located outside the midplane derived from 2D simulations. Note the clear asymmetry between top and bottom due to the action of gravity. It is this asymmetry which produces the positive (negative) α effect in the northern (southern) galactic hemispheres. The positive vertical gradient of the turbulence intensity would lead to opposite signs. Courtesy of O. Gressel.

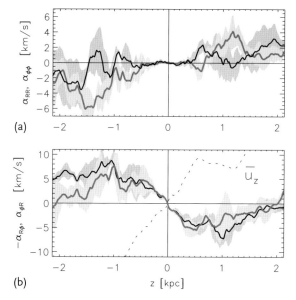

Figure 4.6 The horizontal components of the α-tensor are all of the same order. The shaded areas give the temporal fluctuations. The α_{RR} dark, $\alpha_{\phi\phi}$ light; both quantities are approximately equal and positive in the upper plane (a). The tensor $-\alpha_{R\phi}$ dark, $\alpha_{\phi R}$ light; the off-diagonal elements also antisymmetric with respect to the equator. The advection velocity is directed inward. The dashed line gives the vertical profile with an upward directed wind compensating the turbulent pumping (b). From Gressel *et al.* (2008). Reproduced with permission © ESO.

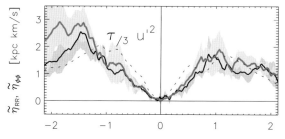

Figure 4.7 The diagonal elements of the eddy diffusivity tensor are of the same order (given in units of 3×10^{26} cm²/s). The diffusivity grows upward following the standard estimation $(1/3)\tau_{\rm corr} u_T^2$ with $\tau_{\rm corr} \simeq 3$ Myr and $u_T \simeq 40$ km/s (dashed line). From Gressel *et al.* (2008). Reproduced with permission © ESO.

explosions. The stronger the field the smaller the α effect. The $\alpha_{\phi\phi}$-component is more strongly quenched in comparison with $\alpha_{\phi R}$. For strong fields $\alpha_{\phi\phi}$ scales as $\bar{B}^{-3}$, while the quenching of the advection velocity only scales as $\bar{B}^{-2}$. While the dependence $\bar{B}^{-3}$ is known from quasi-linear (SOCA) calculations (see Figure 3.19), the weaker feedback of the magnetic field on the advection velocity $\alpha_{\phi R}$ could be delicate for a dynamo model. It may lead to the puzzling situation that under the

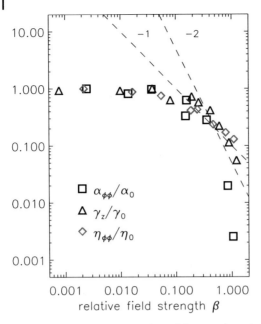

Figure 4.8 The magnetic quenching of the α effect (squares), the diamagnetic pumping (triangles) and the eddy viscosity (diamonds) in the simulations of the SN-driven interstellar turbulence. The α-quenching is stronger than $1/\beta^2$, while the eddy viscosity for large β is less strongly suppressed than $1/\beta$. After Gressel, Bendre, and Elstner (2013).

influence of a mean magnetic field of order of the equipartition value the diamagnetic advection term may dominate the α effect.

Gressel, Bendre, and Elstner (2013) also simulated the magnetic suppression of the α-tensor and the eddy diffusivity. In particular, the $\alpha_{\phi\phi}$-component, the advection term $\alpha_{\phi R}$ and the eddy viscosity, which produce azimuthal values of the EMF have been computed for various amplitudes of the magnetic background field $B_R(z)$ and $B_\phi(z)$. At first glance one finds the magnetic η-quenching with $1/\beta$ as the weakest for large β (Figure 4.8). On the other hand, the α-term and the pumping term are more massively suppressed for large β. Note that the magnetic influence of the generated large-scale dynamo fields on the turbulence does not suppress the eddy diffusivity stronger than the α effect. Moreover, the suppression of the α effect even appears to be stronger than the suppression of η_T^2, that is the α quenching results as stronger than β^{-2}. If this would not be the case, then a magnetically saturated $\alpha\Omega$-dynamo model could not exist.

4.2.3
The Advection Problem

A very special problem for the galactic dynamo theory is formulated by the 'advection problem.' Disk dynamos are highly sensitive to the amplitude of the vertical advection velocity which in cylindrical coordinates equals $\gamma_z = \alpha_{\phi R}$. It is not created by the basic rotation. If the rotation, therefore, is too slow, the α effect becomes too small compared with the own diamagnetic pumping and cannot work. The longer the correlation time the smaller the value of $\hat{\gamma} = \alpha_{R\phi}/\alpha_{\phi\phi}$. Many of the α effect computations on the basis of SN explosions lead to $\hat{\gamma}$-values exceeding 10, for example Ferrière (1992a, $\hat{\gamma} \simeq 50$) and Ziegler, Yorke, and Kaisig (1996, $\hat{\gamma} \simeq 30$). If the phenomenon of networked SN is used as the source of interstellar turbulence (with an increase of the correlation time from 2 to 16 Myr), Ferrière (1996) reached $\hat{\gamma} \simeq 15$. The cause of this complication for galactic dynamos is the slow galactic rotation compared with the correlation time of the interstellar turbulence. The typical values $\tau_{\rm corr} \simeq 10$ Myr and $\Omega = 10^{-15}$ s^{-1} lead to $\Omega^* \simeq 0.6$. However, with the SOCA-expressions for rotating turbulence

$$\alpha_{\phi\phi} = \alpha_{RR} = -\frac{2}{15}\tau_{\rm corr}\Omega^* u_T^2 \left(\frac{d\log\rho}{dz} + \frac{2}{3}\frac{d\log u_T}{dz}\right) \qquad (4.9)$$

and

$$\alpha_{\phi R} = -\alpha_{R\phi} = -\frac{1}{3}\tau_{\rm corr} u_T^2 \frac{d\log u_T}{dz}, \qquad (4.10)$$

$\alpha_{\phi\phi}$ becomes positive in the northern hemisphere if and only if the density stratification term dominates the (opposite) turbulence stratification term. The advection velocity $\alpha_{\phi R}$ becomes negative if the turbulence intensity increases upwards – as it does. The strong halo turbulence pumps the magnetic flux downward toward the equatorial midplane. It follows

$$\hat{\gamma} \simeq \frac{1}{\Omega^*}\frac{H_\rho}{H_u} \qquad (4.11)$$

with the Hs as the scale-heights of the density and the rms-velocity of the interstellar turbulence, respectively. Only for $H_\rho < H_u$ one finds rather small values ($\hat{\gamma} \lesssim 1$), in contrast to the earlier simulations but in agreement with the results given in Figure 4.6. Equation (4.9), taken for a density scale-height of 300 pc and with $u_T \simeq 10$ km/s, yields an α effect of order 1 km/s, in rather perfect agreement with Figure 4.6. Note the contrast of this value to the 50 m/s for isolated SN explosions.

The tensor component $\alpha_{\phi R}$, which is antisymmetric (for galaxies but not for stars, see Rüdiger and Hollerbach, 2004, p. 229) and describes an axial pumping (which is directed upwards for positive $\alpha_{\phi R}$) is shown in Figure 4.6b. It scales linearly with z until $z \simeq \pm 1$, where its amplitude approaches the amplitude of $\alpha_{\phi\phi}$, hence $\hat{\gamma} \simeq 1$, in agreement with the estimate (4.11).

All in all, the advection problem becomes much weaker or even vanishes if both density stratification and turbulence stratification are included in the calculations. The value of $\hat{\gamma}$ is only large if the turbulence stratification dominates the density stratification. Both Figure 4.6 and relation (4.11) demonstrate that for realistic vertical gradients the ratio $\hat{\gamma}$ is of order unity. Even this rather weak equatorward pumping is compensated in the simulations by a vertical wind $\bar{U}_z$ of approximately the same magnitude which plays an important role to solve the advection problem.

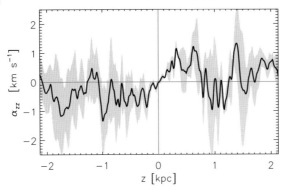

Figure 4.9 The component α_{zz} is much smaller than the $\alpha_{\phi\phi}$. Courtesy of O. Gressel.

The situation is more complicated with the zz-component of the α-tensor, for which the expressions often show (i) opposite sign in comparison to α_{RR} and (ii) a much weaker contribution of the density stratification. It is thus not clear which of both stratifications wins, that is not even the sign of this component is obvious. Estimations show that the influence of the turbulence stratification should dominate, that is

$$\alpha_{zz} = \frac{8}{15} \tau_{\text{corr}} \Omega^* u_T^2 \frac{d \log u_T}{dz} , \tag{4.12}$$

hence $\alpha_{zz}/\alpha_{\phi R} \simeq -\Omega^*$. The α_{zz}, therefore, should be *positive* in the upper disk plane for interstellar turbulence with positive vertical gradient. It becomes negative for negative gradient when the advection velocity $\alpha_{\phi R}$ becomes positive. The numerical value (4.12) is an upper limit which will still be reduced by the density stratification. One must emphasize that the calculations for single SN explosions by Ferrière (1992a,b), Kaisig, Rüdiger, and Yorke (1993) and Ziegler (1996) provided *negative* α_{zz} in the upper disk layer. It is not yet understood how combinations of density stratifications and turbulence intensity stratification determine the sign of α_{zz}. In all such simulations one finds $|\alpha_{zz}| < |\alpha_{\phi\phi}|$ (see Figure 4.9).

4.3
Dynamo Models

A galaxy might be modeled by an axisymmetric disk-like structure with axisymmetric functions α, η_T and $\bar{U}$. While η_T and $\bar{U}$ are taken as symmetric with respect to the galactic plane, the components of the α-tensor are antisymmetric. On the other hand, the eigensolutions of the linear induction equation in a domain with such symmetries are either symmetric (S) or antisymmetric (A) with respect to the midplane, and depend on the azimuth ϕ according to $e^{im\phi}$. The field modes are again denoted by Am or Sm.

4.3.1
Linear Models

The dynamo equation is often studied in a 1D approximation (Parker, 1971). The integration region extends only in the z-direction, where the remaining radial and azimuthal components of the magnetic field depend on z only, so that from the divergence condition one has $\bar{B}_z = 0$. Normalizing time with the diffusion time H^2/η_T and vertical distances with H yields

$$\frac{\partial A}{\partial t} = C_\alpha \hat{\alpha}(z) \Psi(B) B + \frac{\partial^2 A}{\partial z^2},$$

$$\frac{\partial B}{\partial t} = -C_\alpha \frac{\partial}{\partial z}\left(\hat{\alpha}(z)\Psi(B)\frac{\partial A}{\partial z}\right) - C_\Omega \frac{\partial A}{\partial z} + \frac{\partial^2 B}{\partial z^2}. \quad (4.13)$$

A (pseudo)vacuum surrounds the disk. Ψ may represent the magnetic quenching of the α effect.

The α effect depends on the magnetic field as well as on the location in the object, expressed simply by $\hat{\alpha}(z) = -\sin 2\pi z$ if the lower and upper boundaries are located at $z = 0$ and $z = 1$. The galactic rotation law leads to negative C_Ω, it is $|C_\Omega| = \Omega_0 H^2/\eta_T$. Positive C_α means positive (negative) α effect above (below) the disk midplane. The magnetic α-quenching has often been expressed by functions such as $\Psi \propto 1/[1 + (B/B_{\max})^2]$, with a cutoff field strength $B_{\max}$ related to the energy of velocity fluctuations or the gas pressure.

The solutions of the system (4.13) are represented in Figure 4.10 by the growth rates which result for arbitrary choices of the C_α and C_Ω. By definition the

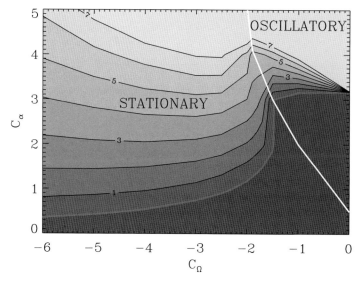

Figure 4.10 The growth rates if time is measured in rotation rates for the 1D disk dynamo. In the dark area the magnetic field decays. Note that in the disk geometry the $\alpha\Omega$-dynamo (lower-left corner) is stationary while the α^2-dynamo (upper-right corner) oscillates – contrary to the spherical dynamo models. Courtesy of R. Arlt.

marginal stability has a zero growth rate. For strong C_Ω the eigenvalue $\mathcal{D} = C_\alpha C_\Omega = -2.08$ for the excitation of steady quadrupoles is found. For small C_Ω, however, the quadrupole with the lowest eigenvalue oscillates. In this (disk!) approximation an α^2-dynamo produces oscillating quadrupoles. For negative α effect (not shown) all the solutions with negative C_Ω lead to oscillating quadrupoles. In this regime the dynamo number $\mathcal{D}$ is 44.1.

The growth times reach values of a few rotation periods close to the line of neutral stability. For galaxies this time is not much smaller than 1 Gyr. For highly supercritical dynamos the growth times are formally smaller than the rotation period.

For $|C_\Omega| > C_\alpha$ the pitch angles are much too low for galaxies. In order to produce large pitch angles one has to fulfill a relation such as $C_\Omega \lesssim C_\alpha$ leaving the stationary $\alpha\Omega$-regime. In consequence, the galactic dynamo can *not* be a standard $\alpha\Omega$-dynamo; it exists close to the limit where the α^2-regime prevails. Obviously, despite their slow rotation galaxies develop a rather high α effect. The condition $C_\Omega \simeq C_\alpha$ leads to $\alpha \simeq \Omega H$, which is of the order of 10 km/s.

Let us now suppose that the interstellar turbulence is driven by a field of random SN explosions under the influence of global rotation. The α effect only occurs in a stratified medium under the combined influence of turbulence and global rotation. Both density and intensity stratification of the turbulence field form the α-sources. The α effect is positive (negative) in the northern (southern) plane, and the turbulent field advection is toward the midplane. The resulting expressions are given by (4.9) and (4.10), multiplied by an appropriate function for the magnetic quenching.

The galactic rotation law can simply be approximated by

$$\Omega = \Omega_0 \left(1 + \frac{R^2}{R_\Omega^2}\right)^{-1/2}. \tag{4.14}$$

We assume $\bar{U}_\phi = R_\Omega \Omega_0 = 100$ km/s as the linear velocity. For the eddy diffusivity we work with the expression (3.25) with $\eta_{ijp} = \eta_T \epsilon_{ijp}$, which should be sufficiently accurate for slow rotation. Both turbulent and microscopic contributions have to be included in the magnetic diffusivity, so that $\eta = \eta_T + \eta_{\text{micro}}$, with $\eta_{\text{micro}} = 1/\mu_0 \sigma$ and σ the microscopic conductivity. In the bulk of the galaxy, for $z < 0.5 H$, the first term will always dominate. If, for greater vertical distances, the gas becomes neutral the second term dominates there.

Characteristic of our approach is the inclusion of the advection term simultaneously with the α effect. It transports magnetic field to regions of lower turbulence intensity. If higher velocities are present in the galactic halo then the field is carried toward the galactic midplane. The advection velocity term does not vanish for vanishing rotation. The galactic dynamo thus possesses a minimum rotation rate for which dynamo generation is possible. The dynamo dies off for $\bar{U}_\phi$ of order 20 km/s (see also Gressel et al., 2008). Note that NGC 4449 with $\bar{U}_\phi \simeq 30$ km/s possesses regular fields with ~ 14 μG amplitudes (Chyży et al., 2000).

One may ask how long the dynamo tolerates an extremely high turbulence-originated advection velocity. Therefore, we again write $\alpha_{R\phi} = \hat{\gamma} \alpha_{\phi\phi}$, and vary the $\hat{\gamma}$ as a parameter. For large negative $\hat{\gamma}$ the upward vertical turbulent transport is strong while for positive $\hat{\gamma}$ the field is advected to the galactic midplane. One finds that the dynamo only exists for $|\hat{\gamma}| < 5$. There is thus not too much freedom in the choice of the value of $\alpha_{R\phi}$. For too strong turbulent field advection the dynamo cannot operate.

The pitch angles of such models are always small ($\lesssim 5°$). The dynamo models yield pitch angles in accordance with

$$\tan p \simeq \sqrt{\frac{C_\alpha}{|C_\Omega|}}. \tag{4.15}$$

It completely vanishes if the field is purely azimuthal. Fields with globally large pitch angles form the observed patterns of galaxies so that the α effect must be strong enough compared with the differential rotation to create considerable radial magnetic field components. Not only the azimuthal field component has been amplified, for example, by differential rotation, but there must exist also an effective amplification mechanism of the poloidal component. Indeed, SN explosions may form a particularly strong driver of the interstellar turbulence which under the influence of the galactic rotation should form high values of the kinetic helicity which leads to the required high values of the α effect.

4.3.2
Nonlinear Dynamo Models

Equation (4.15) is only true within the linear dynamo theory. If in the nonlinear regime the feedback of the induced field suppresses mainly the helicity, then the resulting α effect becomes much smaller and the pitch angle is also reduced. Within the concept of α-quenching as the basic nonlinear feedback the pitch angle problem cannot be solved.

The model F by Elstner, Gressel, and Rüdiger (2009) demonstrates this idea. A disk of radius $R_0 = 5$ kpc and a height of 2 kpc rotates with the rotation law (4.14) with $R_\Omega = 1$ kpc and the α effect formulated as

$$\alpha = \alpha_0 \sin \frac{\pi z}{H_\alpha} \left(1 + \frac{B^2}{\bar{B}_{eq}^2}\right)^{-1} \tag{4.16}$$

with $H_\alpha = 1.5$ kpc. The initial values are $C_\alpha = 5$ and $C_\Omega = 100$. The resulting kinematic pitch angle works out to $p_0 = 16°$ which is very close to $p = 13°$ according to (4.15). In the nonlinear regime, however, the α effect is reduced by the magnetic suppression, and the pitch angle is thereby reduced to smaller values. Hence, standard $\alpha\Omega$-dynamos with magnetic α-quenching cannot solve the pitch angle puzzle. The observed large pitch angles indicate that the galactic dynamo saturates by a different nonlinearity than by the standard α-quenching.

An improved model contains two more field transport effects in the vertical direction. At first, if the intensity of the interstellar turbulence has a minimum at the galactic midplane and grows outward the diamagnetic effect will transport the fields toward the midplane. This effect is also quenched by the induced magnetic field. The nonlinear dynamo will thus not saturate by the diamagnetic effect. The only hope to saturate the dynamo without α-quenching is to include the flux loss by the vertical wind $\bar{U}_z \propto z$, which is a result of the density-stratified turbulence driven by SN explosions (see Gressel et al., 2009). This wind blows opposite to the diamagnetic effect and is *not* magnetically suppressed. The field losses by the galactic wind saturate the dynamo with a growth time of 250 Myr at the level of the equipartition field. The nonlinear model A of Elstner, Gressel, and Rüdiger (2009) shows indeed a *final* pitch angle of $p = 16°$.

The vertical wind of about 5 km/s is a result of the simulations (Gressel et al., 2008) which provided the α-tensor described in Section 4.2.2. At the same time a large-scale magnetic field developed with a growth time of about 250 Myr (Figure 4.11). The ratio of the radial and azimuthal field components takes the (large) value of 0.2, and the energy of the fluctuating field dominates that of the mean field, that is $q \simeq 6$.

Hence, in the box with SN-driven interstellar turbulence a dynamo operates growing by orders of magnitude and which is not a classical $\alpha\Omega$-dynamo. The resulting pitch angle p is larger and q is smaller than the related values of the Sun. The effective magnetic Reynolds number is thus smaller for galaxies than for the Sun, but it is nevertheless larger than unity. The galactic dynamo operates in the high-conductivity regime and, indeed, the magnetic Reynolds number in the simulation is around 500. A comparison with Figure 3.4 shows that these values fit well together. From Figure 4.11 we can also take that the (vertically averaged) toroidal field exceeds the (vertically averaged) poloidal field only by a factor of 5, leading to the large pitch angles. Note that the simulation started with very small field amplitudes. The fields grow by three orders of magnitudes over a time interval of 1.5 Gyr,

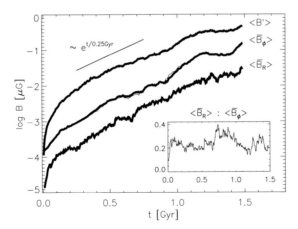

Figure 4.11 Box simulation of a SN-driven galactic dynamo with a growth rate of 250 Myr by Gressel et al. (2008). Note the initial field to be smaller by three orders of magnitude. The fluctuations dominate and the pitch angle is large. Reproduced with permission © ESO.

Table 4.2 Dynamo models with vertical wind for various SFR. For rigid rotation (last row) there is no dynamo. From Elstner and Gressel (2012).

R_Ω [kpc]	SFR	B_ϕ/B_{eq}	B_R/B_{eq}	p [deg]	τ_{gr} [Myr]
1	0.4	1.3	0.25	12	180
1	1	1.1	0.25	14	160
1	2.6	0.95	0.25	17	250
5	1	–	–		∞

that is six growth times. This behavior cannot be due to the linear amplification by differential rotation.

It can also be shown with this numerical model that a minimum rotation exists where the dynamo remains marginal. The critical rotation rate is 25 km s^{-1} kpc^{-1}, corresponding to a rotation velocity of 100 km/s at a radius of 4 kpc. The model demonstrates the correctness of the concept of the mean-field dynamo theory on the basis of externally driven (interstellar) turbulence.

With fixed density stratification Elstner and Gressel (2012) probed the influence of the rotation law and the star formation rate (SFR) with mean-field models where the shape of the electromotive force has been parameterized by means of the simulations. For fixed SFR the rotation law is varied and for fixed R_Ω the SFR is varied (Table 4.2). All models with strong differential rotation ($R_\Omega = 1$ kpc) operate as dynamos with large pitch angles. The calculation with $R_\Omega = 5$ kpc shows, they are not α^2-dynamos. With too weak differential rotation no dynamo is possible.

It is also interesting that the large value of B_R does not depend on the star formation rate. On the other hand, the azimuthal component is anticorrelated with the SFR, so that the pitch angle grows with growing SFR. The same is true with respect to the growth time of the dynamo-induced fields. One is confronted with the puzzling situation that for high SFR the pitch angle grows but the dynamo is more and more suppressed. One must note, however, that pitch angles of 35° as observed by Chyzy and Buta (2008) for the ring-shaped galaxy NGC 4736 can hardly be explained with such models.

4.4
Magnetic Instabilities

4.4.1
The Seed Field Problem

In order to form a quadrupolar geometry of the dynamo-induced fields the seed field must already have a (small) quadrupolar component. Ruzmaikin, Shukurov, and Sokoloff (1988) discuss the possibility of magnetic field amplification by a collapsing protogalactic cloud. The geometry of such fields remains antisymmetric with respect to the galactic midplane, so they cannot serve as a seed field for quadrupoles. A battery effect, caused by the early collapse of the material to the galactic midplane, can also not generate such an initial field. The battery effect forms electrical currents with *equatorial* symmetry, so that the corresponding magnetic fields are always antisymmetric with respect to the galactic midplane (Krause and Beck, 1998). The battery effect cannot provide any quadrupolar seed field – not even of the smallest amplitude. There is no clear route from the formation of a single galaxy to the present-day large-scale, dynamo-generated quadrupolar magnetic fields of a few µG amplitude. The typical seed field strength in the early Universe is given as 10^{-18} G (Hanasz and Lesch, 1997).

Poezd, Shukurov, and Sokoloff (1993) and Beck *et al.* (1994) suggest considering small-scale dynamo action to produce the necessary seed fields. The turnover time of such fluctuations with $\ell_{corr} \simeq 100$ pc and $u_{rms} \simeq 10$ km/s reaches 10 Myr. The magnetic amplitude for balancing kinetic and magnetic energy is not more than 3 µG. Beck *et al.* (1996) favor a value of 10^{-8} G as the mean component of such a fluctuating field. Its dipolar part will decay, and its quadrupolar part (if it exists) will be amplified. On the other hand, Deiss *et al.* (1997) and Thierbach, Klein, and Wielebinski (2003) report a few microgauss magnetic field strength in the intracluster medium of Coma. Even higher field amplitudes are given for the intergalactic medium for two radio source samples by Kronberg *et al.* (2001) and Carilli and Taylor (2002).

The magnetorotational instability (MRI) is one of the candidates to solve this problem. As shown below the instability that results from the interaction of a vertical magnetic field (antisymmetric with respect to the galactic midplane) and the rotation law $\Omega \propto R^{-1}$ leads to the excitation of axisymmetric modes with *quadrupolar* magnetic field geometry. With the microscopic diffusivity $\eta \simeq 10^7$ cm^2/s of the interstellar gas the resulting minimum field for the MRI is only $\sim 10^{-25}$ G – much smaller than the abovementioned 10^{-18} G.

Not only axial fields but also strong axial currents can become unstable, even without rotation. Here the rotation plays a suppressing role if it is rigid. Differential rotation, however, allows instability even in the case that the toroidal field is current-free (see Section 5.6, AMRI). The kink-type Tayler instability, MRI and AMRI are the main subjects of the Chapters 5 and 6. In the following we shall demonstrate possible applications of the theory to galaxies.

4.4.2
Magnetorotational Instability

The first question arises whether the MRI also exists in galaxies, with their differential rotation given by (4.14) and their aspect ratio $R_\Omega/H \simeq 5$ with H as the disk half-thickness. The *maximal* amplitude of an unstable vertical magnetic field B_0 is given on the right-hand branch in Figure 4.12a, and according to Ha $\lesssim 0.4 \overline{\mathrm{Rm}}$ with

$$\overline{\mathrm{Rm}} = \frac{\mathrm{Rm}}{\sqrt{\mathrm{Pm}}} = \frac{\Omega H^2}{\sqrt{\nu \eta}}, \quad \mathrm{Ha} = \frac{B_0 H}{\sqrt{\mu_0 \rho \nu \eta}} \qquad (4.17)$$

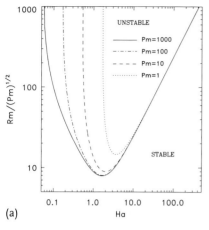 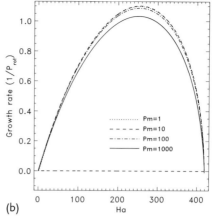

Figure 4.12 Stability map for a galactic rotation law against quadrupolar perturbations under the influence of a vertical magnetic field with the Hartmann number Ha and for high magnetic Prandtl numbers (a). The growth rates in units of the rotation time of the rigidly rotating core. $\overline{Rm} = 1000$ (b). From Kitchatinov and Rüdiger (2004). Reproduced with permission © ESO.

yields $B_0 \lesssim 0.5\sqrt{\mu_0\rho}\,\Omega\, H$. With the characteristic values $\Omega = 10^{-15}\,\text{s}^{-1}$ and $\rho = 10^{-24}\,\text{g/cm}^3$ one finds critical field strengths of order microgauss, that is $B_0 \lesssim H/200\,\text{pc}$ (in microgauss). Sellwood and Balbus (1999) concluded that up to amplitudes of a few microgauss galaxies might be MRI-unstable. If so, the existence of interstellar turbulence could be explained without any stellar activity or SN explosions – in other words, stellar winds and SN explosions would merely perturb the MRI. The observations of the H I disk in NGC 1058, with its uniform distribution of interstellar turbulence without any star formation activity could confirm this idea. In NGC 4414 one also finds ordered, large-scale magnetic fields and turbulence without any strong star formation activity.

For a linear, global MRI model the magnetic Prandtl number should be larger than unity due to the high values of the microscopic (Spitzer, 1962) viscosity in young galaxies ($\nu \sim 10^{18}\,\text{cm}^2/\text{s}$, $\eta \sim 10^7\,\text{cm}^2/\text{s}$, see Kulsrud and Anderson, 1992). The neutral stability lines for Pm $\leq 10^3$ are given in Figure 4.12a. One finds that

- for large Pm the MRI scales with $\overline{Rm}$,
- the strong-field branch leads to $\Omega_A/\Omega \simeq 0.4$,
- for large Pm the total minimum of the Reynolds number $\overline{Rm}$ is 8,
- quadrupolar solutions are preferred.

The *minimum* magnetic field which allows MRI for large Pm follows from S = 1.7 (Figure 4.12a) with the Lundquist number $S = \text{Ha}\cdot\sqrt{\text{Pm}}$. Accordingly, with the microscopic diffusivity, the minimum axial magnetic field for which the MRI starts is $B_\text{min} \simeq 10^{-25}$ G. Sigl, Olinto, and Jedamzik (1997) estimate the primordial magnetic field as up to 10^{-20} G on a scale of 10 Mpc. Much higher magnetic amplitudes for cosmological fields are reported by Widrow (2002). Banerjee and Jedamzik (2003)

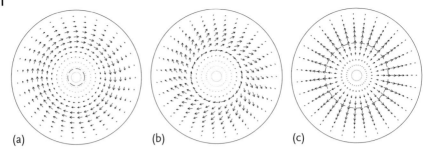

Figure 4.13 The field geometry in the midplane for the quadrupole mode with $\overline{\mathrm{Rm}} = 1000$. At the weak-field branch (marginal) (a) the mode with the maximum growth rate (b), at the strong-field branch (marginal) (c). Pm = 1. From Kitchatinov and Rüdiger (2004). Reproduced with permission © ESO.

present numerical models for the magnetic field evolution from the cosmic past to the present. They favor the surprisingly large cluster fields to be of primordial origin.

In Figure 4.12b the growth rates for $\overline{\mathrm{Rm}} = 1000$ are given in units of the rotation time of the inner core. Between the left- and right-hand branches the growth time is basically the rotation time (for all magnetic Prandtl numbers). For a rotation time of 70 Myr the amplification factor after 1 Gyr is thus 10^6. In 1 Gyr the MRI can amplify a seed field of 10^{-12} G to 1 μG.

The results for the geometry of the induced magnetic field are also interesting. Not only does the perturbation mode with the lowest eigenvalue possess quadrupolar geometry, field lines of the pattern with the maximal growth rates also possess pitch angles of $\lesssim 45°$. For marginal instability the pitch angles are zero at the weak-field branch and almost 90° at the large-field branch, but they have medium values for the solutions with the maximum growth rate. Figure 4.13 demonstrates this situation for $\overline{\mathrm{Rm}} = 1000$ and Ha $\lesssim 400$.

It is worth mentioning that the main properties of the MRI for Pm ≥ 1 also follow from the dispersion relation for marginal instability. The solution of the dispersion relation for galaxies with $\mathrm{d}\log\Omega/\mathrm{d}\log R = -1$ yields

$$\overline{\mathrm{Rm}}^2 = \frac{\mathrm{Pm}(1 + \mathrm{Ha}^2)^2}{2(\mathrm{PmHa}^2 - 1)} \tag{4.18}$$

(here $\overline{\mathrm{Rm}} = \Omega_0/\bar{\eta}k^2$, Ha $= \Omega_A/k^2\bar{\eta}$, $\Omega_A = kB_0/\sqrt{\mu_0\rho}$, $\bar{\eta} = \sqrt{\nu\eta}$). The minimum rotation for MRI proves to be $\overline{\mathrm{Rm}} = \sqrt{2 + 2/\mathrm{Pm}}$, which only varies between $\sqrt{2}$ and 2 for Pm ≥ 1. The corresponding magnetic field is given by Ha $= \sqrt{1 + 2/\mathrm{Pm}}$, which only varies between $\sqrt{3}$ and 1 for Pm $\gtrsim 1$. The Pm-dependence of the coordinates of the minima of the stability curve – expressed by means of Ha and $\overline{\mathrm{Rm}}$ – is strikingly weak.

Nonlinear MHD simulations of Dziourkevitch, Elstner, and Rüdiger (2004) indicate the occurrence of flow and field fluctuations due to MRI for two runs with uniform density. The characteristic values are given in Table 4.3. The initial value of the vertical field varies by a factor of 4. The typical wavelength ℓ_{corr} of the

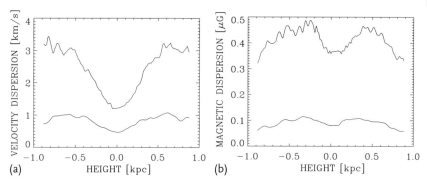

Figure 4.14 The turbulent (a) and the magnetic (b) intensities have a minimum at the disk midplane. The same holds for u_z and b_z (dashed). From Dziourkevitch, Elstner, and Rüdiger (2004). Reproduced with permission © ESO.

Table 4.3 Disk instability for different $\bar{B}_z = B_0$ and identical rotation laws ($R_\Omega = 2$ kpc, $\Omega_0 = 100$ km s^{-1} kpc^{-1}). $\rho = 5 \times 10^{-25}$ g/cm^3.

B_0 [μG]	Ω_A/Ω	ℓ_{corr} [pc]	u_{rms} [km/s]	b_{rms} [μG]	q
0.11	0.002	130	2.3	1.0	83
0.22	0.005	220	8.3	2.1	109
0.44	0.009	400	17	4.3	96
7	0.14	–	–	–	–

resulting turbulence intensity u_{rms} also grows with the magnetic field. It appears to be no problem to produce the observed values for the turbulence intensity of a few km/s. Both turbulence and magnetic intensity show minima at the galactic midplane (Figure 4.14) also as observed. As it must, for too strong initial vertical fields ($\Omega_A/\Omega = 0.14$) the disk proves to be stable.

The Fourier spectrum of the magnetic fluctuations leads to the critical wavelengths as given in Table 4.3. The slope of the spectrum is steeper than a Kolmogorov spectrum. In both cases the energy of the magnetic fluctuations approaches the energy of the kinetic fluctuations. The ratio $q = \langle b^2 \rangle / B_0^2$ is of order 100.

Elstner, Gressel, and Rüdiger (2009) applied a uniform vertical field of 0.1 μG to a galactic disk model with a z-dependent density and a rotation velocity of $\bar{U}_\varphi = 200$ km/s. The midplane density is 10^{-24} g/cm^3 and the vertical density scale is $H_\rho = 300$ pc. The disk reaches in radius from 1 to 4 kpc, with reflecting boundary conditions. At the vertical boundaries at a height of ±5 kpc pseudovacuum has been assumed, with $b_R = b_\varphi = 0$ (and also reflection for the flow). The vertical flux is then conserved, but not the radial one. The field starts to grow exponentially at the inner boundary. As long as the growing field components are still weak the growth time is rather short (50 Myr), but later it is much longer (800 Myr). During the saturation process the energy ratio q grows from $q \simeq 30$ to $q \simeq 100$, while the pitch angle remains constant, that is $B_R/B_\phi \simeq 0.1$ corresponding to $p = 6°$.

4.4.3
Tayler Instability

Observations of the galactic fields reveal the mode with $m = 0$ almost always as dominant but not as singular. Often a *spectrum* exists including magnetic modes with $m = 1$ and $m = 2$ (see Fletcher, 2010). It appears to be complicated to explain this phenomenon by the standard dynamo theory and/or by the standard MRI. On the other hand, nonaxisymmetric modes are the preferred eigensolutions of the current-driven Tayler instability (TI) or the azimuthal magnetorotational instability. These instabilities are discussed in Chapter 6 mostly on the basis of detailed calculation of Taylor–Couette flow models. Here only a short summary is given with the main attention to the excitation of nonaxisymmetric instability modes.

Indeed, the concept of MRI is consistent for galaxies if the background magnetic field would only possess an axial field component and if the rotation law does not depend on the vertical coordinate z. Galaxies, however, possess magnetic fields with toroidal and poloidal components of the same order of magnitude. They rotate with a characteristic velocity $\bar{U}_\phi$ of $\gtrsim 100$ km/s which is almost constant outside a small core which rotates as a solid body. The stability analysis is thus complicated by the helical structure of the background field. If the toroidal field component exceeds a critical value the system becomes unstable against the Tayler instability or – for weak electric currents – against the azimuthal version of MRI. One finds, however, with the characteristic parameters of galaxies the small value (4.2) for the normalized Alfvén frequency of the toroidal field component, indicating strong rotational suppression of the TI. We shall see that by the action of *differential* rotation the modes with low m easily survive the rotation-induced stabilization (see Section 6.5).

Here, in cylindrical geometry the linear stability of a prescribed helical magnetic field $(0, \beta R/R_{out}) B_0$ is studied, where the radial coordinate R runs between $r_{in} R_{out}$ and R_{out} (here $r_{in} = 0.5$). Large $|\beta|$ describe dominant toroidal fields while small $|\beta|$ describe dominant axial fields. The ratio $\mu_B = B_{out}/B_{in}$ of the outer and the inner toroidal field for the given r_{in} is fixed as $\mu_B = 2$, which is the profile of the azimuthal magnetic field which is due to a uniform axial electric current (Tayler, 1957).

The stability of such a field is probed in the following (i) without rotation, (ii) with rigid rotation and (iii) with the quasi-galactic rotation law

$$\Omega = \frac{\Omega_{in}}{6}\left(2 + \frac{R_{out}^2}{R^2}\right). \tag{4.19}$$

It is the rotation law of a Taylor–Couette flow with cylinders that approach the galactic rotation $\bar{U}_\phi = \text{const}$ for $r_{in} = 0.5$ with $\Omega_{out} = \Omega_{in}/2$. The perturbations are developed into normal modes and the solutions are optimized with respect to the axial wave number k. The radial boundary conditions are $u = 0$ and perfect

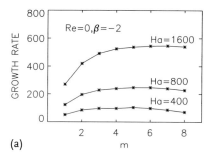
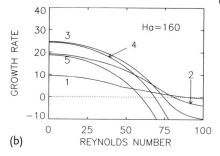

Figure 4.15 The growth rate in units of the dissipation frequency for various Hartmann numbers of the toroidal field, and axial fields of the same order as the toroidal field. No rotation. Note the quadratical dependence of ω_{gr} on the Hartmann number. Higher m are preferentially excited (a). The suppression of TI by rigid rotation. The curves are marked with their azimuthal mode number m. The $m = 1$ mode survives the best (b). $\mu_B = 2$, $\beta = -2$, $Pm = 1$. Reproduced with permission © ESO.

conductors inside the inner and outside the outer cylinder. All calculations are done for $\beta = -2$ (both components of the same order (see Bonanno and Urpin, 2011). The global rotation and the toroidal field are represented by the Reynolds number and the Hartmann number

$$\mathrm{Re} = \frac{U_\phi R_{\mathrm{in}}}{\nu}, \quad \mathrm{Ha} = \frac{B_{\mathrm{out}} R_{\mathrm{in}}}{\sqrt{\mu_0 \rho \nu \eta}}, \tag{4.20}$$

(here taken with the *outer* value of the toroidal magnetic field). We start with $\mathrm{Re}_m = 0$. Figure 4.15a shows the growth rates for supercritical magnetic fields for various mode numbers m. The instability is dominated by higher mode numbers with $m > 1$ if the axial field is strong enough. The mode number with the highest growth rate grows with growing axial field component B_0, that is for smaller $|\beta|$. For given β the characteristic mode number m also grows with growing Ha. Hence, the value of the m with the largest growth rate measures the helicity β of the cylindrical background field. It can easily be of order 10 for fields with $B_z \simeq B_\phi$, but this is only realized for very large Hartmann numbers, that is for very strong fields.

Global rigid rotation, however, changes this picture. It mainly quenches the instability, but the modes with higher m are quenched fastest (Figure 4.15b). For the largest Reynolds numbers only the mode with $m = 1$ survives. In this graph the higher-order modes with $m > 2$ already disappeared for $\Omega_A/\Omega < 2$, where only the $m = 1$ mode still exists[2]. With the axial magnetic field component included one finds fields with $\Omega_A/\Omega \lesssim 1.6$ as stable even against perturbations with $m = 1$, so that the value (4.2) for galaxies is by far too small for TI.

This picture changes again if differential rotation comes into play (Figure 4.16). Then the modes with small m are re-animated below the critical values for Ω_A/Ω known from Figure 4.15b. For large Reynolds number the growth rates grow with the Reynolds number so that finally $\omega_{gr} \simeq \Omega$. We find that under the presence

---

2) For dominant toroidal fields this mode is only stabilized by much faster rotation so that by (4.2) such toroidal fields could become unstable in galaxies.

of differential rotation toroidal fields with (4.2) are indeed unstable against nonaxisymmetric perturbation modes with $m \gtrsim 1$.

However, for small $|\beta|$ the axisymmetric mode returns. This is due to the existence of the standard MRI for the axial component of the field. As the dashed line in Figure 4.16 shows, for rather small fields ($\Omega_A/\Omega < 0.27$) it even possesses the largest growth rates. This figure also demonstrates that for nearly uniform toroidal fields ($\mu_B = 1$) almost the same situation exists as with uniform electric current discussed (in more detail) in Section 6.5.2.

To approach the nonlinear regime Elstner, Bonanno, and Rüdiger (2008) simulated the instability of toroidal fields in a flat isothermal *ring* with periodic boundary conditions in the vertical direction. The minimum wave number is thus fixed. The radial profile of the toroidal field can be (i) rather flat or (ii) highly concentrated as a ring. Tests with current-free fields, that is $B_\phi \sim 1/R$, always provide stability, also for rigid rotation. Without rotation both the uniform field and the ring-like field become unstable, where during the initial growth and in the saturation the $m = 1$ mode dominates. They differ, however, in the normalized growth time $\tau = \tau_{gr}\Omega_A$. For rapid rigid rotation all sorts of toroidal fields prove to be stable.

In Table 4.4 the rotation velocity and the magnetic background field are normalized with the Alfvén rotation Ω_A or the Alfvén speed $V_A = R_{out}\Omega_A$. The fastest ro-

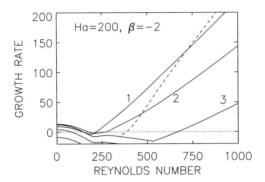

Figure 4.16 As in Figure 4.15 but with differential rotation. Homogeneous field ($\mu_B = 1$). For rapid rotation the modes with small m are re-animated by AMRI. The dashed curve demonstrates the appearance of the axisymmetric mode excited by standard MRI. $\beta = -2$, $\mu_\Omega = 0.5$, Pm $= 1$. From Rüdiger, Schultz, and Elstner (2011). Reproduced with permission © ESO.

Table 4.4 The influence of the (rigid) rotation on TI for ring-like toroidal fields in thin disks of uniform density. The almost constant growth time τ_{gr} is given in units of the Alfvén time $1/\Omega_A$.

Ω	B_0	Ω_A/Ω	τ_{gr}	u_{rms}	b_{rms}	q
0	4	∞	0.07	0.9	0.9	0.05
1	4	4	0.07	0.8	0.9	0.05
2	4	2	0.08	0.6	0.8	0.04

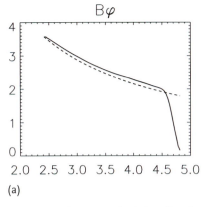

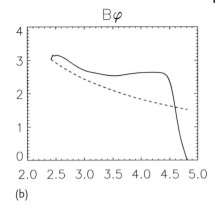

Figure 4.17 The saturated radial profiles of the toroidal field (with a peaked initial profile) is an almost current-free profile, with no rotation (a). For slow rotation the profiles exhibit a reduced mean electric current for $\Omega_A/\Omega = 4$ (b). Dashed lines represent a current-free field, that is $R\bar{B}_\phi = $ const.

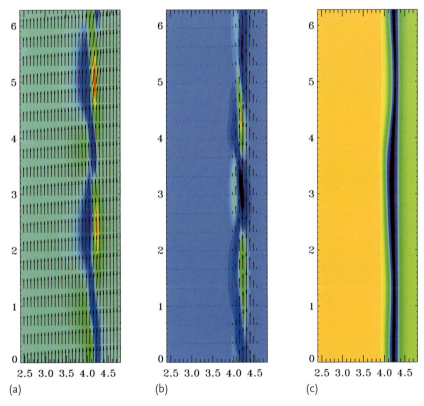

Figure 4.18 The instability pattern during the exponential growth phase in the R–ϕ plane for velocity (a), magnetic field (b) and density (c). $\Omega_A/\Omega = 4$ (slow rotation). Note that the leading mode is here of the type with $m = 2$. From Elstner, Bonanno, and Rüdiger (2008).

tation reaches the value $\Omega_A/\Omega = 2$, which after (4.2) is still too large by a factor of ten for applications to galaxies. Without rotation the ring profile decays after about two Alfvén crossing times and evolves to a current-free R^{-1} profile (Figure 4.17a) which is stable so that the $m = 1$ pattern disappears. A possible explanation is that the TI develops turbulence with high eddy diffusivities which reduce the corresponding electric current.

The saturation of TI due to the resulting turbulence via the eddy diffusivity seems plausible. Simulations of rotating disks show how the electric current is reduced (Figure 4.17). The field amplitude sinks for faster rotation but the ring structure of the field survives. The numbers in Table 4.4 suggest that the rotation suppresses the instability (last two lines).

The numbers of the first two lines in Table 4.4 indicate the TI as magnetic dominated: the turbulence never exceeds the intensity of the magnetic fluctuations. Both quantities u_{rms} and b_{rms} grow proportionately to the amplitude of the toroidal background field. Note, however, that the energy of the magnetic fluctuations is surprisingly small compared with the energy of the background field ($q = O(10^{-2})$).

The geometry of the TI-induced patterns during the exponential growth for the slow-rotation model with $\Omega_A/\Omega = 4$ is presented in Figure 4.18. Because of the ring-like structure of the background field the leading mode for velocity, magnetic field and density is of the type $m = 2$ existing only at the radius of the initially defined magnetic ring. Figure 4.17b demonstrates that during the course of the saturation process the maximum of the toroidal field migrates inward in order to reduce the applied mean-field current.

5
The Magnetorotational Instability (MRI)

5.1
Taylor–Couette Flows

The Taylor–Couette (TC) flow between concentric rotating cylinders (Figure 5.1a) is a classical problem of hydrodynamic stability (Couette, 1890; Taylor, 1923). For viscous flows in the absence of any longitudinal pressure gradient the general form of the rotation law in the container is

$$\Omega(R) = a_\Omega + \frac{b_\Omega}{R^2}, \tag{5.1}$$

where a_Ω and b_Ω are two constants related to the angular velocities $\Omega_{\rm in}$ and $\Omega_{\rm out}$ with which the inner and outer cylinders rotate (and we will only be interested in the case where $\Omega_{\rm in}$ and $\Omega_{\rm out}$ are both positive). With $R_{\rm in}$ and $R_{\rm out}$ ($R_{\rm out} > R_{\rm in}$) being the radii of the two cylinders, one obtains the coefficients

$$a_\Omega = \frac{\mu_\Omega - r_{\rm in}^2}{1 - r_{\rm in}^2} \Omega_{\rm in}, \quad b_\Omega = \frac{1 - \mu_\Omega}{1 - r_{\rm in}^2} \Omega_{\rm in} R_{\rm in}^2, \tag{5.2}$$

where $\mu_\Omega = \Omega_{\rm out}/\Omega_{\rm in}$ and $r_{\rm in} = R_{\rm in}/R_{\rm out}$. According to the Rayleigh criterion the ideal flow is stable whenever the specific angular momentum increases outwards, that is

$$\mu_\Omega \geq r_{\rm in}^2. \tag{5.3}$$

Viscosity, however, has a stabilizing effect, so that a flow with $\mu_\Omega < r_{\rm in}^2$ becomes unstable only if the Reynolds number of the inner rotation exceeds some critical value, or in other words, if the inner cylinder rotates sufficiently rapidly. The classical flow with stationary outer cylinder is given by $\mu_\Omega = 0$. We will consider primarily the case $r_{\rm in} = 0.5$, that is $R_{\rm out} = 2 R_{\rm in}$. There are then three more characteristic values of μ_Ω. The 'Rayleigh line' of marginal centrifugal instability indicating uniform angular momentum density ($R^2 \Omega = $ const) exists for $\mu_\Omega = 0.25$, a pseudo-Kepler rotation of the cylinders ('cylinders are rotating like planets') is given by $\mu_\Omega = 0.35$, and $\mu_\Omega = 0.5$ mimics a quasi-galactic rotation law with $\bar{U}_\phi = $ const. With his pioneering work of 1959, Velikhov was the first who demonstrated that (ideal) TC flows

Magnetic Processes in Astrophysics, First Edition. G. Rüdiger, L.L. Kitchatinov, and R. Hollerbach.
© 2013 WILEY-VCH Verlag GmbH & Co. KGaA. Published 2013 by WILEY-VCH Verlag GmbH & Co. KGaA.

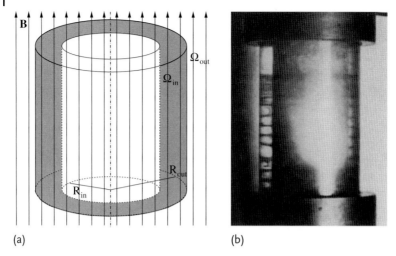

Figure 5.1 Taylor vortices at the onset of the hydrodynamic instability (a). Note that the aspect ratio of the vortices is close to unity. From Koschmieder (1993). The cylindrical geometry of the magnetic Taylor–Couette flow with a liquid metal between the cylinders (b).

of conducting fluids under the presence of an axial magnetic background field become linearly unstable for all $\mu_\Omega < 1$ (magnetorotational instability, MRI). With the Hall effect included even flows with $\mu_\Omega > 1$ can be destabilized. Moreover, we have verified that stable TC flows fulfilling the condition $r_{\rm in}^2 \leq \mu_\Omega < 1$ can even be destabilized by toroidal fields which are totally current-free in the gap between the two cylinders (azimuthal magnetorotational instability, AMRI). On the other hand, Tayler (1957) and Pitts and Tayler (1985) have shown that TC flows which are subject to a strong enough axial electrical current become unstable without and with (rigid) rotation (Tayler instability, TI).

A particular challenge for understanding hydrodynamic TC flows is the experimental finding of Withjack and Chen (1974) that nonaxisymmetric disturbances appear for stable rotation laws with (5.3) if the fluid is stably density-stratified in the axial direction (stratorotational instability, SRI).

We have computed that for dissipative Taylor–Couette flows with (5.3) the MRI should also be visible for rotating liquid sodium as the conducting fluid, but only for Reynolds numbers of order 10^6. For a possible experimental realization of this finding it is important to know up to what Reynolds numbers hydrodynamically linearly stable Taylor–Couette flows are also nonlinearly stable. According to (5.3) the flow should be maximally stable for stationary inner cylinder. Richard and Zahn (1999) focused attention on the experimental results of Wendt (1933), who found a *nonlinear* instability for this case for Reynolds numbers of order 10^4–10^5. However, later experiments suggested that the results of Wendt were due to imperfections in the container, and with a more precise setup the flow remains laminar for the same order of the Reynolds number (Schultz-Grunow, 1959).

There are several present-day experiments with Reynolds numbers of order 10^6 with various gaps between the cylinders and various aspect ratios $\Gamma = H/(R_{\text{out}} - R_{\text{in}})$ (with H as the height of the container). The Princeton experiment (Ji et al., 2006; Burin, Schartman, and Ji, 2010; Schartman, Ji, and Burin, 2009; Schartman et al., 2012) has the smallest aspect ratio ($\Gamma \simeq 2$), and no direct torque measurements, but the most precisely controlled endplates, which are split into several independently rotating rings. The resulting flux of angular momentum in the form

$$\beta_{\text{vis}} = \frac{Q_{R\phi}}{R_{\text{in}}^2 \Omega_{\text{in}}^2} \tag{5.4}$$

(the β-viscosity, Lynden-Bell and Pringle, 1974; Huré, Richard, and Zahn, 2001) only reaches values of $O(10^{-6})$.

In contrast, experiments in Maryland and Twente (Paoletti and Lathrop, 2011; van Gils et al., 2011; Huisman et al., 2012; Paoletti et al., 2012), and also Cottbus (Merbold, Brauckmann, and Egbers, 2013), have considerably greater aspect ratios, and also direct torque measurements, but no split-ring endplates, and hence potentially greater end-effects. The measured torques (and hence angular momentum transport) are significantly greater than the results inferred in the Princeton experiment. The normalization $J = \nu^2 G$ for the flux of angular momentum represents the observed (eddy) viscosity. G is a normalized torque. Traditionally, one writes $G \propto \text{Re}^\alpha$, so that for $\alpha = 1$ the eddy viscosity simply scales with the molecular viscosity and $\nu_T/\nu = O(1)$. The measurements, however, show that $\alpha \simeq 1.78$ for $\text{Re} = 10^5$–10^6. As $J \propto \nu^{2-\alpha}$ or $\nu_T/\nu \propto \nu^{1-\alpha}$, one finds that an eddy viscosity exists which scales as $\nu^{0.22}$ for $\nu \to 0$ or $\nu_T/\nu \propto 1/\nu^{0.78}$. The maximum β_{vis}-viscosity according to the definition (5.4), which also reads as $\nu_T/\nu \simeq \beta_{\text{vis}}\text{Re}$ scales as $\beta_{\text{vis}} \propto \text{Re}^{\alpha-2}$. The numerical value for β_{vis} which can be taken from Figure 5.2

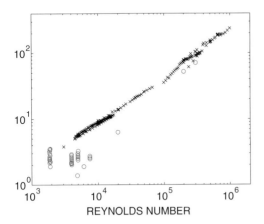

Figure 5.2 The normalized torque G vs. Reynolds number by the experiments of Merbold, Brauckmann, and Egbers (2013) before and beyond the Rayleigh limit. The crosses mark the reference values for the unstable rotation law with resting outer cylinder but the circles represent an experiment with $\mu_\Omega = 0.251$.

is of order 10^{-5}, which exceeds the Princeton values (Ji *et al.*, 2006) by an order of magnitude.

One of the critical differences between the Princeton experiment and the experiments elsewhere concerns the issue of the endplates, and how much influence the resulting Ekman circulations have. Clearly in a setup as short as in Princeton, end-effects are potentially important throughout the entire container, making the use of split-ring endplates crucial. If split-ring endplates are used though, the Princeton results indicate that they are very effective at minimizing end-effects. In contrast, in the much taller setups elsewhere, it is less clear how important end-effects are, and whether split-ring endplates are necessary. Hollerbach and Fournier (2004) argued that because of the Taylor–Proudman theorem end-effects could be important even in very tall cylinders, and that a large number of split rings would be necessary to produce a smooth rotation profile in the interior, rather than a step-like profile consisting of a series of Stewartson layers. The numerical calculations by Avila *et al.* (2008) also indicate that 'end-effects' could in fact occur in the middle of the container, as far away from the ends as it is possible to go. Avila (2012) did further calculations designed specifically to reproduce some of the existing setups, and suggests that end-effects are sufficiently strong to generate unwanted instabilities (also Schartman *et al.*, 2012).

At the moment therefore it should probably be considered an open question whether flows that are linearly Rayleigh-stable are also nonlinearly stable, or whether finite amplitude instabilities exist that can generate turbulence and enhanced angular momentum transport. The astrophysical discussion of such conjectured nonlinear instabilities is thus likely to continue (e.g., Richard and Zahn, 1999; Lesur and Longaretti, 2005; Balbus, 2011; Paoletti *et al.*, 2012). It would certainly be of interest also to do further experiments combining the best aspects of all of the setups to date, that is, tall cylinders, split-ring endplates, and direct torque measurements. Doing all that at Reynolds numbers exceeding 10^6 is of course rather challenging. Nevertheless, in the following sections a series of existing experiments can be described which already realized almost all of the mentioned instabilities in the laboratory.

5.2
The Stratorotational Instability (SRI)

According to (5.3) the rotation law $\Omega \propto R^{-2}$ plays an exceptional role. It separates the unstable flow region from the stable flow region and thereby represents the Rayleigh limit. Withjack and Chen (1974) experimentally found nonaxisymmetric disturbances in density-stratified Taylor–Couette flows even beyond the Rayleigh limit. With their wide-gap container ($r_{in} = 0.2$) they found the stability curve *crossing* the Rayleigh limit. The resulting experimental instability line, however, is rather steep (see Figure 8 in Withjack and Chen, 1974) and never crosses the line $\bar{U}_\phi = $ const of quasi-galactic rotation. It has also been shown theoretically that the combination of stably stratified (along the rotation axis) and centrifugally stable differential rotation can lead to a flow which is unstable against nonaxisymmetric perturbations (Molemaker, McWilliams, and Yavneh, 2001; Yavneh, McWilliams, and Molemaker, 2001; Dubrulle et al., 2005; Shalybkov and Rüdiger, 2005; Umurhan, 2006).

The equations for incompressible stratified fluids are

$$\frac{\partial U}{\partial t} + (U \cdot \nabla)U = -\frac{1}{\rho}\nabla P + g + \nu \Delta U, \qquad \frac{\partial \rho}{\partial t} + (U \cdot \nabla)\rho = 0 \qquad (5.5)$$

and $\operatorname{div} U = 0$ where U is the velocity, P is the pressure, g is the gravitational acceleration (supposed as vertical and constant), and ν is the kinematic viscosity. In the presence of a vertical density gradient ($\rho = \rho(z)$) it is clear that the system (5.5) allows the angular velocity profile (5.1) only in the limit of slow rotation and small stratification, that is $R\Omega^2/g \ll 1$ and $|d \log \rho / d \log z| \ll 1$.

We are interested in the stability properties of the basic state

$$\bar{U} = (0, R\Omega(R), 0), \quad \bar{P} = P_0(R) + P_1(R, z), \quad \bar{\rho} = \rho_0 + \rho_1(z), \qquad (5.6)$$

where ρ_0 is the uniform reference density, $\bar{P}$ is the pressure with $|P_1/P_0| \ll 1$ and also for the density holds $|\rho_1/\rho_0| \ll 1$. The linear stability problem is considered for the perturbed state of U, ρ and P. Using the above conditions the linearized equation system (5.5) takes the Boussinesq form with the coefficients depending only on the radial coordinate. Hence, a normal mode expansion $F = F(R)\exp(i(m\phi + kz + \omega t))$ of the solution can be used with F as any of the disturbed quantities. After a normalization the system

$$\frac{d^2 u_R}{dR^2} + \frac{1}{R}\frac{du_R}{dR} - \frac{u_R}{R^2} - \left(k^2 + \frac{m^2}{R^2}\right)u_R - 2i\frac{m}{R}u_\phi - i\text{Re}(\omega + m\Omega)u_R$$

$$+ 2\text{Re}\,\Omega\, u_\phi - \text{Re}\frac{dp}{dR} = 0,$$

$$\frac{d^2 u_\phi}{dR^2} + \frac{1}{R}\frac{du_\phi}{dR} - \frac{u_\phi}{R^2} - \left(k^2 + \frac{m^2}{R^2}\right)u_\phi + 2i\frac{m}{R}u_R$$

$$- i\text{Re}(\omega + m\Omega)u_\phi - i\text{Re}\frac{m}{R}p - \frac{\text{Re}}{R}\frac{d}{dR}(R^2\Omega)u_R = 0,$$

$$\frac{d^2 u_z}{dR^2} + \frac{1}{R}\frac{du_z}{dR} - \left(k^2 + \frac{m^2}{R^2}\right)u_z - i\text{Re}(\omega + m\Omega)u_z - i\text{Re}\,kp - \text{Re}\rho' = 0,$$

$$\frac{du_R}{dR} + \frac{u_R}{R} + i\frac{m}{R}u_\phi + iku_z = 0, \quad i(\omega + m\Omega)\rho' - N^2 u_z = 0 \qquad (5.7)$$

results where the same symbols are used for the normalized quantities.

The dimensionless numbers of the problem are the Reynolds number Re and the buoyancy frequency N

$$\text{Re} = \frac{\Omega_{\text{in}} R_0^2}{\nu}, \quad N^2 = -\frac{g}{\rho_0}\frac{d\rho_1}{dz}. \qquad (5.8)$$

We used $R_0 = \sqrt{R_{\text{in}}(R_{\text{out}} - R_{\text{in}})}$ as the unit of length, ν/R_0 as the unit of the perturbation velocity, Ω_{in} as the unit of the frequencies ω, N, and Ω, $R_0\Omega^2$ as the unit of g; ρ_0 is the density unit. It is convenient to describe the influences of the basic rotation with its normalized value, that is by the dimensionless Froude number Fr = Ω_{in}/N. Always no-slip boundary conditions for the velocity and no-flux conditions for the temperature on the walls are used (see Figure 5.3a, for the geometry of the container). Slippery boundary conditions have been adopted at the endplates to avoid Ekman circulations.

According to the Rayleigh condition the nonstratified Taylor–Couette flow with the rotation law (5.1) is stable against axisymmetric disturbances for (5.3). This condition has been extended to stratified fluids by Ooyama (1966). It is known that the axisymmetric mode is the most unstable mode for dissipative Taylor–Couette flows for $\mu_\Omega \geq 0$. The SRI, however, exists beyond the Rayleigh line but it is basically nonaxisymmetric, that is the unstable modes possess azimuthal mode numbers of $m > 0$. Our prediction that the instability for a small-gap container ($r_{in} = 0.78$) exists even beyond the pseudo-Kepler line has experimentally been confirmed by Le Bars and Le Gal (2007, see Figure 5.4).

In Figure 5.3b the marginal stability line which separates stable and unstable regions is calculated as a function of N^2/Ω_{in}^2 for $m = 1$ disturbances and for a container with $r_{in} = 0.5$ and with $\mu_\Omega = 0.3$. The latter value lies beyond the Rayleigh limit where nonstratified flows are linearly stable. An absolute-minimum critical Reynolds number exists for the Froude number Fr $\simeq 1.4$. The SRI is revealed as a delicate balance of buoyancy and rotation. Both the frequencies must approximately be equal otherwise one of the two stabilities dominates and stabilizes the system.

The next question is how this balance depends on the rotation law and on the geometry of the container. In Figure 5.5 containers are considered with different gap sizes. The computations also concern various density stratifications. For any

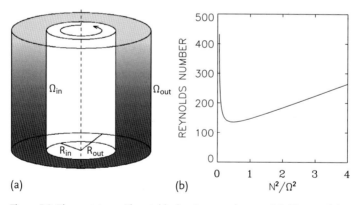

Figure 5.3 The container with a stable density (or temperature) stratification where the rotation of the cylinders mimics the rotation laws with Ω decreasing outwards ($\Omega_{in} > \Omega_{out}$) (a). The marginal stability line for the $m = 1$ perturbations in a container with $r_{in} = 0.5$ and $\mu_\Omega = 0.3$. The instability is suppressed for both too weak and too strong stratifications. The minimum Reynolds number occurs at a Froude number of 1.4 depending on the rotation law (b). Reproduced with permission © ESO.

5.2 The Stratorotational Instability (SRI)

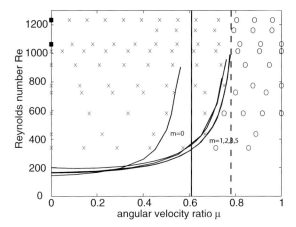

Figure 5.4 The stability lines for $m = 0, 1, 2, 3$ of Shalybkov and Rüdiger (2005) in comparison with experimental results for a flow with an axial salt stratification. The crosses stand for the observation of unstable flows while the circles denote stability. The very left curve for $m = 0$ does not cross the Rayleigh line while the other curves for nonaxisymmetric perturbations indicate instability up to the limit for quasi-galactic rotation laws (dashed), see Figure 5.5 ($r_{in} = 0.78$). From Le Bars and Le Gal (2007). Copyright 2007 by The American Physical Society.

gap size three vertical lines are given. The left line is the line for the Rayleigh limit where $\Omega_{out} = \Omega_{in} R_{in}^2 / R_{out}^2$. The right line mimics the rotation law $\Omega \propto 1/R$ which is typical for galaxies ($\Omega_{out} = \Omega_{in} R_{in}/R_{out}$). The central line represents the rotation law $\Omega \propto R^{-1.5}$, that is $\mu_\Omega = r_{in}^{1.5}$, which may be called the pseudo-Kepler limit. The main result of the Figure 5.5 is that the pseudo-Kepler rotation is *always unstable*. This, however, is not true for the more flat quasi-galactic rotation which for large gaps becomes stable. From Figure 5.5c we realize that the quasi-galactic rotation law with $\Omega \propto 1/R$ can be too flat for the possible existence of SRI in galaxies.

The SRI can also be realized with a model where stratification is due to a temperature gradient which is positive, that is the cylindrical gap is heated from above (Gellert and Rüdiger, 2009b). This configuration is perfectly stable for rigid rotation. If the outer cylinder rotates sufficiently slow, the flow without temperature gradient becomes unstable due to the centrifugal instability. Beyond the Rayleigh limit, when the outer cylinder rotates more rapidly, the unstratified flow is hydrodynamically stable. The buoyancy frequency is now $N^2 = \alpha g dT/dz$. Here α is the coefficient of volume expansion and T is the temperature.

For the numerical simulations the Froude number Fr, the cylinder height and the Prandtl number $\mathrm{Pr} = \nu/\chi$, with χ as the heat diffusivity, must be fixed. For too strong or too weak stratification there is no instability. Rotation frequency and buoyancy frequency of the same order appears to be a good choice to probe the existence of SRI. Figure 5.6 gives the spectrum of the unstable modes for the two Froude numbers 0.7 and 1.4. The spectra drastically differ. For larger buoyancy the spectrum peaks at $m = 5$ while for faster rotation the spectrum peaks at $m = 3$.

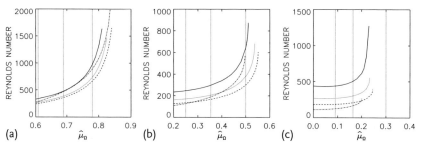

Figure 5.5 The marginal stability lines for the $m = 1$ modes for small-gap (a, $r_{in} = 0.78$), for medium gap $r_{in} = 0.5$ (b) and for wide gap $r_{in} = 0.3$ (c). The vertical dotted lines denote the Rayleigh limit, the pseudo-Kepler rotation and the quasi-galactic rotation. The curves represent different density stratifications with Fr = 2.2 (dot-dashed), Fr = 1 (dashed), Fr = 0.7 (dotted), Fr = 0.5 (solid, maximum stratification). For narrow gaps instability occurs even for rather flat rotation laws beyond the quasi-galactic rotation but not for too wide gaps. From Rüdiger and Shalybkov (2009). Reproduced with permission © ESO.

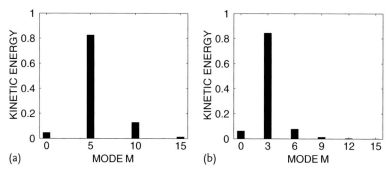

Figure 5.6 The azimuthal Fourier modes for a small gap container. Fr = 0.7 (a). Fr = 1.4 (b). Note the dominance of medium mode numbers with $m > 0$. Re = 500, $r_{in} = 0.78$, Pr = 7. From Gellert and Rüdiger (2009b).

The existence of the SRI due to temperature stratification has been demonstrated by an experiment operated at the University of Cottbus by Harlander *et al.* (2013), see Figure 5.7. The main question is whether there is a rotational quenching of the SRI as indicated by numerical simulations and the above argumentation. The container has an aspect ratio of 10 and a gap of $r_{in} = 0.52$. The temperature difference between top and bottom varies from 4 to 9 K. The typical spiral pattern of SRI is indeed visible. The experiment provides instability for fast rotation (slightly) beyond the Kepler limit (Figure 5.8) but not as clear as predicted by Figure 5.5.

5.2.1
The Angular Momentum Transport

Based on the idea that angular momentum transport is mainly enhanced via an increase of viscosity, Shakura and Sunyaev (1973) introduced in a parameterized

5.2 The Stratorotational Instability (SRI) | 193

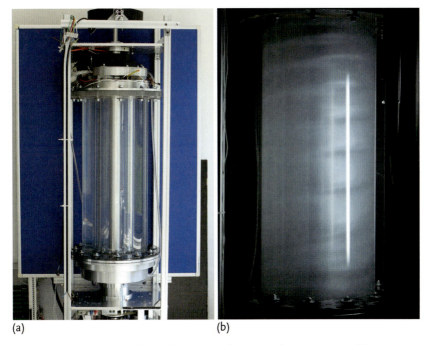

(a) (b)

Figure 5.7 SRI experiment of Harlander *et al.* (2013) from the BTU Cottbus in a water container heated from above with a centrifugally stable differential rotation ($\mu_\Omega = 0.27$). The outer radius is 14.5 cm and for the inner radius is $r_{in} = 0.52$, the height of the container is 70 cm. Re = 1100, Fr = 1.4, Pr = 7. Courtesy of U. Harlander and C. Egbers.

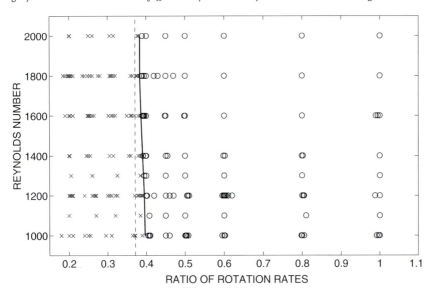

Figure 5.8 The gray crosses indicate the domain of instability. The line of marginal instability in the Cottbus SRI experiment is slightly beyond the pseudo-Kepler limit (dashed).

model for an accretion disk the coefficient

$$\alpha_{SS} = \frac{Q_{R\phi}}{H^2 \Omega_{\mathrm{in}}^2}, \qquad (5.9)$$

with the density scale height H of the disk. On the other hand, taking the normalized Reynolds stress in the form (5.4) for the SRI, leads to the values of β_{vis} given by Figure 5.9. Figure 5.9a shows a dependence of the β-value on the Reynolds number such as $\beta_{\mathrm{vis}} \simeq 10^{-5}\sqrt{\mathrm{Re}}$. For cosmical applications one has to mention that the β-viscosity works with the radius as the normalization. For thin disks this makes a big difference to the α_{SS} which are normalized with the vertical height. By definition it is

$$\alpha_{SS} = \beta_{\mathrm{vis}} \left(\frac{R_{\mathrm{in}}}{H}\right)^2 \approx 10^{-2}\sqrt{\mathrm{Re}}, \qquad (5.10)$$

so that in these units the angular momentum transport by SRI for high Reynolds number becomes surprisingly strong. A saturation for high Reynolds number must exist in the profile shown in Figure 5.9a as otherwise the transport grows too strong. Already with the maximum value shown in Figure 5.9a the resulting α_{SS} reaches values of 0.1–1. Such a saturation tendency is indeed suggested by Figure 5.9a.

From the following astrophysical argumentation one even expects much smaller values of β_{vis} for higher Reynolds numbers. In the center of the accretion theory one finds the angular momentum transport by viscous dissipation (von Weizsäcker, 1943). The general diffusion equation that governs the evolution of the distribution of the column density

$$\Sigma(R) = \int_{-\infty}^{\infty} \rho(R, z)\,dz \qquad (5.11)$$

in a thin Kepler disk is

$$\frac{\partial \Sigma}{\partial t} = \frac{3}{R}\frac{\partial}{\partial R}\left[\sqrt{R}\frac{\partial}{\partial R}(\sqrt{R}\nu\Sigma)\right] \qquad (5.12)$$

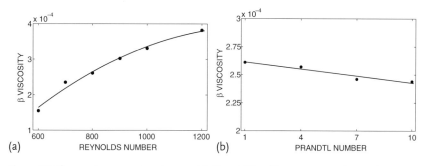

Figure 5.9 The angular momentum transport in the middle of the gap averaged over z in terms of the β-viscosity. Pr = 1, Reynolds number varies (a). Re = 800, Prandtl number varies (b). Small gap ($r_{\mathrm{in}} = 0.78$, pseudo-Kepler rotation). Fr = 1.4. From Gellert and Rüdiger (2009b).

(von Weizsäcker, 1948; Lüst, 1952; Pringle, 1981), which results from conservation of mass and angular momentum after elimination of the radial (accretion) flow. Hence, the diffusion of the mass (or, what is almost the same, the angular momentum) is subject to the characteristic timescale $\tau_{\mathrm{diff}} = R^2/\nu$ which means that $\tau_{\mathrm{diff}} \simeq 1/\beta_{\mathrm{vis}}\Omega$. The diffusion time at 1 AU is thus $\tau_{\mathrm{diff}} \simeq 1/(2\pi\beta_{\mathrm{vis}})$ in years which is only 1000 years if the maximum value of β_{vis} is taken from Figure 5.9a, which is too short by orders of magnitude compared with the expected 1 Myr.

As shown in Figure 5.9b the Prandtl number dependence of the SRI effect is only weak. Pr $= 1$ provides the maximum angular momentum transport. One can indeed expect that the angular momentum transport sinks if one diffusion coefficient dominates the other.

A principal solution of the high-β_{vis} puzzle focuses on the basic difference of TC flows and Kepler disks. Note that in the system (5.7) the first two equations decouple from the rest if the pressure is small or even vanishing – as in Kepler disks. The resulting planar inertia-Coriolis modes excite sound waves in the axial direction but this is not an exponential instability. Just in thin Kepler disks does the pressure become very small due to the balance of the gravity and the supersonic rotation so that the SRI in thin Kepler disks might maximally exist as a secondary instability (Shtemler et al., 2010). The appearance of SRI modes in simulated density-stratified thin Kepler disks has indeed never been reported.

5.2.2
Electromotive Force by Magnetized SRI

Stratorotational instability under the influence of a global magnetic field is interesting for the mean-field electrodynamics in turbulent fields. With the characteristics of the flow the pseudoscalar $\boldsymbol{g} \cdot \boldsymbol{\Omega}$ can be formed so that the existence of small-scale helicity $\langle \boldsymbol{u} \cdot \mathrm{curl}\, \boldsymbol{u} \rangle$ (also a pseudoscalar) can be expected. The combination of stable differential rotation, stable density stratification and stable magnetic field should thus be one of the most simple constructions to realize the α effect in simulations and/or experiments. If the applied magnetic fields is a current-free toroidal field then the ϕ-component of the turbulent EMF $\langle \boldsymbol{u} \times \boldsymbol{b} \rangle$ represents the $\alpha_{\phi\phi}$-component of the tensor. If the applied magnetic field is axial then it is the α_{zz}-component. Figure 5.10 shows the magnetic suppression of the SRI by azimuthal current-free magnetic fields. Without any magnetic influence the model becomes unstable against perturbations with $m = 1$ for Re $= 261$. The Hartmann numbers are taken with the maximal amplitudes of the magnetic field components. The magnetic suppression is given for the standard value Pm $= 1$ but also for Pm $= 10^{-5}$. The differences of the stability curves for the two magnetic Prandtl numbers are not too big to avoid experiments in the laboratory.

For azimuthal external fields first results for $\alpha_{\phi\phi}$ have been obtained. The non-axisymmetric components of flow and field form the fluctuations, while the axisymmetric components are considered as the mean quantities. The averaging procedure is simply the integration over the azimuth ϕ. It is standard to express the

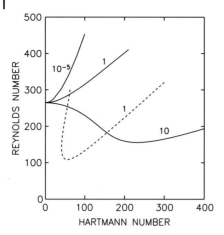

Figure 5.10 The magnetic suppression of SRI by current-free azimuthal fields for various magnetic Prandtl numbers. It is $m = 1$, $\mu_B = 0.5$, $\mu_\Omega = 0.3$, Fr = 0.5. The dashed curve is for homogeneous fluids with Pm = 1. From Rüdiger and Shalybkov (2009). Reproduced with permission © ESO.

turbulence-induced electromotive force as

$$\mathcal{E} = \langle \boldsymbol{u} \times \boldsymbol{b} \rangle = \alpha \circ \bar{\boldsymbol{B}} - \eta_T \, \mathrm{curl}\, \bar{\boldsymbol{B}} \tag{5.13}$$

with the alpha-tensor α_{ij} and the scalar eddy diffusivity η_T. In cylindrical geometry, a mean magnetic field $\bar{\boldsymbol{B}}$ only has a ϕ-component and the mean current curl $\bar{\boldsymbol{B}}$ only has a z-component. Hence, $\mathcal{E}_\phi = \alpha_{\phi\phi} \bar{B}_\phi$ and $\mathcal{E}_z = -\eta_T \,\mathrm{curl}_z\, \bar{\boldsymbol{B}} = 0$. The latter relation only holds if the mean magnetic field is current-free in the considered domain.

Let us consider the α effect in the frame of a linear theory where all functions are free to one and the same arbitrary parameter. This free parameter disappears if the second-order quantities such as $\mathcal{E}_\phi$ are normalized with another second-order quantity such as for the correlation function

$$f^\alpha = \frac{\langle \boldsymbol{u} \times \boldsymbol{b} \rangle_\phi}{\mathrm{MAX}\left(\sqrt{\langle \boldsymbol{u}^2 \rangle \langle \boldsymbol{b}^2 \rangle}\right)}. \tag{5.14}$$

All the correlations along the radius R are normalized with one and the same parameter. Hence, this function no longer contains the arbitrary factor of the eigenfunctions, and by definition it is smaller than unity. The only question is the sign of this quantity. Note that the B_ϕ has been given as positive in the model setup where rotation axis and density gradient are antiparallel.

Figures 5.11 present the correlation function f^α without and with density stratification. Note the influence of the density stratification. The resulting correlations change their sign at a certain radius within the container. Averaged over the radius, the f^α vanishes for $N = 0$ but not for the presence of a finite density stratification. The density gradient is a preferred direction in the system aligned with rotation axis

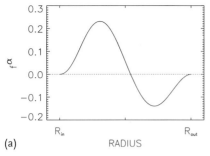
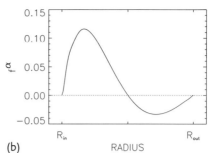

Figure 5.11 The correlation function (5.14) for a container with conducting cylinders in which a vacuum field $B_\phi \propto 1/R$ exists. Density stratification is zero (a) and nonzero (b, Fr = 0.5). $r_{in} = 0.5$, $\mu_\Omega = 0.35$ (pseudo-Kepler), Pm = 1. From Rüdiger and Shalybkov (2009). Reproduced with permission © ESO.

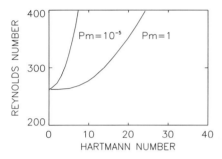

Figure 5.12 The magnetic suppression of SRI by homogeneous axial fields for various magnetic Prandtl numbers. The SRI is stabilized for small Pm, there is no subcritical excitation for Pm ≲ 1. It is $m = 1$, $\mu_\Omega = 0.3$, Fr = 0.5.

and allows the formation of large-scale helicity. In a uniform fluid the correlations are vanishing in the radial average. With an axial density stratification the calculations model the polar region of a rotating sphere or disk. Figure 5.11 demonstrates the $\alpha_{\phi\phi}$ at the northern pole as *positive* for negative density gradients.

It is also possible to compute the α_{zz}-component by application of a large-scale axial field. The magnetic suppression of the SRI by axial fields seems to be more massive than that for toroidal fields (Figure 5.12). The SRI under the influence of an axial field could be interesting to measure the α effect along the density stratification. Figure 5.12 shows that for small magnetic Prandtl numbers axial fields up to Hartmann numbers of order ten can be applied without destruction of the SRI. First calculations showed the resulting α_{zz} much smaller than $\alpha_{\phi\phi}$ and even negative. If this was true then the very elementary system of SRI drastically demonstrates the restricted meaning of a scalar as the representation of the α-tensor.

5.3
The Standard Magnetorotational Instability (SMRI)

If the fluid is electrically conducting and an axial magnetic field is applied, then the critical Reynolds number for the onset of axisymmetric Taylor rolls grows with growing strength of the magnetic field (Donnelly and Ozima, 1960; Chandrasekhar, 1961). These results were obtained for very small magnetic Prandtl numbers. Theory and observation are in nearly perfect agreement, but there is no indication of any magnetic-induced instability such as MRI. Ji, Goodman, and Kageyama (2001) supposed that this absence of MRI is due to the use of the small magnetic Prandtl number limit. The magnetic Prandtl number is really very small under laboratory conditions (Table 5.1). Within the small-gap approximation but with Pm as a free parameter Kurzweg (1963) found that for *weak* magnetic fields and sufficiently large magnetic Prandtl number the critical Taylor number becomes smaller than in the hydrodynamic case. If the field is not too strong it can play a destabilizing role and can lead to MRI via the Lorentz force. For ideal magnetic Taylor–Couette flows this was first discovered by Velikhov (1959). In the MHD regime the Rayleigh criterion for stability (5.3) changes to $\mu_\Omega \geq 1$, that is flows with superrotation are stable in the MHD regime. Velikhov found a positive growth rate along the Rayleigh line (i.e., $a_\Omega = 0$) of order $\sqrt{\Omega_{in}\Omega_{out}}$ and a critical wave number for marginal instability of $k \leq \sqrt{\Omega_{in}\Omega_{out}}/V_A$, with $V_A = B_0/\sqrt{\mu_0\rho}$ the Alfvén velocity of the given axial field. The stabilizing effect of an applied toroidal field decreasing outwards has also been suggested[1]. These results were derived via a dispersion relation of second order in the Fourier frequency ω. Instability is possible but only if V_A is smaller than the shear $|R^2 d\Omega/dR|$ multiplied by some positive factor, that is the instability is stabilized by too strong fields. Chandrasekhar (1960) confirmed these results.

The ideal hydrodynamic flow is stable if its angular momentum increases with radius, but the hydromagnetic flow is only stable if the angular velocity itself increases with radius. This remains true also for dissipative fluids. The MRI decreases the critical Reynolds number for weak magnetic field strengths for hydrodynam-

Table 5.1 Parameters of the liquid metals suitable for MHD experiments taken from Chandrasekhar (1961) and Noguchi et al. (2002). It is $\bar{\eta} = \sqrt{\nu\eta}$.

	ρ [g/cm^3]	ν [cm^2/s]	η [cm^2/s]	$\bar{\eta}$ [cm^2/s]	Pm
Mercury	5.4	1.1×10^{-3}	7600	2.9	1.4×10^{-7}
Gallium	6.0	3.2×10^{-3}	2060	2.6	1.5×10^{-6}
Galinstan (GaInSn)	6.4	3.4×10^{-3}	2428	2.9	1.4×10^{-6}
Sodium	0.92	7.1×10^{-3}	810	2.4	0.88×10^{-5}

[1] Azimuthal fields were later considered for axisymmetric perturbations by Knobloch (1992, 1996) and Pringle (1996).

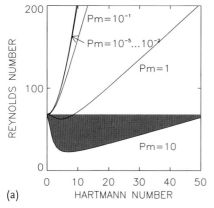 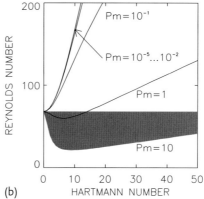

Figure 5.13 Marginal stability lines for axisymmetric modes with resting outer cylinder of conducting material (a) and vacuum (b). The shaded areas denote subcritical excitations of unstable axisymmetric modes by the external magnetic field. It only appears for Pm ≳ 1. $r_{\rm in} = 0.5$, $\mu_\Omega = 0$. From Rüdiger, Schultz, and Shalybkov (2003). Copyright 2003 by The American Physical Society.

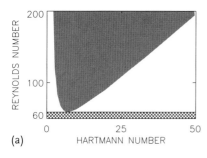 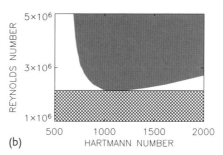

Figure 5.14 The instability domain for axisymmetric modes in containers with rotating outer cylinder of conducting material for Pm = 1 (a) and Pm = 10^{-5} (b). $r_{\rm in} = 0.5$, $\mu_\Omega = 0.33$.

ically unstable flow (Figure 5.13) and it destabilizes the hydrodynamically stable flow for $r_{\rm in}^2 < \mu_\Omega < 1$ (Figure 5.14). It depends only on the amplitude of the external axial magnetic field and not on its sign.

As we shall demonstrate, the critical Reynolds numbers vary as 1/Pm for hydrodynamically stable flows, so that the magnetic Reynolds number Rm = Pm · Re controls the instability. According to the numbers given in Table 5.1 the magnetic diffusivity of fluid metals is at least 1000 cm²/s (very similar to the solar plasma); it is therefore not easy to reach magnetic Reynolds numbers of the required $O(10)$. Already the first studies revealed the high-Rm problem for any realization of the MRI in the laboratory (Rüdiger and Zhang, 2001; Ji, Goodman, and Kageyama, 2001).

If the Hall effect is included in the induction equation then the new situation appears that the Taylor–Couette flow can become unstable for *any* ratio of the angular velocities of the inner and the outer cylinder. This new instability, however, does not exist for both signs of the axial magnetic field B_0. For positive shear $d\Omega/dR$ the Hall instability exists for negative fields and for negative shear it exists for positive fields. For negative shear, of course, the Hall instability combines with the MRI (see Section 5.5).

5.3.1
The Equations

A viscous electrically conducting incompressible fluid between two rotating infinite cylinders in the presence of a uniform axial magnetic field B_0 admits the basic solution $B_z = B_0$, $U_\phi = a_\Omega R + b_\Omega/R$, and all other vector components vanish. We are interested in the stability of this solution. By developing the disturbances $\boldsymbol{u}$ and $\boldsymbol{b}$ into normal modes, the solutions of the linearized MHD equations are considered in the form

$$\boldsymbol{u} = \boldsymbol{u}(R)e^{i(\omega t+kz+m\phi)}, \quad \boldsymbol{b} = \boldsymbol{b}(R)e^{i(\omega t+kz+m\phi)}. \tag{5.15}$$

The Navier–Stokes equation is used in its standard form, (2.15). To normalize lengths, again $R_0 = \sqrt{R_{\rm in}(R_{\rm out} - R_{\rm in})}$ is used as the unit of distances, η/R_0 as the unit of the perturbed velocity, ν/R_0^2 as the unit of frequencies, B_0 as the unit of the magnetic field fluctuations, R_0^{-1} as the unit of the wave number and $\Omega_{\rm in}$ as the unit of Ω. The dimensionless numbers of the problem are then the magnetic Prandtl number Pm, the Hartmann number Ha and the Reynolds number Re,

$$\text{Pm} = \frac{\nu}{\eta}, \quad \text{Ha} = \frac{B_0 R_0}{\sqrt{\mu_0 \rho \bar{\eta}}}, \quad \text{Re} = \frac{\Omega_{\rm in} R_0^2}{\nu} \tag{5.16}$$

with the geometric average of the diffusivities, $\bar{\eta} = \sqrt{\nu \eta}$. We shall see that in most cases, where no solution for $B_0 = 0$ exists, the magnetic Reynolds number $\text{Rm} = \text{Pm}\,\text{Re}$ and the Lundquist number $S = \sqrt{\text{Pm}}\,\text{Ha}$ are better representations of the characteristic eigenvalues. There are also exceptions from this rule when the stability/instability of rather steep rotation laws under the presence of toroidal fields is considered (see Section 5.3). The ratio S/Rm which is free of diffusivity values can be expressed by the Alfvén frequency $\Omega_A = V_A/R_0$ normalized with the inner rotation rate, that is

$$\frac{\Omega_A}{\Omega_{\rm in}} = \frac{S}{\text{Rm}}, \tag{5.17}$$

which is the inverse of the magnetic Mach number Mm.

Using the same symbols for normalized quantities as before, the equations can be written as a system of ten equations of first order. After elimination of both pressure fluctuations and the fluctuations of the vertical magnetic field, b_z, the linearized equations are

$$\frac{du_R}{dR} + \frac{u_R}{R} + \frac{im}{R}u_\phi + iku_z = 0, \tag{5.18}$$

$$\frac{d^2 u_\phi}{dR^2} + \frac{1}{R}\frac{du_\phi}{dR} - \frac{u_\phi}{R^2} - \left(\frac{m^2}{R^2} + k^2\right) u_\phi - i(m\mathrm{Re}\Omega + \omega) u_\phi$$

$$+ \frac{2im}{R^2} u_R - \mathrm{Re}\frac{1}{R}\frac{d}{dR}(R^2\Omega) u_R - \frac{m}{k}\left[\frac{1}{R}\frac{d^2 u_z}{dR^2} + \frac{1}{R^2}\frac{du_z}{dR}\right.$$

$$\left. - \left(\frac{m^2}{R^2} + k^2\right)\frac{u_z}{R} - i(m\mathrm{Re}\Omega + \omega)\frac{u_z}{R}\right]$$

$$+ \frac{m}{k}\mathrm{Ha}^2\left(\frac{1}{R}\frac{db_R}{dR} + \frac{b_R}{R^2}\right) + \frac{i}{k}\mathrm{Ha}^2\left(\frac{m^2}{R^2} + k^2\right) b_\phi = 0, \quad (5.19)$$

$$\frac{d^3 u_z}{dR^3} + \frac{1}{R}\frac{d^2 u_z}{dR^2} - \frac{1}{R^2}\frac{du_z}{dR} - \left(\frac{m^2}{R^2} + k^2\right)\frac{du_z}{dR} + \frac{2m^2}{R^3} u_z - i(m\mathrm{Re}\Omega + \omega)\frac{du_z}{dR}$$

$$- im\mathrm{Re}\frac{d\Omega}{dR} u_z - \mathrm{Ha}^2\left(\frac{d^2 b_R}{dR^2} + \frac{1}{R}\frac{db_R}{dR} - \frac{b_R}{R^2} - k^2 b_R + \frac{im}{R}\frac{db_\phi}{dR} - \frac{im}{R^2} b_\phi\right)$$

$$- ik\left[\frac{d^2 u_R}{dR^2} + \frac{1}{R}\frac{du_R}{dR} - \frac{u_R}{R^2} - \left(k^2 + \frac{m^2}{R^2}\right) u_R\right]$$

$$- k(m\mathrm{Re}\Omega + \omega) u_R - 2\frac{km}{R^2} u_\phi - 2ik\mathrm{Re}\Omega u_\phi = 0, \quad (5.20)$$

$$\frac{d^2 b_R}{dR^2} + \frac{1}{R}\frac{db_R}{dR} - \frac{b_R}{R^2} - \left(\frac{m^2}{R^2} + k^2\right) b_R - \frac{2im}{R^2} b_\phi - i\mathrm{Pm}(m\mathrm{Re}\Omega + \omega) b_R + ik u_R = 0, \quad (5.21)$$

$$\frac{d^2 b_\phi}{dR^2} + \frac{1}{R}\frac{db_\phi}{dR} - \frac{b_\phi}{R^2} - \left(\frac{m^2}{R^2} + k^2\right) b_\phi + \frac{2im}{R^2} b_R$$

$$- i\mathrm{Pm}(m\mathrm{Re}\Omega + \omega) b_\phi + ik u_\phi + \mathrm{Pm}\mathrm{Re} R\frac{d\Omega}{dR} b_R = 0 \quad (5.22)$$

(after elimination of the pressure, Shalybkov, Rüdiger, and Schultz, 2002). The rotation law $\Omega = \Omega(R)$ is normalized with $\Omega_{\mathrm{in}} = \Omega(R_{\mathrm{in}})$.

An appropriate set of ten boundary conditions is needed to solve the system. We always use the no-slip conditions for the velocity $u_R = u_\phi = u_z = 0$, and for the perfectly conducting walls $db_\phi/dR + b_\phi/R = b_R = 0$. The boundary conditions are valid at both R_{in} and R_{out}. For insulating walls the magnetic boundary conditions are different at R_{in} and R_{out}, that is for $R = R_{\mathrm{in}}$

$$b_R + i\frac{b_z}{I_m(kR)}\left[\frac{m}{kR} I_m(kR) + I_{m+1}(kR)\right] = 0, \quad (5.23)$$

and for $R = R_{\mathrm{out}}$

$$b_R + i\frac{b_z}{K_m(kR)}\left[\frac{m}{kR} K_m(kR) - K_{m+1}(kR)\right] = 0, \quad (5.24)$$

where I_m and K_m are the modified Bessel functions. The condition for the toroidal field at both locations is $k_R b_\phi = m b_z$.

The homogeneous set of linear equations with the boundary conditions determines the eigenvalue problem for given value of Pm. The real part of ω, $\Re(\omega)$, describes a drift of the pattern along the azimuth. For a fixed Hartmann number, a fixed Prandtl number and a given vertical wave number one finds the eigenvalues Re and $\Re(\omega)$. For a certain wave number a minimum of the Reynolds number exists which is the desired critical Reynolds number.

Figure 5.13 shows the stability lines for axisymmetric modes for containers with both conducting and insulating walls with resting outer cylinder, for fluids of various magnetic Prandtl number. Only the vicinity of the classical hydrodynamic solution with Re = 68 (for r_{in} = 0.5) is shown. There is a strong difference of the bifurcation lines for Pm $\gtrsim$ 1 and Pm < 1. For fluids with low viscosity the magnetic field only suppresses the instability, so that all the critical Reynolds numbers exceed the value 68, by more and more for increasingly strong magnetic fields.

For materials with high magnetic Prandtl number, however, the critical Reynolds numbers are smaller than Re = 68. The magnetic field with sufficiently small Hartmann numbers supports the instability rather than suppresses it. This effect becomes more effective for increasing Pm, but again it vanishes for stronger magnetic fields. Obviously, the MRI exists when the fields can be considered as frozen-in and/or enough viscosity prevents the validity of the Taylor–Proudman theorem.

In order to find a minimum in the Re profile the magnetic Prandtl number must exceed some critical value, Pm_{min}, for hydrodynamically unstable flow. The critical magnetic Prandtl numbers lie in the narrow interval 0.25–1.75 for all μ_Ω and r_{in}. Thus, for liquid metals the MRI in the form described by Kurzweg (1963) *cannot* be observed by experiments with hydrodynamically unstable flows.

5.3.1.1 The Rayleigh Limit

There is a universal scaling with Pm for the special case $\mu_\Omega = r_{in}^2$, that is with $a_\Omega = 0$ in (5.1). Then the terms with $d(R^2\Omega)/dR$ vanish in (5.18) to (5.22) and for $m = \omega = 0$ one finds that the quantities u_R, u_z, b_R and b_z scale as $Pm^{-1/2}$, while u_ϕ, B_ϕ, k and Ha scale as Pm^0. The Reynolds number for the axisymmetric modes then clearly scales as Re $\propto Pm^{-1/2}$. This scaling does not depend on the boundary conditions, as these also comply with the relations for $m = 0$. On the other hand, for $a_\Omega \gtrsim 0$ one finds the much steeper scaling Re $\propto Pm^{-1}$ (see Figure 5.15), resulting in the relation Rm $\simeq$ const for the magnetic Reynolds number, and Ha $\propto Pm^{-1/2}$, resulting in S $\simeq$ const for the Lundquist number S. For small magnetic Prandtl numbers the exact value of the microscopic viscosity is thus not relevant for the excitation of the instability. In consequence, however, the corresponding Reynolds numbers for the MRI at and beyond the Rayleigh limit differ by at least two orders of magnitude. In the vicinity of $\mu_\Omega = r_{in}^2$ there is a vertical jump in the Reynolds numbers from 10^4 to 10^6 in an extremely small interval of the abscissa. This sharp transition only exists for very small values of Pm. It is thus clear that experiments with liquid metals with $\mu_\Omega = r_{in}^2$ are not possible. Even the smallest deviation from the condition $\mu_\Omega = r_{in}^2$ would drastically change the excitation condition (Rüdiger, Schultz, and Shalybkov, 2003).

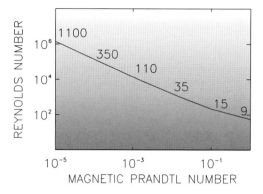

Figure 5.15 The critical Reynolds numbers vs. magnetic Prandtl number marked with those Hartmann numbers where the Reynolds number is minimum. There is a clear scaling of the Reynolds number with Pm^{-1}. Conducting walls, $r_{in} = 0.5$, $\mu_\Omega = 0.33$. From Rüdiger (2004).

5.3.1.2 Pseudo-Kepler Rotation

Another situation occurs if the outer cylinder rotates so fast that the rotation law no longer fulfills the Rayleigh criterion (5.3) and a solution for Ha = 0 cannot exist. The nonmagnetic eigenvalue along the vertical Re-axis moves to infinity, but a minimum remains for finite values of Ha. Figure 5.14 presents the results for $Pm = 1$ and $Pm = 10^{-5}$. There are always minima of Re for certain Hartmann numbers. The minima and the critical Hartmann numbers increase for decreasing magnetic Prandtl numbers. For $r_{in} = 0.5$ and $\mu_\Omega = 0.33$ the critical Reynolds numbers together with the critical Hartmann numbers are plotted in Figure 5.15. Obviously, the coordinates of the absolute minima for experiments with rotating outer cylinder, for $Pm \leq 1$, are basically characterized by magnetic Reynolds numbers of $O(10)$, very similar to the values of the successful dynamo experiments (Stieglitz and Müller, 2001; Gailitis et al., 2001).

From the numbers given in Table 5.2 one easily finds that a container with an outer radius of 10 cm (and an inner radius of 5 cm) filled with liquid sodium requires a rotation of 90 Hz in order to produce the MRI. The required magnetic field is about 3 kG. For insulating walls the necessary rotation rates are lower but the magnetic field amplitudes are higher. Obviously, the influence of the boundary conditions is not small.

Table 5.2 Coordinates of the absolute minima in Figure 5.14b for rotating outer cylinder with $\mu_\Omega = 0.33$, $r_{in} = 0.5$ and $Pm = 10^{-5}$.

	Conducting walls	Insulating walls
Reynolds number, Re	2.13×10^6	1.42×10^6
Mag. Reynolds number, Rm	21.3	14.2
Hartmann number, Ha	1100	1400
Lundquist number, S	3.47	4.42

Figure 5.15 shows that the value Rm = 21 holds for all values for Pm ≲ 0.1. The same is true for the Lundquist number S = 3.5. For Pm ≲ 0.1 the standard MRI scales with Rm and S almost independent of the actual value of Pm. Consequently, the SMRI does not exist at all in the quasi-static approximation with Pm = 0 as defined by Roberts (1967), Zikanov and Thess (1998) and Youd and Barenghi (2006). The viscosity must be finite to obtain the MRI but in the eigenvalues Rm and S it does not appear. It is interesting that the nonaxisymmetric instability of (weak) differential rotation under the presence of *azimuthal* toroidal fields shows the same scalings and restrictions (see Figure 5.34).

5.3.2
Nonaxisymmetric Modes

According to Cowling's theorem only nonaxisymmetric magnetic fields can be maintained by a dynamo process. It is thus important to know the excitation conditions of nonaxisymmetric modes. We start with the results for containers with outer cylinders at rest and for Pm = 10^{-5} (Figure 5.16). There is the linear Rayleigh instability even without magnetic fields. Hydrodynamic solutions for $m = 0$ (Re = 68), $m = 1$ (Re = 75) and $m = 2$ (Re = 127) are known. The axisymmetric mode possesses the lowest Reynolds number. This is not always true in the MHD regime: we find a crossover of the instability lines for axisymmetric and nonaxisymmetric modes, but only in containers with conducting walls. This is also true for containers with rotating outer cylinder (Figure 5.17).

A basic difference between the two sorts of boundary conditions is thus the existence of crossovers of the instability lines for $m = 0$ and $m = 1$ for conducting walls. For both resting and rotating outer cylinders Hartmann numbers exist above which the nonaxisymmetric mode possesses a lower Reynolds number than the ax-

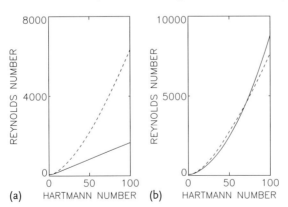

Figure 5.16 Resting outer cylinder: stability limits for axisymmetric (solid lines) and nonaxisymmetric instability modes with $m = 1$ (dashed). Insulating cylinders (a), and conducting cylinders (b). Pm = 10^{-5}, $r_{in} = 0.5$, $\mu_\Omega = 0$. From Rüdiger, Schultz, and Shalybkov (2003). Copyright 2003 by The American Physical Society.

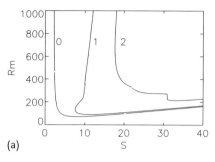 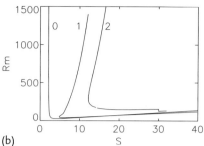

Figure 5.17 Stability map for $m = 0$, $m = 1$ and $m = 2$ in a pseudo-Kepler flow. $Pm = 1$ (a), $Pm = 0.01$ (b). For stronger magnetic fields the $m = 1$ mode is preferred. $r_{in} = 0.5$, $\mu_\Omega = 0.35$, conducting walls. From Gellert, Rüdiger, and Schultz (2012). Reproduced with permission © ESO.

isymmetric mode. The occurrence of nonaxisymmetric solutions as the preferred modes is a rather general phenomenon for containers with conducting walls.

Moreover, there is a very basic difference between the axisymmetric and the nonaxisymmetric modes. For $m = 0$ and $S \gtrsim 1$ always *one* critical Rm exists above which the MRI is excited for all larger magnetic Reynolds numbers. For $S > S_{MIN}$ (where $S_{MIN} \simeq 1$ is the smallest possible Lundquist number) there are always *two* critical Reynolds numbers between which the nonaxisymmetric MRI modes exist. All the nonaxisymmetric modes, therefore, cannot survive too rapid rotation. The differential rotation excites the MRI but – if too strong – it suppresses its nonaxisymmetric parts.

The $m = 1$ mode in Figure 5.17 needs increasingly strong background fields for increasingly rapid rotation. As an estimate on the basis of (5.17) one finds that the nonaxisymmetric instability only exists in the domain

$$0.013 < \frac{\Omega_A}{\Omega_{in}} < 0.25, \tag{5.25}$$

formed thus by rather weak fields. These limits hardly depend on the magnetic Prandtl number (Figure 5.17b). The curves of marginal instability of the $m = 1$ mode in the S–Rm plane are almost identical for $0.01 \leq Pm \leq 1$ (Gellert, Rüdiger, and Schultz, 2012). At the strong-field branch the marginal-stability curves for $m = 0$ and $m = 1$ cross for $S \simeq 18$. For weaker fields the mode with the smallest Reynolds number is always axisymmetric, but for stronger fields the critical Reynolds numbers for $m = 1$ are smaller than those for $m = 0$. For such fields and for large enough Reynolds numbers the MRI sets in as a nonaxisymmetric flow pattern. When the magnetic Reynolds number reaches the upper value of the marginal instability of the $m = 1$ mode the solution again becomes axisymmetric (see Figure 5.22).

Figure 5.18 shows for the two different rotation laws (from Rayleigh limit to Kepler limit) the critical Reynolds numbers for various magnetic Prandtl numbers. The general rule is that the excitation of nonaxisymmetric modes is much more complicated than the excitation of axisymmetric modes. For $Pm \simeq 1$ the differ-

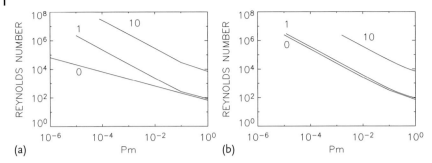

Figure 5.18 Stability map for rotation at the Rayleigh limit (a) and with a quasi-Kepler rotation law (b) for the modes with $m = 0$, $m = 1$ and $m = 10$. $r_{in} = 0.5$, conducting walls. From Rüdiger and Schultz (2008).

ences between $m = 0$ and $m = 1$ are not as big but for small Pm and steep rotation law the differences become huge (Figure 5.18a). It is hard to imagine that nonaxisymmetric modes in MRI experiments with fluid conductors and close to the Rayleigh limit can be excited.

5.3.3
Wave Numbers

The unstable magnetic Taylor–Couette flow forms axisymmetric Taylor vortices. With our normalizations the vertical extent δz of the vortex is given by the expression

$$\frac{\delta z}{R_{out} - R_{in}} = \frac{\pi}{k}\sqrt{\frac{r_{in}}{1 - r_{in}}}. \tag{5.26}$$

The dimensionless vertical wave numbers k, which belong to the optimized critical Reynolds numbers are given in Figure 5.19. For the nonmagnetic Taylor vortices we have $\delta z \simeq R_{out} - R_{in}$, independently of gap size and boundary conditions. The cell therefore has the same vertical extent as it has in radius (see Koschmieder, 1993).

The influence of strong magnetic fields on turbulence mostly consists of suppression and deformation. The deformation consists in a prolongation of the cell structure in the vertical direction, so that δz is expected to become larger (the wave number becomes smaller) for increasing magnetic field. This is true for $Pm \gtrsim 1$, but for smaller Pm the vertical cell size has a minimum for an intermediate value of the magnetic field. The cell size is minimum for the critical Reynolds number for all calculated examples of hydrodynamically stable flows with a conducting boundary. For containers with insulating walls the cell size simply grows with increasing magnetic field. For experiments with the critical Reynolds numbers the vertical cell size is generally 2–3 times larger than the radial one. Also, the smaller the magnetic Prandtl number the longer the cells in the vertical direction.

The influence of boundary conditions on the cell size disappears, of course, for sufficiently wide gaps. For the small and medium gaps, however, one finds for the

5.3 The Standard Magnetorotational Instability (SMRI)

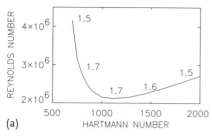
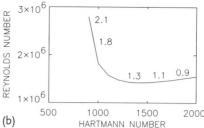

Figure 5.19 The stability lines for $\mu_\Omega = 0.33$ (no centrifugal instability) marked with the wave numbers for which the Reynolds number is minimum. Conducting walls (a) and insulating walls (b). The variation of the wave numbers with the magnetic field differs in the two cases. Note the minimum Reynolds number for insulating walls as smaller than for conducting walls (see Table 5.2). $r_{in} = 0.5$, $Pm = 10^{-5}$. From Rüdiger and Shalybkov (2002). Copyright 2002 by The American Physical Society.

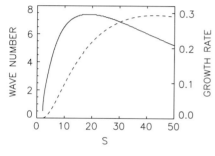

Figure 5.20 Wave number (solid) and growth rate (dashed) for supercritical rotation rate (Rm = 500). Wave numbers are normalized with $1/R_0$ and growth rates are normalized with Ω_{in}. Both quantities have maxima between the two critical Lundquist numbers for marginal instability. Perfectly conducting boundaries. $\mu_\Omega = 0.35$, $m = 0$. Rm = 160, S = 13, Pm = 1. From Gellert, Rüdiger, and Schultz (2012). Reproduced with permission © ESO.

smallest Reynolds number the cells are vertically more elongated for containers with insulating walls.

The behavior of the wave number for supercritical rotation (beyond the limit of marginal instability) forms a surprise. Figure 5.20 shows the wave numbers and the growth rates for the axisymmetric mode for fixed magnetic Reynolds number. There are two critical Lundquist numbers where the growth rates vanish and where the wave numbers are rather small ($\lesssim \pi$) so that the ring-like cells are oblong for marginal instability. Between the two limits, however, the wave number grows representing very flat rings (in opposition to the Taylor–Proudman theorem). The maximum wave number for Rm = 500 is shown in the graph. It basically exceeds the value π. Such flat rings for supercritical excitation can indeed be observed in the simulations (see Figure 5.22c).

Another general remark about nonaxisymmetric modes follows: only the solutions with those wave numbers k are of interest for which the Reynolds numbers are minimized. All solutions with other k have of course higher Reynolds num-

Figure 5.21 The simultaneous existence of the two modes with k and $-k$ in the nonlinear regime for pseudo-Kepler flow. The whole pattern drifts in the direction of rotation. Rm = 160, S = 13. Pm = 1.

bers. One can then show that the solutions with a certain positive k are always accompanied by a solution with $-k$ but with the same Reynolds number and drift frequency (for given Ha and m). As the pitch angle of the resulting spirals is given by $b_z/b_\phi \propto -m/k$ it is clear that the two solutions have opposite pitch angles so that it is always a combination of a left spiral and a right spiral drifting in the same direction. One finds the system as invariant against the simultaneous transformation $k \to -k$, $u_z \to -u_z$, $b_R \to -b_R$ and $b_\phi \to -b_\phi$. Hence, if a solution is known for a certain k, then a modified solution always exists for $-k$. The standard MRI in cylindrical geometry always produces the same number of left and right spirals. The total helicity (kinetic and magnetic) is thus vanishing (Figure 5.21). The system is also invariant against the simultaneous transformation $m, k, \Re(\omega) \to -m, -k, -\Re(\omega)$ after conjugation which, however, describes the identical situation. It is thus enough to consider only the solutions with $m \geq 0$.

5.3.4
Nonlinear Simulations

With a 3D spectral MHD code for cylindrical geometry nonlinear simulations of global MRI are also possible unless too small Pm. The cylinder can be regarded with a periodicity in the vertical coordinate. The MHD Fourier spectral element code described by Fournier et al. (2004) and Gellert, Rüdiger, and Fournier (2007) is used. In this approach the solution is expanded in M Fourier modes in the azimuthal direction. The resulting decomposition is a collection of meridional problems each of which is solved using a Legendre spectral element method (Deville,

Fischer, and Mund, 2002). Either $M = 8$ or $M = 16$ Fourier modes are used, two or three elements in radius, and twelve or eighteen elements in the axial direction. The polynomial order is varied between $N = 8$ and $N = 16$. With a semi-implicit approach consisting of second-order backward differentiation and third-order Adams–Bashforth for the nonlinear forcing terms, time stepping is done with second-order accuracy. At the inner and outer wall no-slip and perfectly conducting boundary conditions (or pseudo-vacuum) are applied (see Gellert, Rüdiger, and Schultz, 2012).

The first simulations are related to the map of marginal instability for $Pm = 1$ for a fixed Lundquist number $S = 13$ (Figure 5.17). Three examples are given here. The first operates with slow rotation, while the others rotate faster or much faster. The marginal instability appears for a minimum magnetic Reynolds number of about 70. The first model lies below the instability domain of $m = 1$, the second one lies inside and the third model again lies outside.

The nonlinear calculations are important as the curves for marginal instability of nonaxisymmetric modes only concern the linear stability behavior of the Kepler flow under the presence of the axial field. It is also possible that the rolls of the $m = 0$ solution become unstable against disturbances with $m > 0$ leading to secondary instabilities with nonaxisymmetric patterns, but here they turn out not to. For given magnetic field amplitude ($S = 13$) simulations with $Rm = 88, 600,$ and 1250 are presented in Figure 5.22. Only the value $Rm = 600$ lies between the lower and the upper limit for nonaxisymmetric instability and, indeed only in this case a nonaxisymmetric (drifting) magnetic pattern results. From the beginning of the simulations the mode with $m = 0$ grows fastest and also the mode with $m = 1$ grows continuously. The other modes come much later so that the complete pattern only occurs after 30 orbits.

The opposite is true for faster rotation (Figure 5.22c). Here only the mode with $m = 0$ grows while all the nonaxisymmetric modes decay. The resulting magnetic pattern is axisymmetric despite the high value of the Reynolds number. A nonaxisymmetric instability of the ring-like magnetic pattern does not appear in the numerical simulations.

Note that in agreement with Figure 5.20 the axisymmetric cells for Reynolds numbers close to the lower instability limit are prolate. They become oblate for faster rotation despite the predictions of the Taylor–Proudman theorem.

The amplitudes of the MRI-induced magnetic fluctuations are also of importance. They easily exceed the strength of the axial background field, but only in the domain where the nonaxisymmetric modes are excited. On the other hand, the amplitudes of the toroidal field components grow with growing Reynolds number. For $Rm = 600$ the maximum b_ϕ exceeds the axial background field already by one order of magnitude.

More information is provided by considering the ratio of the induced magnetic energy to the energy of the applied axial field, that is $q = \langle b^2 \rangle / B_0^2$. Figure 5.23 shows the dependence of this quantity for standard MRI. As for driven turbulence there is a clear relation $q \propto Rm$ (see (3.22)) with a slight dependence of the coefficient on the magnetic field B_0. The viscosity does not appear. The magnetic

210 | 5 The Magnetorotational Instability (MRI)

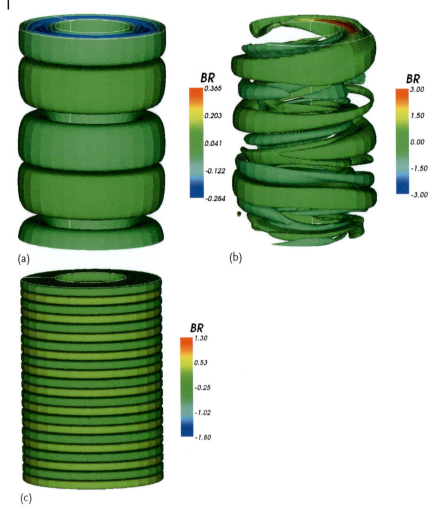

Figure 5.22 The radial component of the magnetic field for S = 13 and for slow rotation (Rm = 88, a), medium rotation (Rm = 600, b) and fast rotation (Rm = 1250, c). For too slow and for too fast rotation the nonlinear instability pattern is purely axisymmetric, and only for medium rotation it is nonaxisymmetric. Conducting walls, $\mu_\Omega = 0.35$, Pm = 1. From Gellert, Rüdiger, and Schultz (2012). Reproduced with permission © ESO.

Reynolds number includes the ratio of the induction by the differential rotation and the dissipation by the magnetic diffusivity. Obviously, the nonlinear SMRI is saturated by diffusion. The higher the magnetic conductivity the stronger the normalized induced magnetic field fluctuations. This relation is basically linear. On the other hand, a magnetic quenching is also visible in Figure 5.23. All curves can be compressed in the relation $q \simeq 7\,\mathrm{Rm}/S^2$, which leads to

$$\Lambda \simeq 7 \tag{5.27}$$

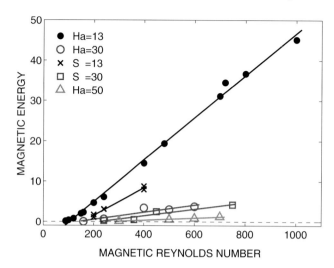

Figure 5.23 The ratio q averaged over the whole container in its dependence on Rm. The models represented by circles are for Pm = 1 while for the others the magnetic Prandtl numbers are smaller (0.1, 0.2, 0.5). Note the dependence on the magnetic Prandtl number as rather weak. Mid-gray: strong fields, light gray: weak fields. $r_{in} = 0.5$, $\mu_\Omega = 0.35$.

for the turbulent Elsasser number

$$\Lambda = \frac{\langle b^2 \rangle}{\mu_0 \rho \Omega \eta}. \tag{5.28}$$

The mean axial magnetic field does not appear in this equation as it only acts in the sense of a catalyst. The uniform axial field must be finite (above the threshold value) but it does not appear in the resulting expression for the energy of the magnetic fluctuations.

5.3.5
The Angular Momentum Transport

The total angular momentum transport in the radial direction is $T_{R\phi} = \langle u_R u_\phi \rangle - \langle b_R b_\phi \rangle / \mu_0 \rho$ which in the usual approximation can be written as

$$T_{R\phi} = -\nu_T R \frac{d\Omega}{dR} = \alpha_{SS} \Omega^2 D^2, \tag{5.29}$$

with $D = R_{out} - R_{in}$ as the gap width of the container. Hence, α_{SS} as the normalized angular momentum transport can be computed from the definition $\alpha_{SS} = T_{R\phi}/\Omega^2 D^2$. If the average procedure concerns the quantity (5.29) by averaging only over the azimuth, one finds that the angular momentum transport is positive everywhere, without any appearance of negative angular momentum transport (Figure 5.24). One also finds that the magnetic-induced angular momentum transport dominates. By averaging over the z coordinate the resulting expression becomes a

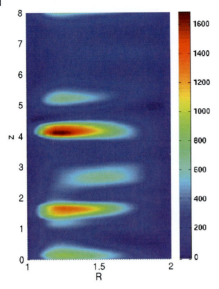

Figure 5.24 The angular momentum transport parameter α_{SS} averaged over the azimuth is everywhere positive, and only slightly reflects a cellular structure. Rm = 400, S = 30, μ_Ω = 0.35, Pm = 1. From Gellert, Rüdiger, and Schultz (2012). Reproduced with permission © ESO.

function of R only. The profile shows a characteristic maximum as it must vanish at the boundaries because of the boundary conditions. The question is how this maximum is related to the mean pressure in the computational domain. The pressure in the container vanishes at the inner cylinder and monotonically increases towards the outer cylinder. The radially averaged pressure P for Kepler rotation between the bounding cylinders is of order $\rho\Omega^2 D^2$ so that indeed the relation (5.29) can be read as

$$\rho T_{R\phi} = \alpha_{SS} P \,. \tag{5.30}$$

This standard equation of the accretion theory holds in our model after averaging over the entire cylinder.

The averaging procedure may thus concern the entire container. The results for the α_{SS} parameter are only valid for this model. They are given in Figure 5.25b, and they can be represented by the *linear* relation

$$\alpha_{SS} = 4.8 \times 10^{-5} S \,, \tag{5.31}$$

valid for all the considered Reynolds numbers and magnetic Prandtl numbers (Pm $\leq$ 1). There is no dependence of α_{SS} on the rotation rate (Figure 5.25a) unless the chosen value of the magnetic Reynolds number lies too close to the boundaries of the instability map. For two examples for Pm = 1 (diamonds) the outer boundary condition has been changed from perfect conductor to pseudovacuum. According to (5.4) and (5.29) one can also understand the α_{SS} as a realization of

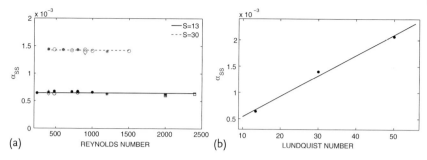

Figure 5.25 The normalized angular momentum transport α_{SS} (averaged over the entire cylinder) in its dependence on the Reynolds number Rm (a) and the Lundquist number S (b). Dots: Pm = 1, circles: Pm = 0.5, stars: Pm = 0.2, square: Pm = 0.1. Diamonds: Pm = 1 and pseudovacuum as outer boundary condition. $r_{in} = 0.5$, $\mu_\Omega = 0.35$. Reproduced with permission © ESO.

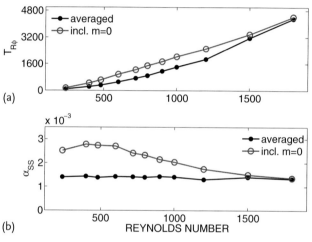

Figure 5.26 The angular momentum transport (a) and the viscosity parameter α_{SS} (b) in its dependence on the Reynolds number. Open circles represent the values if the contribution of the axisymmetric mode with $m = 0$ is added. Averaging is done over the entire container. S = 30, Pm = 1. From Gellert, Rüdiger, and Schultz (2012). Reproduced with permission © ESO.

the β viscosity in the sense of Duschl, Strittmatter, and Biermann (2000) and Huré, Richard, and Zahn (2001).

The angular momentum transport shown in Figure 5.26 (dark color) is only due to the nonaxisymmetric modes with $m > 0$. Only these modes have been defined as the 'fluctuations' in the definitions of the random functions of flow and field, and the well-defined averaging procedure is considered as the integration over ϕ. Only these modes in the simulations are close to forming turbulence.

The inclusion of the angular momentum transport by the axisymmetric ('channel') modes gives a slight modification of this picture. The amplification of the angular momentum transport by the channel mode is characterized by a factor of two, which for slow rotation gives an increase of α_{SS} (Figure 5.26). The total angular momentum transport expressed by α_{SS} due to axisymmetric and nonaxisymmetric modes finally yields a weak decrease for increasing Reynolds number.

The simulations do not indicate a decay of the angular momentum transport of the MRI for small magnetic Prandtl number as argued by Fromang et al. (2007). As is also known from the bifurcation maps of the standard MRI, the actual value of the magnetic Prandtl number is not very important if the rotation and the magnetic background field are normalized without use of the microscopic viscosity. Expressed by α_{SS} the results are of a striking simplicity. The Lundquist number gives the only influence, represented by the linear relation (5.31). The numerical value of α_{SS} taken at (say) $S \simeq 100$ (1 G for protoplanetary disks at 1 AU) is of order 10^{-3}. This value brings to mind the numerical results of Brandenburg et al. (1996) obtained with shearing box simulations. Higher values of α_{SS} require stronger fields.

5.4
Diffusive Kepler Disks

So far, the wave number was an optimized free parameter in an axially unbounded cylindrical container. In order to find the influence of a fixed vertical boundary a rotating disk of constant thickness, $2H$, threaded by a uniform axial magnetic field has been considered. The rotation axis is normal to the disk plane and the angular velocity approaches a Kepler profile

$$\Omega = \frac{\Omega_0}{\sqrt{1 + (R/R_\Omega)^3}} \tag{5.32}$$

for $R > R_\Omega$. The aspect ratio of the disk may be fixed to $R_\Omega/H = 5$. The equations are linearized around this rotation flow and a uniform axial magnetic field of amplitude B_0. Pseudovacuum conditions for the magnetic field, $b_R = b_\phi = 0$, are applied on the disk surfaces at $z = \pm H$. The kinetic and magnetic modes are developed in Fourier modes in axial and azimuthal direction with the standard notation with m and l. For fixed numbers m and l, and a given symmetry type, the eigenvalue problem is defined. The remaining three free parameters are used in the notation (5.16), but with the disk height H as the characteristic length scale.

Figure 5.27 shows that for small Pm the stability map in the S–Rm plane does not depend on Pm, that is the viscosity of the fluid once again does not play a role. There is a minimum value of S of order unity below which no instability exists. The minimum grows slightly with growing magnetic Prandtl number. This is also true for the minimum magnetic Reynolds number Rm. Much more dramatic, however, is the positive slope of the weak-field limit of the nonaxisymmetric modes (Figure 5.28a). There is always an upper limit of the rotation frequency above which the nonaxisymmetric MRI modes are no longer unstable in the linear theory. The positive slope of this weak-field branch is almost parallel to the positive slope of

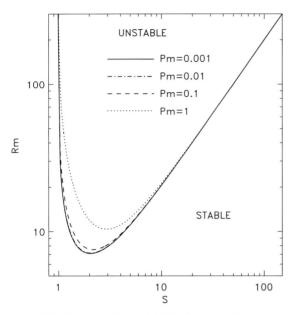

Figure 5.27 The lines of neutral stability for the axisymmetric quadrupole-type modes of a Kepler disk with small magnetic Prandtl numbers. Note the very weak Prandtl number dependence of the eigenvalues. The viscosity of the medium has almost no influence on the excitation of the MRI (and also not on the growth rates). From Kitchatinov and Rüdiger (2010). Reproduced with permission © ESO.

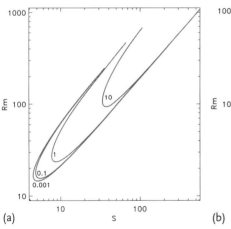
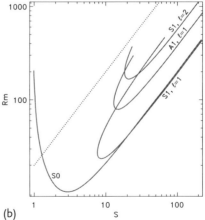

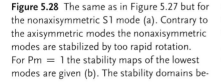

Figure 5.28 The same as in Figure 5.27 but for the nonaxisymmetric S1 mode (a). Contrary to the axisymmetric modes the nonaxisymmetric modes are stabilized by too rapid rotation. For Pm = 1 the stability maps of the lowest modes are given (b). The stability domains become increasingly narrow the smaller the axial wavelengths are. The dotted line represents the relation $\Omega_A/\Omega_0 = 0.05$. From Kitchatinov and Rüdiger (2010). Reproduced with permission © ESO.

the strong-field branch. The excitation of nonaxisymmetric magnetic modes also requires the existence of stronger fields. The value of the plasma $\beta = \mathrm{Rm}^2/S^2$ (or equivalently, $(\Omega_0/\Omega_A)^2$) must be smaller than a hundred as an instability condition. For larger plasma-β the linear MRI only produces axisymmetric rolls.

For the strong-field branch the relation between field and rotation is $\Omega_A/\Omega_0 \simeq 0.5$ similar to the slope of the curves in Figure 5.14. Hence, an upper limit of the field exists for the MRI which from Figure 5.28 does not differ for axisymmetric and nonaxisymmetric modes. One must have $\Omega_A/\Omega_0 < 0.5$ to allow MRI in any form. The positive slope on the upper limit of the instability curve also exists for higher l modes (Figure 5.28b). The plasma β of shearing box simulations (Fromang et al., 2007), however, should not be too high. Otherwise, for too large plasma-β the nonaxisymmetric modes will not appear. As a possible upper limit the value $\Omega_A/\Omega_0 \simeq 0.05$ is marked in Figure 5.28b. Expressed with the Alfvén frequency the excitation condition for nonaxisymmetric modes here is

$$0.05 \lesssim \frac{\Omega_A}{\Omega_0} \lesssim 0.5 . \tag{5.33}$$

Another global analysis (in vertical dimension) of the stability of thin Kepler disks is by Shtemler, Mond, and Liverts (2011) and Liverts, Shtemler, and Mond (2012) who consider an ideal fluid. By the thin-disk approximation and/or the equilibrium of centrifugal acceleration and gravity the horizontal components of the MHD equations decouple from the vertical one. The eigensolution of the horizontal equations is called as Alfvén–Coriolis wave, which becomes unstable after the calculations if $\Omega_A/\Omega \lesssim 1.2$. In contrast to (5.33) the rotation rate Ω here depends on the radius so that one must be careful with a direct comparison of the results but the numerical difference of the results is rather small.

5.5
MRI with Hall Effect

The influence of the Hall effect on the linear marginal stability of an MHD Taylor–Couette flow in the presence of an axial uniform magnetic field is now considered. The basic induction equation is given by (2.64) where now β_H denotes the Hall coefficient (see Section 2.7). Introducing another dimensionless quantity, which only includes material parameters rather than the magnetic field, we write

$$\beta_0 = \frac{R_B}{S}, \tag{5.34}$$

with $S > 0$ by definition (see (2.67)). The parameter β_0 has both signs, depending on the orientation of the magnetic fields in relation to the rotation axis. The amplitude of β_0 should be smaller than unity. Since it scales as R_{out}^{-1} it is inversely proportional to the size of the container. The boundary conditions are not influenced by the Hall effect.

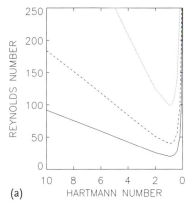 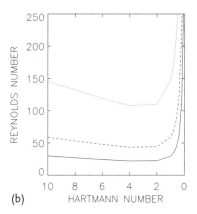

Figure 5.29 The axisymmetric mode of the Shear–Hall instability for positive shear and for conducting boundaries (a) and vacuum boundary conditions (b). The curves are for $\mu_\Omega = 1.2$ (dotted), $\mu_\Omega = 1.5$ (dashed) and $\mu_\Omega = 2$ (solid). $\beta_0 = -1$, Pm = 1. From Rüdiger and Shalybkov (2004). Copyright 2004 by The American Physical Society.

We first describe an instability which exists only for cool fluids with Hall effect in the presence of differential rotation (Shear–Hall instability, SHI). It is even able to destabilize flows with positive shear $d\Omega/dR$, for which no other instability is known. Figure 5.29 illustrates this phenomenon for both conducting and nonconducting boundary conditions for a container with $r_{\rm in} = 0.5$. The flow is unstable only for negative fields, that is $R_B < 0$, which simply means that the rotation axis and magnetic field have opposite orientations (see (2.66)).

Containers with negative shear are more complicated, as such flows can already be unstable even without the Hall effect. The black lines in Figure 5.30 give the lines of marginal instability without Hall effect ($\beta_0 = 0$) for axisymmetric and nonaxisymmetric MRI modes. The Lundquist number represents the magnetic field, and the magnetic Reynolds number represents the rotation amplitude of the assumed pseudo-Kepler flow. These curves are known from Section 5.3.1. Note again that the instability domain for the nonaxisymmetric mode is much small-

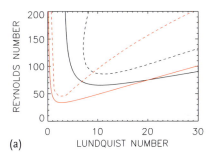 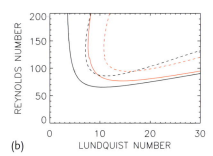

Figure 5.30 Instability maps for pseudo-Kepler flow with vacuum boundary conditions. Black lines: $\beta_0 = 0$ (MRI, no Hall effect). Red lines represent the influence of the Hall effect. $\beta_0 = 1$ (a), $\beta_0 = -0.1$ (b). Solid lines: $m = 0$, dashed lines: $m = 1$. Pm = 1.

er than that for the axisymmetric mode. It is increased, however, for fluids with Hall effect if both the magnetic axis and the rotation axis have the same orientation (see (2.66)). This case is realized in Figure 5.30a for the rather strong Hall effect $\beta_0 = 1$. The minimum Lundquist number for instability is much smaller than for MRI. Moreover, the increase of the instability domain also happens for the nonaxisymmetric modes, so that the Hall-MRI produces flow patterns which are much more complex than the rings which are formed by the axisymmetric modes dominating without the Hall effect.

Magnetic fields antiparallel to the rotation axis stabilize the MRI by means of the Hall effect. In Figure 5.30b the small value $\beta_0 = -0.1$ already gives a drastic stabilization mainly of the nonaxisymmetric modes. The characteristic Lundquist number for instability strongly increases for smaller β_0, for example for $\beta_0 = -1$. Obviously, the stabilization by negative β_0 appears to be more effective than the destabilization by positive β_0. The important meaning of the latter, however, is that nonaxisymmetric modes become unstable even for rather weak fields. An interesting application of this effect has recently been given by Cheung and Cameron (2012) and Pandey and Wardle (2012) for the magnetohydrodynamics of flux tubes in the solar atmosphere.

Note that the critical magnetic Reynolds number is also reduced by the Hall effect, but this reduction remains small. It seems unlikely that the critical Reynolds number of the Kepler disks can be massively reduced by orders of magnitudes by the Hall effect. One only finds a reduction by a factor of two as realized in Figure 5.30a for rather strong positive Hall effect. Whether the bifurcation lines for the $m = 1$ mode show a similar positive slope of both branches as it appears for nonaxisymmetric MRI modes is not finally apparent from the given results.

5.6
The Azimuthal MRI (AMRI)

The current-free toroidal field $B_\phi \propto 1/R$ is considered which is a stable profile. For axisymmetric disturbances there is a necessary and sufficient criterion for stability, that is

$$\frac{\mathrm{d}}{\mathrm{d}R}\left(\frac{B_\phi}{R}\right)^2 < 0 \tag{5.35}$$

(Michael, 1954; Velikhov, 1959), which under the presence of rotation takes the form

$$\frac{1}{R^3}\frac{\mathrm{d}}{\mathrm{d}R}(R^2\Omega)^2 - \frac{R}{\mu_0\rho}\frac{1}{\mathrm{d}R}\left(\frac{B_\phi}{R}\right)^2 > 0 \tag{5.36}$$

(Michael, 1954). Obviously, for ideal fluids *any* rotation law can be destabilized by toroidal fields with radial profiles which do not fulfill (5.35). Both the Kepler law or the galactic rotation with $\Omega \propto 1/R$ can become unstable under the presence of particular toroidal fields.

If both flow and field are stable against axisymmetric disturbances, then also both together should be stable. This statement is not true, however, if nonaxisymmetric perturbations are allowed. In this case the stability condition (5.35) for nonaxisymmetric perturbations becomes

$$\frac{\mathrm{d}}{\mathrm{d}R}(R B_\phi^2) < 0 \tag{5.37}$$

(Vandakurov, 1972; Tayler, 1973) which also indicates current-free fields with $B_\phi \propto 1/R$ as stable. A new question is, however, what happens with the stability of this field against nonaxisymmetric perturbations under the presence of differential rotation, where the rotation law is stable according to the (Rayleigh) criterion (5.36) for $B_\phi = 0$. The surprising answer is that indeed the interaction of a stable field and a stable rotation law can lead to an instability (Ogilvie and Pringle, 1996; Rüdiger et al., 2007a) which we here named as azimuthal magnetorotational instability (AMRI).

5.6.1
The Equations

Consider a viscous, electrically conducting, incompressible fluid between two rotating infinite cylinders in the presence of an azimuthal magnetic field. The equations of the problem are

$$\frac{\partial \boldsymbol{U}}{\partial t} + (\boldsymbol{U} \cdot \nabla)\boldsymbol{U} = -\frac{1}{\rho}\nabla P + \nu \Delta \boldsymbol{U} + \frac{1}{\mu_0 \rho}\,\mathrm{curl}\,\boldsymbol{B} \times \boldsymbol{B},$$

$$\frac{\partial \boldsymbol{B}}{\partial t} = \mathrm{curl}(\boldsymbol{U} \times \boldsymbol{B}) + \eta \Delta \boldsymbol{B} \tag{5.38}$$

and $\mathrm{div}\,\boldsymbol{U} = \mathrm{div}\,\boldsymbol{B} = 0$, where $\boldsymbol{U}$ is the velocity, $\boldsymbol{B}$ the magnetic field, P the pressure, ν the kinematic viscosity and η the magnetic diffusivity. Equation (5.38) provides the basic state ('background') solution $U_R = U_z = B_R = B_z = 0$ and for the azimuthal components

$$U_\phi = R\Omega = a_\Omega R + \frac{b_\Omega}{R}, \quad B_\phi = a_B R + \frac{b_B}{R}, \tag{5.39}$$

where a_Ω and b_Ω are given by (5.2) and

$$a_B = \frac{B_\mathrm{in}}{R_\mathrm{in}}\frac{r_\mathrm{in}(\mu_B - r_\mathrm{in})}{1 - r_\mathrm{in}^2}, \quad b_B = B_\mathrm{in} R_\mathrm{in} \frac{1 - \mu_B r_\mathrm{in}}{1 - r_\mathrm{in}^2}. \tag{5.40}$$

The term $a_B R$ corresponds to a uniform axial current everywhere within $R < R_\mathrm{out}$, and b_B/R corresponds to an additional current but only within $R < R_\mathrm{in}$. For the following it is useful to define the two parameters

$$\mu_\Omega = \frac{\Omega_\mathrm{out}}{\Omega_\mathrm{in}}, \quad \mu_B = \frac{B_\mathrm{out}}{B_\mathrm{in}}, \tag{5.41}$$

measuring the variation in Ω and B_ϕ across the gap. We are interested in the linear stability of the solution (5.39) with $a_B = 0$ so that the considered toroidal field is current-free in the fluid between the cylinders (Figure 5.31). Fields with positive (negative) b_B are due to axial currents parallel (antiparallel) with the rotation axis. The condition $a_B = 0$ is always fulfilled for fields with $\mu_B = r_{\rm in}$. The full state of the flow is given by u_R, $R\Omega + u_\phi$, u_z, b_R, $B_\phi + b_\phi$, b_z. The Hartmann number of the *inner* value of the toroidal field

$$\mathrm{Ha} = \frac{B_{\rm in} R_0}{\sqrt{\mu_0 \rho \nu \eta}} \qquad (5.42)$$

and the Reynolds number (5.16)$_3$ are the dimensionless numbers of the problem, where again $R_0 = \sqrt{R_{\rm in}(R_{\rm out} - R_{\rm in})}$ is taken as the unit of length. We have used R_0^{-1} as the unit of the wave number, η/R_0 as the unit of the velocity fluctuations, $\Omega_{\rm in}$ as the unit of frequencies, and $B_{\rm in}$ as the unit of the magnetic field fluctuations. The general equations for TC flows with toroidal fields after normalization are

$$\frac{d^2 u_R}{dR^2} + \frac{1}{R}\frac{du_R}{dR} - \frac{u_R}{R^2} - \left(k^2 + \frac{m^2}{R^2}\right)u_R - 2\mathrm{i}\frac{m}{R}u_\phi - \mathrm{i}\,\mathrm{Re}(\omega + m\Omega)u_R$$
$$+ 2\mathrm{Re}\,\Omega u_\phi - \frac{dp}{dR} + \mathrm{i}\frac{m}{R}\mathrm{Ha}^2 B_\phi b_R - 2\mathrm{Ha}^2 \frac{B_\phi}{R} b_\phi = 0,$$

$$\frac{d^2 u_\phi}{dR^2} + \frac{1}{R}\frac{du_\phi}{dR} - \frac{u_\phi}{R^2} - \left(k^2 + \frac{m^2}{R^2}\right)u_\phi + 2\mathrm{i}\frac{m}{R}u_R - \mathrm{i}\,\mathrm{Re}(\omega + m\Omega)u_\phi$$
$$- \mathrm{i}\frac{m}{R}p - \frac{\mathrm{Re}}{R}\frac{d}{dR}(R^2\Omega)u_R + \frac{\mathrm{Ha}^2}{R}\frac{d}{dR}(B_\phi R)b_R + \mathrm{i}\frac{m}{R}\mathrm{Ha}^2 B_\phi b_\phi = 0,$$

$$\frac{d^2 u_z}{dR^2} + \frac{1}{R}\frac{du_z}{dR} - \left(k^2 + \frac{m^2}{R^2}\right)u_z - \mathrm{i}\,\mathrm{Re}(\omega + m\Omega)u_z - \mathrm{i}k\,p$$
$$+ \mathrm{i}\frac{m}{R}\mathrm{Ha}^2 B_\phi b_z = 0,$$

$$\frac{du_R}{dR} + \frac{u_R}{R} + \mathrm{i}\frac{m}{R}u_\phi + \mathrm{i}k u_z = 0, \qquad (5.43)$$

and

$$\frac{d^2 b_R}{dR^2} + \frac{1}{R}\frac{db_R}{dR} - \frac{b_R}{R^2} - \left(k^2 + \frac{m^2}{R^2}\right)b_R - 2\mathrm{i}\frac{m}{R^2}b_\phi$$
$$- \mathrm{i}\,\mathrm{Pm}\,\mathrm{Re}(\omega + m\Omega)b_R + \mathrm{i}\frac{m}{R}B_\phi u_R = 0,$$

$$\frac{d^2 b_\phi}{dR^2} + \frac{1}{R}\frac{db_\phi}{dR} - \frac{b_\phi}{R^2} - \left(k^2 + \frac{m^2}{R^2}\right)b_\phi + 2\mathrm{i}\frac{m}{R^2}b_R - \mathrm{i}\,\mathrm{Pm}\,\mathrm{Re}(\omega + m\Omega)b_\phi$$
$$+ \mathrm{Pm}\,\mathrm{Re}\,R\frac{d\Omega}{dR}b_R - R\frac{d}{dR}\left(\frac{B_\phi}{R}\right)u_R + \mathrm{i}\frac{m}{R}B_\phi u_\phi = 0,$$

$$\frac{db_R}{dR} + \frac{b_R}{R} + \mathrm{i}\frac{m}{R}b_\phi + \mathrm{i}k b_z = 0, \qquad (5.44)$$

with the boundary conditions described in Section 5.3. Here we shall consider only perfectly conducting and no-slip boundaries. The rotation law is flat, that is $\mu_\Omega =$

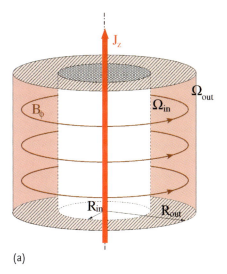

(a) (b)

Figure 5.31 An electric current flows (only) through the axis of the differentially rotating container producing the toroidal field $B_\phi \propto 1/R$ in the fluid between the cylinders (a). The container for the AMRI experiment (b) in the Helmholtz-Zentrum Dresden-Rossendorf (similar design as for the PROMISE, see below). $R_{in} = 4$ cm, $R_{out} = 8$ cm, Γ = height/gap = 10.

0.5 for $r_{in} = 0.5$ (quasi-galactic) and B_ϕ is current-free between the cylinders, that is $\mu_B = 0.5$. Again the wave number is varied until the Reynolds number for a given Hartmann number reaches its minimum. The resulting wave number belongs to the most unstable mode. Both the background flow and magnetic field are normalized with their values at $R = R_{in}$, hence $\hat{\Omega} = \Omega/\Omega_{in}$, $\hat{B}_\phi = B_\phi/B_{in}$ (and the hats are then dropped).

The equation system has a characteristic symmetry. If at the same time with $m \to -m$ for fixed k also the eigenvalue $i\omega$, the flow $\boldsymbol{u}$ and the field $\boldsymbol{b}$ are transformed to their complex conjugate values then the equation system remains unchanged. That means that $m = 1$ and $m = -1$ form a solution with the same drift $\Re(\omega)/m$ and the same Reynolds number for the same Hartmann number.

A constant phase of a nonaxisymmetric pattern drifts with the velocities

$$\left.\frac{\partial z}{\partial t}\right|_\phi = -\frac{\Re(\omega)}{k}, \quad \left.\frac{\partial \phi}{\partial t}\right|_z = -\frac{\Re(\omega)}{m}. \tag{5.45}$$

The first relation describes the phase velocity of the modes in the axial direction, the second one in the azimuthal direction (only existing for nonaxisymmetric modes). The wave is traveling upwards if the real part of the eigenfrequency is negative.

At a fixed time the phase relations can also be written as $\partial z/\partial \phi = -m/k$. The wave numbers k and m are both real numbers, and without loss of generality we can take one of them (say k) to be positive. Then m must be allowed to have both signs. Negative m describe right-hand spirals and positive m describe left-hand spirals. The resulting solution has a combined structure as shown in Figure 5.32 but

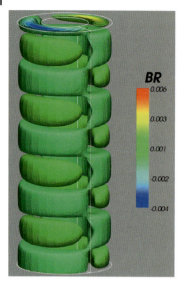

Figure 5.32 Isolines of the radial field component for AMRI very close to the Rayleigh limit ($\mu_\Omega = 0.255$). The entire pattern rotates with (nearly) the same rate as the outer cylinder does. As no large-scale currents flow through the fluid conductor a magnetorotational instability is shown rather than a current-driven instability. Without shear the magnetic field itself is stable for all amplitudes. Ha = 80, Re = 1000, $r_{in} = 0.5$, Pm = 0.01, conducting boundaries.

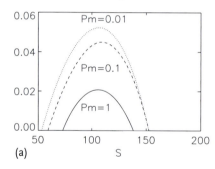
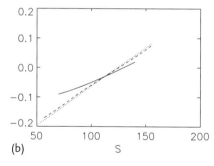

Figure 5.33 The real and the imaginary parts of the $i\omega/\Omega_{in}$ in the unstable domains of Figure 5.34. The growth rate normalized with Ω_{in} for Rm = 200 for various magnetic Prandtl numbers (a). The negative drift rate in the resting system (solid: Pm = 1, dashed: Pm = 0.1, dotted: Pm = 0.01). The azimuthal drift of the modes for small Pm does not depend on the viscosity (b). $\mu_\Omega = 0.5$, $\mu_B = 0.5$. From Rüdiger et al. (2007a).

both spirals drift in the same positive ϕ-direction. The drift rate for flat rotation laws is only small but for small Pm it does not depend on the magnetic Prandtl number (Figure 5.33b). As the drift is negative it is a weak eastward migration in the stationary system, so that in the corotating system it is a westward migration, or in other words, the pattern rotates slower than the cylinders. One finds, however, close to the Rayleigh limit another situation (see below).

5.6.2
The Instability Map

Figure 5.34 shows the stability curve for $m = \pm 1$. The bifurcation line for a fixed Pm in the S–Rm-plane has the characteristic form of a tilted cone. The various cones for Pm $\lesssim$ 1 converge for Pm $\to$ 0 so that for small Pm the characteristic values scale with Rm and S. For a given Pm there is an absolute minimum for the Lundquist number. The minimum Lundquist number is about 40 (for Pm $\to$ 0). For larger values S there are always two critical Reynolds numbers limiting the instability domain. Again, the rotation can be too slow or too fast for the existence of AMRI. This is why the shear in the rotation law can be too large for the existence of nonaxisymmetric magnetic modes. The lower ('slow rotation') branch behaves like S $\simeq$ Rm while the upper branch ('fast rotation') approximately follows the relation S $\simeq$ Rm/3. The oblique domain of the instability is thus described as

$$0.3 \lesssim \frac{\Omega_A}{\Omega} \lesssim 1, \tag{5.46}$$

showing the AMRI as an instability of medium fields. The magnetic Mach number of rotation, Mm $= \Omega/\Omega_A$, is large ('superAlfvénic'). It is only small for the Tayler instability which is considered in the next chapter.

For large magnetic Prandtl numbers the scaling switches from S to Ha and from Rm to $\overline{\mathrm{Rm}} = \sqrt{\mathrm{Pm}\mathrm{Re}}$ just like for the SMRI. Also in this case the ratio between the scaling factors can be expressed by Ω_A/Ω. We shall demonstrate below that (surprisingly) for sufficiently steep rotation laws close to the Rayleigh limit the scaling switches from Rm to Re, which leads to the possibility to realize the AMRI in the laboratory.

The formal reason for the existence of the axial and the azimuthal MRI is that the linearized induction term curl($u \times \bar{B}$) in (5.38) produces expressions like $(\bar{B} \cdot \nabla) u$ coupling the flow fluctuations to the field fluctuation. This term exists for axisymmetric modes only for axial $\bar{B}$ while for azimuthal $\bar{B}$ a possible unstable mode

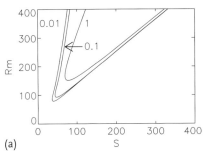
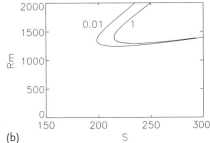
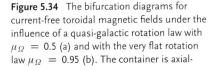

Figure 5.34 The bifurcation diagrams for current-free toroidal magnetic fields under the influence of a quasi-galactic rotation law with $\mu_\Omega = 0.5$ (a) and with the very flat rotation law $\mu_\Omega = 0.95$ (b). The container is axially periodic, with perfectly conducting walls. $m = \pm 1$, $r_{in} = 0.5$. The curves are marked with Pm. For small Pm the AMRI scales here with Rm and S. $r_{in} = 0.5$, $\mu_B = 0.5$. Copyright © 2007 RAS.

must be nonaxisymmetric. One can easily see that for toroidal background fields the equation for the radial magnetic component only couples for $m \neq 0$ to the other components, otherwise it decays. The radial magnetic field b_R is needed to transfer the energy of the differential rotation into the magnetic part of the instability. Also the nonlinear numerical simulations reveal the AMRI as a basically three-dimensional phenomenon.

The growth rates of AMRI are only small (Figure 5.33a). One finds that they grow for decreasing magnetic Prandtl number and for increasing Reynolds numbers, but a saturation seems to exist for Pm $\to$ 0 and for large Rm. It seems to be typical for AMRI that its growth rates scale as only a (small) fraction of the rotation rate.

As the energy for the AMRI exclusively comes from the differential rotation, the question arises as to what happens for very flat rotation profiles. Figure 5.34b gives the results for the profile $\mu_\Omega = 0.95$. Scaling again with Rm and S, the differences to the case of $\mu_\Omega = 0.5$ are that for flat rotation laws the magnetic Reynolds number and also the magnetic fields must be much higher (factor 10) to excite the AMRI. The system requires not only faster rotation in order to get more energy from the differential rotation but also the magnetic field must be higher to translate this energy into the fluctuations. Note that the shear $1 - \mu_\Omega$ also differs by a factor of 10 in the two cases.

5.6.3
Different Scalings with Pm

For small magnetic Prandtl numbers a dramatic difference of the eigensolutions of AMRI exists for steep and flat rotation laws. As shown in Figure 5.35 there is a sharp transition in the scaling of the solutions between the rotation law R^{-2} (Rayleigh limit, left y-axis) and $R^{-3/2}$ (Kepler rotation, right y-axis). For $\mu_\Omega = 0.25 - 0.31$ the solutions for Pm $\to$ 0 are characterized by their critical Reynolds number so that Rm scales with Pm, becoming very small for small Pm. The critical Reynolds number close to the Rayleigh limit is only of order 10^3 so that an experimental realization must be possible of this (nonaxisymmetric) instability.

The scaling with Re is in opposition to the scaling known from SMRI which in general scales for small Pm with the magnetic Reynolds number of the rotation and the Lundquist number of the field. In this case the corresponding Reynolds number sharply increases like 1/Pm, reaching very high numbers for small Pm. The same is true for the AMRI for rotation laws with $\mu_\Omega > 0.31$. The resulting Rm for small Pm are of order 50 for insulating cylinders while Figure 5.34 for conducting cylinders provides the somewhat higher value of about 80.

5.6.4
Nonlinear Results

We have to probe the AMRI in the nonlinear regime. The regular Taylor–Couette flow as an initial setup with $\mu_\Omega = 0.5$ for $r_{\rm in} = 0.5$ and white noise for the initial magnetic field are used. For the external toroidal field $B_\phi \propto 1/R$ is applied, which

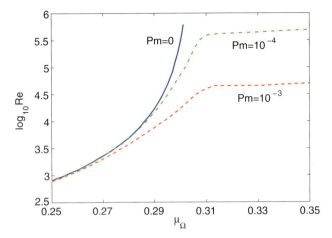

Figure 5.35 The minimum Reynolds numbers for various rotation laws and magnetic Prandtl numbers. For rotation laws close to the Rayleigh limit ($\mu_\Omega = 0.25$) the influence of (small) Pm disappears so that the solutions even exist for Pm = 0. For flatter rotation profiles ($\mu_\Omega > 0.31$) the condition Rm = const holds. Insulating boundaries, $r_{in} = 0.5$. From Hollerbach, Teeluck, and Rüdiger (2010). Copyright 2010 by The American Physical Society.

proves to be unstable. The instability limits found with the nonlinear MHD code for Pm = 1 perfectly agree with the lines of marginal instability of the linear theory which can thus be considered as theory and simulations (Figure 5.36).

The code is also used to demonstrate the nonaxisymmetric structure of AMRI by means of the radial magnetic field for a rather steep rotation law very close to the Rayleigh limit (Figure 5.32). One only needs very small Reynolds ($\simeq 1000$) and Hartmann ($\simeq 80$) numbers to excite the instability. These values comply with the results of the linear theory very close to the Rayleigh limit. The influence of the magnetic Prandtl number is small. The magnetic pattern in Figure 5.32 appears as nearly circular in the meridional plane. The azimuthal drift is directed eastward with $\omega/\Omega_{in} \simeq 0.25$, that is the patterns (almost) corotate with the outer cylinder.

In the simulations only the Fourier mode $m = 1$ grows exponentially. Already after roughly one rotation, when nonlinear effects become important, higher modes also appear. The mode spectrum drops very quickly with increasing m. More than 99% of the energy is stored within the first three modes. After only a few rotations all excited modes saturate to a stationary spectrum. The spectral distribution of the total magnetic energy of the instability for various azimuthal mode numbers m for S = 110 is shown in Figure 5.37. The distribution of the magnetic energy in the entire container is spread in the modes with $m = 0, 1, 2, \ldots$ The smallest and the highest possible Reynolds number within the tilted instability cone are probed. Almost all energy is concentrated at the nonaxisymmetric mode $m = 1$. The energy of the axisymmetric mode ($m = 0$) exceeds the energy in the higher modes ($m \geq 2$). The nonlinearity, therefore, produces an inverse cascade: the magnetic energy starts from the injection at $m = 1$ to lower modes rather than to high-

5 The Magnetorotational Instability (MRI)

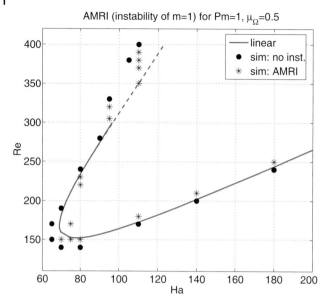

Figure 5.36 Nonlinear simulation of the instability domain of the nonaxisymmetric AMRI for $\mu_\Omega = 0.5$ and perfectly conducting walls. The solid line presents the results of the linear theory. Note the small discrepancy between both curves for fast rotation. $r_{in} = 0.5$, $\mu_B = 0.5$, Pm = 1. From Gellert, Rüdiger, and Fournier (2007).

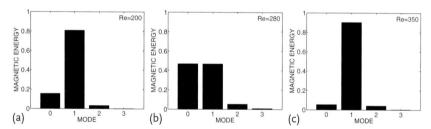

Figure 5.37 The magnetic energy (integrated over the entire volume) spectrum for $S = 110$ and Pm = 1 in dependence on the Reynolds number; Rm = 200 (a), Rm = 280 (b), Rm = 300 (c). Rm = 280 describes the highly developed instability while the other plots belong to the two branches of marginal instability. $\mu_\Omega = 0.5$. From Gellert, Rüdiger, and Fournier (2007).

er modes. This might be a consequence of the differential rotation, which favors axisymmetric magnetic fields over nonaxisymmetric ones. In the center of the instability strip (Figure 5.37b) the energies of the modes with $m = 0$ and $m = 1$ are almost equal and only a rather small part remains for $m = 2$. It is an open question whether the $m = 0$ energy can ever exceed the energy in the $m = 1$ mode.

Let us ask whether the most unstable mode with $m = 1$ also dominates in very narrow gaps between the cylinders. In Section 2.4 it is shown for current instability in spherical geometry that for magnetic field belts concentrated into a narrow ring (in radius) Fourier modes with $m \simeq 5$ appeared as the most unstable pertur-

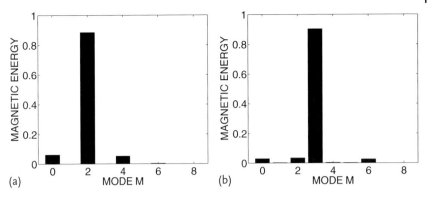

Figure 5.38 The excitation of AMRI modes with $m > 1$. The same as in Figure 5.37 but for narrow gaps between the cylinders. $r_{in} = \mu_B = 0.83$ (a). $r_{in} = \mu_B = 0.9$ (b). $Pm = 1$.

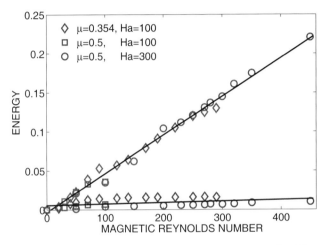

Figure 5.39 The magnetic energy q (upper line) and the kinetic energy (lower line) of the AMRI perturbations normalized in both cases with $\bar{B}_{in}^2$ averaged over the entire container. Perfectly conducting walls, $r_{in} = 0.5$. $Pm = 1$. As abscissa the difference $Rm - Rm_{crit}$ is used. The fluctuation energies are not in equipartition with the background field. Courtesy of M. Gellert.

bations. One can also demonstrate for narrow gaps between the cylinders that the AMRI is realized by Fourier modes with higher m (Figure 5.38). The first (nonlinear) example with $r_{in} = \mu_B = 0.83$ provides the mode $m = 2$ as the dominant mode while for even thinner gaps (Figure 5.38b) the mode with $m = 3$ is the most unstable one.

Another question is how the ratio q of the energy of the magnetic perturbations and the energy of the vacuum field $\bar{B}_\phi$ scales. The above calculations always lead to rather small amplitudes of the normalized magnetic perturbations. Indeed, the numerical results collected in Figure 5.39 show small values of q, but again the results scale as $q \propto Rm$. The circles in Figure 5.39 belong to various Hartmann numbers and to various magnetic Prandtl numbers between 0.01 and 1.0. A magnetic quenching of q as it exists for the standard MRI (Figure 5.23) is not visible.

The results are of astonishing simplicity: the induced fields gain their energy from the differential rotation, and they saturate by the action of the Ohmic dissipation. Contrary to the standard MRI the Elsasser number Λ for the magnetic small-scale fluctuations is no longer independent of the mean magnetic field. One finds $q \simeq 5 \times 10^{-4}$Rm from Figure 5.39, which can be translated into

$$\Lambda = 5 \times 10^{-4} \, S^2 \qquad (5.47)$$

for the Elsasser number, which may exceed unity, but only for systems with $S > 50$. This is one of the basic differences with the standard MRI.

Figure 5.39 also shows the kinetic energy normalized with the large-scale magnetic field. One finds the magnetic fluctuations as much more intensive than the kinetic fluctuations. As the magnetic fluctuations transport angular momentum but do not mix chemicals, heat and magnetic flux of a conducting fluid which is subject to AMRI should have large effective magnetic Prandtl numbers and Schmidt numbers (see Section 6.3.3). It might be true that the magnetic instabilities discussed here (and in the following) may form the long sought after solution of the open question why in radiative stellar zones much more angular momentum is transported than chemicals (lithium, beryllium, etc.) are mixed.

Note also that Figure 5.39 contains a difference of the curves for magnetic and for kinetic fluctuations. While for the magnetic energy the value μ_Ω of the rotational shear does not play an important role, it seems to play a particular role for the kinetic energy as all the symbols for pseudo-Kepler flow are located above the symbols for quasi-galactic flow. Obviously, the energy of the magnetic field fluctuations dominates the energy of the kinetic fluctuations – and both are *not* in equipartition with the magnetic background field.

5.6.5
The AMRI Experiment

For the small magnetic Prandtl numbers of liquid metals as given in Table 5.1, it is hopeless to plan laboratory experiments to obtain the AMRI for pseudo-Kepler or quasi-galactic rotation laws. Attempts to obtain this nonaxisymmetric MRI in a laboratory experiment would be even more difficult than for the axisymmetric SMRI in an axial field. The required rotation rates would be basically greater, Re $\gtrsim O(10^8)$ (for gallium), with all the difficulties that entails for too fast rotation (see Hollerbach and Fournier, 2004). To produce an azimuthal field with $S \approx O(100)$, would also require a current along the central axis of order of 10^3 kA.

As the instability pattern is a drifting spiral with $m = 1$ the signal which is simplest to measure would be the drift rate $\Re(\omega)$ in the form of an oscillation frequency of the field and/or the flow perturbation at a fixed azimuth in the laboratory system. Indeed, the switch of the scaling from Rm to Re and S to Ha for rotation profiles close to the Rayleigh limit as demonstrated by Figure 5.35 gives the possibility of laboratory experiments. Test calculations of the eigenvalues for Pm $= 10^{-6}$, for

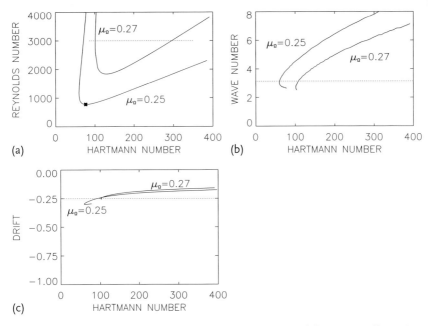

Figure 5.40 The instability map (a), the wave number (b) and the drift frequency (c) $\Re(\omega)/\Omega_{in}$ for μ_Ω at the Rayleigh value and slightly beyond ($\mu_\Omega = 0.25$ and $\mu_\Omega = 0.27$). Perfect-conducting cylinders, $r_{in} = 0.5$, Pm $= 10^{-6}$.

$\mu_\Omega = r_{in}^2$ (the Rayleigh limit) and for different choices of r_{in} lead to the result that the Hartmann number takes its minimum at Ha $\simeq 80$ for $r_{in} = 0.5$. The corresponding Reynolds number has the value of $\simeq 800$. It should be no problem to realize these values in a Taylor–Couette system. In order to express that value of Ha in terms of the axial current which produces the azimuthal field via $I_{axis} = 5 R_{in} B_{in}$ (where I, R and B are measured in ampere, centimeter and gauss) one finds

$$\text{Ha} = \frac{1}{5}\sqrt{\frac{1-r_{in}}{r_{in}}}\frac{I_{axis}}{\sqrt{\mu_0\rho\nu\eta}}. \tag{5.48}$$

For galinstan $\sqrt{\mu_0\rho\nu\eta} \simeq 25.8$ and Ha $= 80$ the critical current becomes $I_{axis} = 10.4$ kA. The value $\sqrt{\mu_0\rho\nu\eta} \simeq 8.2$ for sodium leads to a current of only 3.2 kA. Slightly stronger currents along the axis of the container are needed to destabilize rotation profiles which are little more distant from the Rayleigh limit (see Figure 5.40a).

Concerning the drift frequency of the marginal solution Figure 5.40c reveals interesting findings. For a rotation law with $\mu_\Omega = r_{in}^2$ we have $\Re(\omega)/\Omega_{in} \simeq -r_{in}^2$ if the solution with the *smallest* Reynolds number is considered. Hence, for rotation laws very close to the Rayleigh limit and optimized with the rotation rate, the instability pattern with $m = 1$ almost exactly corotates with Ω_{out} of the outer cylinder. Also for $\mu_\Omega = 0.27$ this striking effect is still visible. As the Reynolds number scales with $(1 - r_{in})/r_{in}$ we have $\Re(\omega) \propto -\nu/R_{out}^2$. It is thus the diffusion time of the entire cylinder which determines the drift velocity.

The AMRI experiment in HZDR uses the same design as used for the Potsdam ROssendorf Magnetic InStability Experiment or PROMISE (see Figure 5.31b). The outer cylinder has an inner radius of 8 cm and the inner cylinder has an outer radius of 4 cm; the height is 40 cm (aspect ratio $\Gamma = 10$). The cylinders are made from copper, so that the boundary conditions can be considered as perfectly conducting. The (insulating) plexiglass endplates are split at $R_{split} = 5.6$ cm. The inner ring (attached to the inner cylinder) reaches in radius from the inner cylinder to 40% of the gap (see Figure 5.56). The outer ring rotates with the rotation rate of the outer cylinder. It is shown in Section 5.8.2 that such a constellation reduces the effects of the Ekman pumping.

Figure 5.41 demonstrates the technical realization of the AMRI in the described experiment. Only the azimuthal Fourier component with odd mode number m of the vertical velocity is measured. The plot shows that a strong jump of the intensity indeed happens from about 10 kA very close to the above estimate. Also the predicted rather strict corotation of this nonaxisymmetric pattern with the outer cylinder (see Figure 5.40c) has been observed. As no electric current flows through the fluid the only energy source for the kinetic energy is the differential rotation. If strong enough the current-free toroidal magnetic field forces the otherwise stable rotation law to become unstable – just in a similar way as it does the standard MRI but with axial fields. The difference is that the fundamental modes of the instability are nonaxisymmetric for AMRI rather than axisymmetric for standard MRI. Note that after the numerical results the amount of energy stored in the nonaxisymmetric modes does not reflect the total energy of all Fourier modes. The fractional energy of the nonaxisymmetric mode to the axisymmetric mode is smaller in the middle

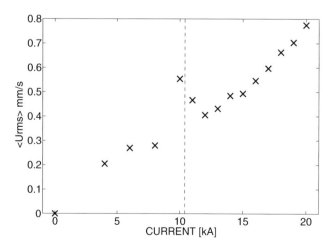

Figure 5.41 The excitation of the nonaxisymmetric ($m = 1$) mode in the AMRI experiment operated in the HZDR with $\mu_\Omega = 0.26$ for a Reynolds number of Re = 2956 (see the dotted line in Figure 5.40a). $r_{in} = 0.5$. Courtesy of M. Seilmayer and F. Stefani.

of the instability domain than at its boundaries (see Figure 5.37) as it seems to be also the case in Figure 5.41.

5.7
Helical Magnetorotational Instability (HMRI)

To study the stability of fields which simultaneously possess axial and azimuthal components under the presence of differential rotation is insofar interesting as the fundamental modes with axial field are axisymmetric while those with azimuthal current-free fields are nonaxisymmetric. The first question concerns the symmetry type of the instability of current-free helical (or better spiral) fields with a preferred handedness. Knobloch (1992, 1996) has shown that possible instabilities of helical background fields – of whatever type – can never be stationary so that a possible axisymmetric mode must travel along the rotation axis. For helical background fields the symmetry of the problem is changed so that z and $-z$ are no longer equivalent. One can hope to utilize this particularity to observe the instability in a laboratory experiment. To this end it would be important to know the oscillation frequency and its dependence on basic parameters.

5.7.1
From AMRI to HMRI

We start with the SMRI exciting both axisymmetric and nonaxisymmetric modes, with the axisymmetric mode being the one with the lowest Reynolds number (Figure 5.42a). For Pm = 1 this overall minimum occurs for Ha $\simeq$ 10 and Re $\simeq$ 80. For sufficiently large Ha the nonaxisymmetric mode with $m = 1$ is actually preferred over the axisymmetric mode. The axisymmetric mode dominates for the globally lowest Re value and the weak-field branch of the instability curves. For the axisym-

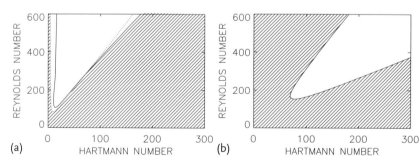

Figure 5.42 SMRI. The axisymmetric mode (dotted) possesses the minimum Reynolds number (a). AMRI. The most unstable mode is nonaxisymmetric. Hartmann number defined with $\bar{B}_z$ (b). The hatched areas are stable. The curves for $m = \pm 1$ are degenerated. Conducting cylinders. $r_{in} = 0.5$, $\mu_\Omega = 0.5$, Pm = 1. Copyright 2010 by The American Physical Society.

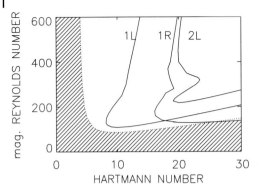

Figure 5.43 Helical magnetorotational instability for $B_{in} = 2B_0$. The mode with the lowest Reynolds number is axisymmetric (dotted). Note that the nonaxisymmetric modes with $m = 1$ (marked by L) and $m = -1$ (marked by R) have different excitation conditions. $r_{in} = 0.5$, $\mu_\Omega = 0.5$, $\beta = 2$, Pm = 1. Copyright 2010 by The American Physical Society.

metric mode the weak-field branch of the line of marginal instability tilts to the left, whereas for the nonaxisymmetric mode it tilts to the right (see Section 5.3.2).

Next the purely azimuthal field without electric currents within the fluid is considered. Figure 5.42b presents results for $\mu_\Omega = 0.5$ showing that for Re $\gtrsim$ 150 and Ha $\gtrsim$ 80 there exists an unstable $m = 1$ mode (AMRI). Both the upper and lower branches of the instability curve tilt to the right.

Figure 5.43 demonstrates that the combination of axial field and toroidal field (with pitch angle of order unity, right-hand helicity) leads to a bifurcation map which for the lowest mode is very similar to that of the SMRI (Figure 5.42a). The lowest mode is axisymmetric for Ha $\lesssim$ 50 and nonaxisymmetric for larger Ha. The weak-field branches are also almost identical. The tilt of the mode $m = 0$ is to the left and the tilt of $|m| > 0$ is to the right. However, the curves for $\pm m$ are different for this helical magnetorotational instability (HMRI) while they are identical for SMRI and AMRI. The instability of spiral background fields distinguishes between left-handed and right-handed nonaxisymmetric modes. In Figure 5.43 only the right-handed mode ($m = -1$, marked by R) becomes for strong magnetic fields the mode with the lowest Reynolds number. The left-handed mode ($m = 1$, marked by L) and the modes for $|m| > 1$ can only be excited for (much) faster rotation. The results are valid for a right-hand combination of azimuthal and axial fields. The weak-field branch of the $m = 0$ mode tilts to the left; all branches for nonaxisymmetric modes tilt to the right. Up to Ha $\approx$ 50 the axisymmetric mode is preferred and for Ha > 50 the right spiral with $m = 1$ is preferred.

Obviously, the (axisymmetric) standard MRI and the (nonaxisymmetric) AMRI both appear if the background field has a helical geometry. The new phenomenon is that the axisymmetric modes shown in Figure 5.43 are oscillatory. The axisymmetric eigenmodes which appear for the globally minimum Reynolds number are stationary for SMRI but they are oscillatory (drifting in z) for HMRI. Both the nonaxisymmetric AMRI and the axisymmetric HMRI show differences of their scaling for Pm ≪ 1 for steep and flat rotation laws:

- For $\mu_\Omega > r_{in}^2$, the magnetic Reynolds number and the Lundquist number are the relevant quantities for SMRI and AMRI. The dependence of the critical Rm and S on the magnetic Prandtl number is very weak, the magnetorotational instabilities scale for Pm ≤ 1 with Rm and S. The consequence is that the kinetic Reynolds number cannot remain finite for Pm = 0. For the standard MRI values of Rm ≃ 10 and S ≃ 5 are characteristic so that for sodium and gallium with their small magnetic Prandtl numbers of 10^{-5} and 10^{-6} very large Reynolds numbers of 10^6 or 10^7 result.
- For rotation laws close to the Rayleigh line AMRI and HMRI scale with Re and Ha for rotation rate and magnetic field, so that the numerical value of the magnetic Reynolds number no longer plays a role (Figure 5.44). Both AMRI and HMRI, therefore, exist at Pm = 0.
- Despite these differences in scalings, from Re and Ha close to the Rayleigh line to Rm and S further beyond it, the HMRI and the SMRI are continuously connected (Hollerbach and Rüdiger, 2005), and similarly for the two different AMRI scalings (Hollerbach, Teeluck, and Rüdiger, 2010). The delicate nature of the transition from one scaling to the other was also explored in considerable detail by Kirillov and Stefani (2010, 2012), who showed how they can be continuously connected and still have the HMRI be a destabilized inertial wave and the SMRI be a destabilized slow magneto-Coriolis wave.

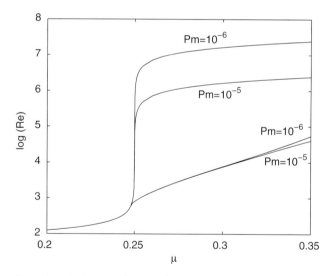

Figure 5.44 SMRI (upper lines) and HMRI (lower lines, $\beta = 2$) between the Rayleigh limit and pseudo-Kepler rotation for conducting boundaries for Pm = 10^{-5} and 10^{-6}. Note the different scaling with Pm of SMRI and HMRI. $r_{in} = 0.5$. From Hollerbach and Rüdiger (2005). Copyright 2005 by The American Physical Society.

- All the presented Reynolds numbers for given Hartmann numbers are optimized with respect to the axial wave number. They form, therefore, the absolute minimum of all excitation conditions. Further conditions on the wave numbers (e.g., a fixed wave number) will increase the corresponding Reynolds number for marginal instability.

Figures 5.45 show for perfectly conducting cylinders how for $Pm = 10^{-5}$ the critical Hartmann number, the corresponding wave number and the drift ('travel') frequency behave just beyond the Rayleigh line. Let β be the pitch angle

$$\beta = \frac{B_{in}}{B_0} \qquad (5.49)$$

between the (inner) toroidal field and the axial field, so the standard MRI is described by $\beta = 0$. The Hartmann numbers in this section are exclusively defined with the *axial* field B_0 (see (5.16)). The only exception is Figure 5.42b where it is not possible. Immediately beyond the Rayleigh line ($\mu_\Omega = 0.25$) the critical Reynolds number jumps to values of 10^6. This is no longer true for finite β. For β of order unity the kinetic Reynolds number for both $Pm = 10^{-5}$ and $Pm = 10^{-6}$ only grows from 10^3 to 10^4 between the steep rotation law of uniform angular momentum (Rayleigh line) and for the flat pseudo-Kepler rotation. Experiments with such Reynolds numbers are possible. For $\beta = 2$ there are rather low values for Re and Ha for values of (say) $\mu_\Omega = 0.27$, which is clearly in the stable area of hydrodynamic TC flow. Reynolds numbers of about 2000 and Hartmann numbers of order 10 (see Table 5.3 for the exact numbers) can actually be realized in the laboratory by use of sodium or gallium as the fluid conductor. It is also clear that for such Reynolds numbers the Taylor–Couette flow remains hydrodynamically stable even in the nonlinear regime.

Figure 5.45 yields the wave numbers and the travel frequencies for the models known from Figure 5.44. The wave number is normalized with the radius R_0 and the frequency is normalized with the inner rotation rate. Note that at the Rayleigh line the Hartmann number and the wave number do not depend on the field parameter β. This, however, is no longer true beyond the Rayleigh limit. Solutions with fixed wave number, therefore, would strongly differ from those given here, where the wave number is optimized to yield the absolute minimum of the Reynolds number. All the normalized wave numbers k are smaller than π so that the traveling cells are elongated in axial direction. The cell elongation is maximal for $\beta = 1$. For variable β the wave number and the drift frequency behave opposite so that the variation of the axial phase velocity is much weaker.

The travel frequency is the lowest frequency in the system. A typical value for medium β is $0.1\Omega_{in}$. One finds from the plots for the phase velocity the numerical value $\omega/k \simeq 0.08$, almost independent of the magnetic field geometry. In physical units it means that $u_z \simeq 0.08 R_{in}\Omega_{in}$, or equivalently, $u_z \simeq 0.08 \text{Re}\,(\nu/R_{in})$. For a Reynolds number of (say) 846 (see Table 5.3) and an inner radius 4 cm of the container and with the viscosity of Table 5.1, one finds $u_z \simeq 0.6$ mm/s as a minimum value. As the drift speed can be precisely measured with Doppler sounding

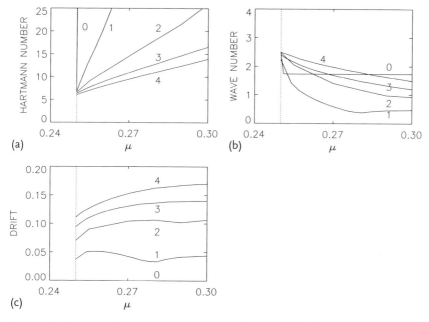

Figure 5.45 The critical Hartmann numbers (a), the corresponding wave numbers (b) and the travel frequencies normalized with the inner rotation rate (c) beyond the Rayleigh limit (here $\mu_\Omega = 0.25$, dotted line), for axisymmetric modes with toroidal fields of order of the axial field. The curves are marked with their β-values. Note that at the Rayleigh line the critical Hartmann numbers and wave numbers hardly depend on β. Conducting boundary conditions, $r_{in} = 0.5$, $\mu_B = 0.5$, $Pm = 10^{-5}$.

Table 5.3 Characteristic values for galinstan experiments ($Pm = 1.4 \times 10^{-6}$) for $r_{in} = 0.5$, $\mu_\Omega = 0.27$ ($R_{in} = 4$ cm, $R_{out} = 8$ cm). Wave number k and travel frequency ω are normalized, B_ϕ denotes the toroidal field at the inner cylinder, the magnetic fields are given in gauss, the current I_z in kA. From Rüdiger et al. (2005b).

β	Re	Ha	k	ω	f_{in} [Hz]	B_0	B_ϕ	I_z	u_z/u_{in}
0	1.02×10^7	1700	1.8	0	318	9460	0	0	0
1	38250	39	0.6	0.04	1.2	215	215	4.30	0.066
2	2386	14.6	1.3	0.10	0.08	80	160	3.20	0.077
3	1160	10.8	1.7	0.13	0.04	59	177	3.54	0.076
4	846	9.5	1.9	0.15	0.03	52	208	4.16	0.079

this prediction for the phase velocity of the traveling modes will be the main target of the experiment PROMISE. It is here of great advantage that the calculated phase velocity shows a very weak dependence on the magnetic field geometry

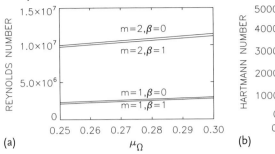
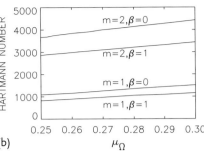

Figure 5.46 The optimized minimal Reynolds numbers (a) and the corresponding Hartmann numbers (b) vs. the shear parameter μ_Ω for the excitation of nonaxisymmetric modes with $m = 1$ and $m = 2$ beyond the Rayleigh limit for $\beta \leq 1$. $r_{in} = 0.5$, $\mu_B = 0.5$, $Pm = 10^{-5}$.

5.7.2
Nonaxisymmetric Modes for small Pm

It is shown in Figures 5.44 and 5.45 for small magnetic Prandtl number how small both the critical Reynolds numbers and Hartmann numbers become beyond the Rayleigh limit if the background field is helical. This is at least true (as shown) for axisymmetric perturbation modes and for not too large amplitudes of the toroidal field compared to the axial field. The question is what happens if one of the two conditions is removed. Of particular interest is here of course the behavior of the nonaxisymmetric modes. To know the excitation conditions is certainly important for the theoretical understanding of the magnetic instabilities but it might also be important for the interpretation of experimental results.

The critical Reynolds numbers and Hartmann numbers for the excitation of the modes with $m = 1$ and $m = 2$ for rotation laws with $\mu_\Omega \geq 0.25$ are plotted in Figure 5.46. For each mode the toroidal field amplitude β varies between 0 and 1. In all cases the excitation of the instability is easier for helical fields ($\beta \neq 0$) but the differences are relatively small. The rescaling of the critical values from Rm to Re which is characteristic for the axisymmetric mode in fluids with small Pm does not appear for the nonaxisymmetric modes. Their typical Reynolds number exceeds 10^6 and their typical Hartmann number exceeds 10^3. It is thus certainly not possible to observe the nonaxisymmetric modes in experiments with fluid metals.

The lines in Figure 5.46 are the curves of marginal instability where the growth rate changes its sign. The calculations show that for values above the lines the perturbation mode grows while it decays below the curves.

5.7.3
Pseudo-Kepler Rotation

The question arises whether the HMRI also exists for more flat rotation profiles such as the Kepler law. Let the cylinders of the container rotate like planets (i.e., $\mu_\Omega = r_{in}^{1.5}$) which for the standard value $r_{in} = 0.5$ provides $\mu_\Omega = 0.35$. Liu, Good-

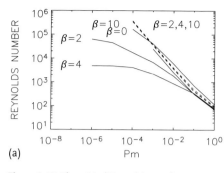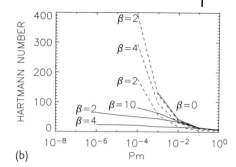

Figure 5.47 The critical Reynolds numbers (a) and Hartmann numbers (b) for pseudo-Kepler rotation between conducting boundaries vs. the magnetic Prandtl number. The solid lines give the axisymmetric mode, the dashed lines are for $|m| = 1$. Only the axisymmetric modes for $\beta = 2$ and $\beta = 4$ scale with Re and Ha. $\mu_\Omega = 0.35$, $r_{in} = 0.5$.

man, and Herron (2006) have shown that in a local approximation with $\nu = \eta = 0$ no solution exists for rotation laws $\Omega \sim R^{-n}$ with $n < 1.66$, thereby excluding the Kepler law with $n = 1.5$. Kirillov and Stefani (2011) and Kirillov, Stefani, and Fukumoto (2012) further demonstrated the validity of the Liu limit for both axisymmetric and nonaxisymmetric instabilities under the mentioned conditions.

Global calculations with conducting boundaries do not confirm this limit for such containers. Figure 5.47 shows that indeed for pseudo-Kepler rotation the HMRI exists but only for not too high β. For small Pm the scaling with Re only exists for $\beta \simeq 2$–4 but no longer for $\beta = 10$. For Pm $= 10^{-5}$ and $\beta = 4$, however, the critical Reynolds number with about 6000 is still rather low in comparison with values $O(10^6)$ which are characteristic for SMRI. It has been shown that this statement holds provided at least one of the boundaries is conducting (Figure 5.48). The open question is for the *minimum* value of n below which the HMRI no longer exists for any value of β. Priede and Gerbeth (2009) and Priede (2011) also suggest that the difference between convective and absolute instability may be important in this limit.

The behavior of the nonaxisymmetric modes vs. β and Pm is also demonstrated by the dashed lines in Figure 5.47. The result is that all $m = 1$ modes scale with Rm and S. The scaling (in the pseudo-Kepler limit) with Re and Ha for small Pm (which allows the experiments) is restricted to the axisymmetric modes.

5.7.4
The Frequencies

Knobloch (1992, 1996) has shown that a possible instability of differential rotation under the influence of a helical background field should exist in the form of a wave traveling in the axial direction. It has then been shown by theory and by numerical simulations that at the Rayleigh limit this drift frequency scales with the viscosity frequency $\omega_\nu = \nu/R_0^2$ with only a slight dependence on the magnetic Prandtl

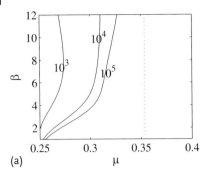
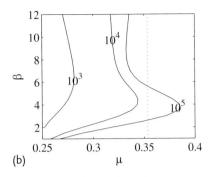

Figure 5.48 The critical Reynolds numbers for various β between Rayleigh limit ($\mu_\Omega = 0.25$) and the pseudo-Kepler law ($\mu_\Omega = 0.35$). Both cylinders insulating (a) and only the outer cylinder insulating (b). In the latter case HMRI exists also for pseudo-Kepler rotation law. $r_{\rm in} = 0.5$, Pm = 0. From Rüdiger and Hollerbach (2007). Copyright © 2007 by The American Physics Society.

number, that is

$$\omega \simeq \omega_\nu \, {\rm Pm}^{-1/4}, \qquad (5.50)$$

without any influence of β (Lakhin and Velikhov, 2007). However, Figure 5.45 shows that this is only true at the Rayleigh line. At some distance from this limit, that is for more flat rotation laws, the influence of β becomes stronger. This influence can be reduced by dividing the growth rates by the critical Hartmann number to consider the ratio (ω/Ω) (Re/Ha) which for pseudo-Kepler rotation leads to

$$\omega \simeq \Omega_{{\rm A},z} \, {\rm Pm}^{-1/4} \, \Phi(\beta) \qquad (5.51)$$

with $\Omega_{{\rm A},z} = B_z/\sqrt{\mu_0 \rho}\, R_0$ as the Alfvén velocity of the *axial* field. The function $\Phi(\beta)$ is odd in β and does strongly depend on the magnetic Prandtl number. It scales with β for Pm of order unity while it scales with $1/\beta$ for very small Pm (Figure 5.49). For Pm of order unity the drift frequency is thus determined by the Alfvén frequency of the *azimuthal* field. For positive definite $\Omega_{{\rm A},z}$ it is the sign of β which fixes the direction of the traveling wave. The traveling frequencies (5.50) and (5.51) vanish for inviscid media, hence HMRI does not exist in ideal MHD (Shtemler, Mond, and Liverts, 2011).

5.8
Laboratory Experiment PROMISE

The simplest idea to realize the MRI in an experiment is based on a Rayleigh-stable flow between differentially rotating cylinders. Such a flow could then be destabilized via the MRI by an externally imposed magnetic field. If the imposed field is purely axial, however, the relevant parameter for the onset of the instability is the magnetic Reynolds number, which must exceed about 10 (Rüdiger and Zhang,

2001; Ji, Goodman, and Kageyama, 2001). The kinetic Reynolds number then exceeds 10^6 because of the small magnetic Prandtl numbers of liquid sodium or gallium. Such large Reynolds numbers are not only difficult to achieve in precisely controlled experiments, but end-effects also become increasingly important and must be dealt with (Hollerbach and Fournier, 2004).

For a combined axial and azimuthal field the relevant parameter slightly beyond the Rayleigh limit is Re, which must only be of order $O(10^3)$ for instability. The main difference of the solutions to those for purely axial imposed fields is that

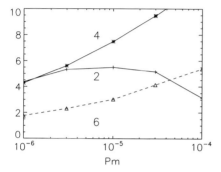

Figure 5.49 Normalized traveling frequency $\hat{\omega} = \omega\mathrm{Pm}^{1/4}\beta/\Omega_{A,z}$ for pseudo-Kepler flows. The curves are marked with their values of β. The convergence for small Pm only occurs for those β-values for which the solutions scale with the Reynolds number and the Hartmann number (i.e., not for $\beta = 6$, see Figure 5.47). $m = 0$, $r_{in} = 0.5$, $\mu_\Omega = 0.35$.

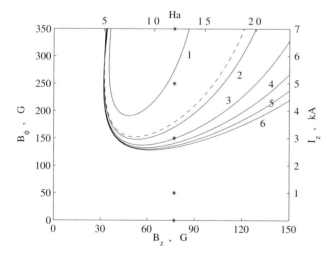

Figure 5.50 The critical Reynolds number (between 1000 and 6000) as a function of B_z and B_ϕ for conducting boundary conditions. On the vertical axes $B_\phi (R_{in})$ in gauss, and in terms of the required axial current. The dashed line is Re = 1775 used in the experiment. The five asterisks correspond to the five experiments in Figure 5.52. $r_{in} = 0.5$, $B_z = 77.2\,\mathrm{G}$, $\mu_\Omega = 0.27$, Pm = 1.4×10^{-6}. From Rüdiger et al. (2006). Reproduced by permission of AAS.

the pattern now drifts along the axis of the cylinders. Figure 5.50 shows the critical Reynolds number for the onset of the instability in an infinite cylinder with perfectly conducting walls as a function of the external fields B_z and B_ϕ. Provided $B_z > 30$ G and $B_\phi \gtrsim 150$ G, Reynolds numbers of order 10^3 are sufficient to excite the traveling waves. As the threshold values of Reynolds number and Hartmann number for insulating boundaries are higher by a factor of almost two the cylinders are better made from highly conducting material (copper).

If B_ϕ is less than 150 G, Re gradually rises, until for $B_\phi = 0$ we would have Re $> O(10^6)$, the known result from the analysis of standard MRI. In contrast, if B_z is less than 30 G, the MRI no longer exists at all (see Figure 5.14).

5.8.1
Experimental Results

The experimental apparatus consists of a cylindrical annulus made of copper, with $R_{in} = 4$ cm, $R_{out} = 8$ cm and height 40 cm (Figure 5.51). The stationary top endplate is made of plexiglass; the bottom copper endplate rotates with the outer cylinder (PROMISE 1, Stefani et al., 2006, 2007). An axial magnetic field is due to a current up to 200 A through an external coil, and an azimuthal magnetic field is imposed by a current up to 7 kA through a water-cooled rod along the central axis. The field within the fluid is current-free. The fluid within the vessel is a GaInSn alloy (see Table 5.1). Only the vertical velocity u_z is measured by two ultrasonic high-focus transducers mounted on opposite sides of the top endplate.

The rotation rates of the inner and outer cylinders were fixed at $\Omega_{in} = 0.377$ s^{-1} and $\Omega_{out} = 0.102$ s^{-1}, so $\mu_\Omega = \Omega_{out}/\Omega_{in} = 0.27$ and Re $= 1775$. The linear

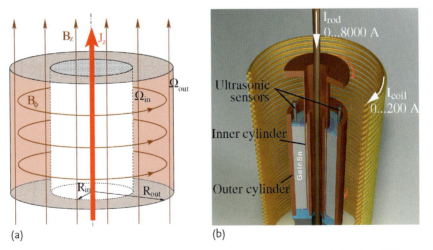

Figure 5.51 The magnetic field geometry for the HMRI experiment. Azimuthal and axial field components form a helical magnetic background field (a). $R_{in} = 4$ cm, $R_{out} = 8$ cm, $\Gamma = 10$. Construction of the Potsdam ROssendorf Magnetic InStability Experiment (PROMISE) (b).

velocity of the inner cylinder is thus fixed to $u_{in} = 1.51$ cm/s. The axial field B_z is fixed to 77.2 G but the azimuthal field B_ϕ is varied (see Figure 5.50).

The waves travel upward, but downward if the sign of one of the field components is changed. The five asterisks in Figure 5.50 correspond to the five plots shown in Figure 5.52. The experimental runs with $I_z \leq 3$ kA should not lead to instability, in contrast to the ones with 5 and 7 kA. The traveling waves are indeed well developed for the 5 and 7 kA runs. Figure 5.52 shows the waves of the axial velocity u_z. The two Ekman vortices induced by the endplates are removed. The predicted travel speeds for unbounded cylinders are of order 1.2 mm/s (for Re = 1775). The experiment shows that 5 and 7 kA yield approximately 0.8 mm/s, almost independent of B_ϕ so that $u_z/u_{in} \simeq 0.052$ (Stefani et al., 2006). This is of the order of the predicted value of 0.079 (see Table 5.3). The discrepancy could be due to the end-effects, by the restricted height of the container with its aspect ratio of only 10. The measurements yield two indications for this suggestion:

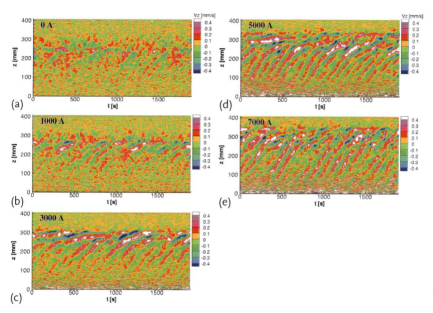

Figure 5.52 Ultrasound measurements of the axial velocity u_z as a function of time (a–e). Time is plotted on the horizontal axis; depth on the vertical axis. The time average is removed to eliminate the two Ekman vortices. The axial current varies from 0 to 7 kA corresponding to $B_\phi(R_{in})$ reaching from 0 to 350 G. The stronger the currents the more coherent are the structures. The axial drift speed is 0.8 mm/s for currents between 5 and 7 kA. Due to the end-effects at the upper boundary no waves exist there. The speed of the waves only depends slightly on the amplitude of the axial current. The data of the five experiments are marked as the five asterisks in Figure 5.50. $B_z = 77$ G. From Rüdiger et al. (2006). Reproduced by permission of AAS.

- The resulting wavelengths are with 5–6 cm, much shorter than the predicted values of 13 cm which cannot find enough space in the container with 40 cm height.
- The power of the end-effects can be observed directly through the difference between top and bottom even for $I_z = 0$ (see Figure 5.52a).

It is thus necessary to discuss the influences of the Ekman–Hartmann layers at the endplates of the container in more detail.

5.8.2
Endplate Effects

At an interface between an incompressible fluid with low viscosity and a rapidly rotating rigid surface an Ekman layer develops with thickness $d_E \propto \sqrt{\nu/\Omega}$. On the other hand, if the fluid is conducting in the vicinity of a solid resting boundary, and under the influence of an external magnetic field $\bar{B}_z$, a Hartmann layer results with the thickness $d_H \simeq \bar{\eta}/V_A$. Typical thickness values taken for PROMISE are $d_E \simeq 1$ mm and $d_H \simeq 0.3$ mm. The Ekman flow velocity is $u_E \simeq \sqrt{\mathrm{Re}}\nu/R_0$

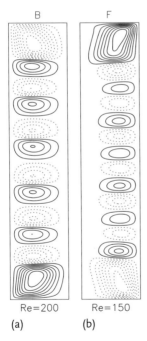

Re=200 Re=150
(a)   (b)

Figure 5.53 Contour lines of stream functions between the cylinders with endplates from conducting material attached to the outer cylinder (a, Re = 200) or attached to the inner cylinder (b, Re = 150). Solid lines denote clockwise flows, dashed lines denote counter-clockwise flows. Note the opposite orientation of the circulations for opposite shears between fluid and lid. Ha = 3, $\mu_\Omega = 0.27$, $\beta = 0$, $\Gamma = 10$, Pm = 0. From Szklarski and Rüdiger (2007). Copyright 2007 by The American Physical Society.

which yields 0.3 mm/s for the considered experiments. When these two effects are combined, an Ekman–Hartmann layer develops (Acheson and Hide, 1973). In the vicinity of a plane which rotates with Ω_{end} the Ekman–Hartmann layer develops if an axial magnetic field exists for $\Omega_{end} \neq \Omega$. If $\Omega_{end} < \Omega$ the Ekman flows are directed radially outwards ('blowing') and when $\Omega_{end} > \Omega$ they are directed towards the axis ('suction'). Gilman and Benton (1968) have shown that in addition to the Ekman flow an electric Hartmann current exists parallel to the direction of the Ekman flow. The current appears due to the vertical shear and is able to influence the flow via a Lorentz force far away from the boundary. This is true in particular for the MHD Taylor–Couette flow in finite containers if the cylinders are covered with rigidly rotating endplates. The Lorentz force due to the radial current and the axial magnetic field modifies the rotation law in the container. Formally the force is equivalent to an azimuthal pressure gradient leading to a Taylor–Dean flow (Chandrasekhar, 1960).

Consider a hydrodynamically stable flow with $\mu_\Omega = 0.27$ with an aspect ratio $\Gamma = 10$ and with rigidly rotating endplates. There are several possibilities to attach the endplates at the cylinders. Indeed, it makes a strong difference whether the endplate is attached at the inner or at the outer cylinder. One finds the hydrodynamically stable flow becoming unstable in both cases. The Lorentz force of the Hartmann current basically modifies the rotation law between the cylinders so that they become Rayleigh-unstable. The instability only disappears for unrealistically high values for the aspect ratio Γ.

Figure 5.53 demonstrates the differences of the flow patterns in the two cases for conducting cylinders. When the plates rotate with $\Omega_{end} = \Omega_{out}$ the Ekman circulation is clockwise and the corresponding Hartman current is positive, that is close to the inner cylinder it leaves the Ekman–Hartmann layer with $I_z > 0$ so that the radial current also has a positive sign. If the endplates are attached to the inner cylinder ($\Omega_{end} = \Omega_{in}$) the Ekman circulation is counterclockwise (Ekman suction) and the corresponding Hartmann current is negative.

For insulating endplates the effect is much weaker (Figure 5.54). One needs higher values of Re and Ha to find the circulation. Again in both cases the patterns are clearly opposite. The circulation in the upper half of the container is clockwise for endplates fixed at the outer cylinder (Figure 5.54a) but it is counterclockwise for endplates fixed at the inner cylinder (Figure 5.54b). These results suggest that the endplate-induced circulations in hydromagnetic TC flows can be minimized by endplates which are split so that the outer part rotates with the outer cylinder and the inner part rotates with the inner cylinder. Using a 2D MHD code for finite containers with Pm $= 0$ and $\Gamma = 10$, Szklarski (2007) found 40% as the optimum width of the inner ring of the two endplates, which should be made with insulating material.

Figure 5.55 shows the differences between rigid-body plates (Figure 5.55a,b) and split-rings endplates (Figure 5.55c,d) both from insulating material. It is $\mu_\Omega = 0.27$ in both cases with Ha $= 10$ and Re $= 1775$. For $\beta = 0$ the flow is stable for infinite containers. Nevertheless, the two insulating solid-body plates produce a traveling wave (Figure 5.55a,c) which could easily lead to wrong physical explanations. Figure 5.55b,d shows that the effect almost completely disappears by use of the insulating and split endplates. At Figure 5.55c HMRI for $\beta = 6$ develops as Re and Ha are supercritical. Note the clear presentation of the traveling wave allowed by the split endplates (Figure 5.55d).

Note also that in finite container simulations such as those by Szklarski (2007) issues such as absolute versus convective instability are incorporated automatically; that is, the computed global eigenmode corresponds to the absolute instability case. The PROMISE experiment, certainly the PROMISE 2 version discussed below, also corresponds to absolute instability.

5.8.3
PROMISE 2

PROMISE 2 (Stefani et al., 2008, 2009) is an improved version of the PROMISE experiment which possesses endplates from plexiglass which are split into two rings where the inner one is attached to the inner cylinder and the outer one to the outer cylinder. Based on the numerical simulations the splitting position is at $R_{split} = 56$ mm, according to the above mentioned 40% rule minimizing the Ekman pumping activity of rigid endplates (Figure 5.56).

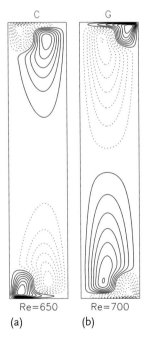

Figure 5.54 Streamlines between the cylinders with endplates from insulating material attached to the outer cylinder (a, Re = 650) or attached to the inner cylinder (b, Re = 700). Again the orientation of the circulations is opposite for opposite shears between fluid and lid. Ha = 10, $\beta = 0$, $\mu_\Omega = 0.27$, $\Gamma = 10$, Pm = 0. Copyright 2007 by The American Physical Society.

A series of experiments is started increasing μ_Ω from $\mu_\Omega = 0$ to $\mu_\Omega = 0.26$, crossing the Rayleigh line beyond which the rotation laws are hydrodynamically

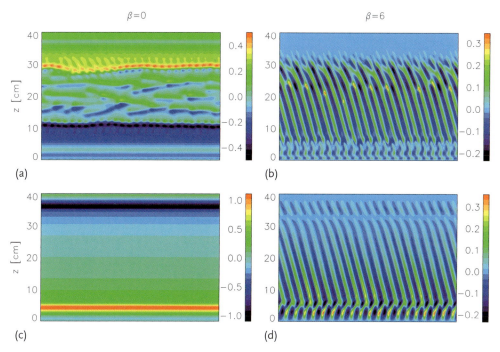

Figure 5.55 The axial velocity in the middle of the gap as a function of time. Rigid lids rotating with $\Omega_{end} = \Omega_{out}$ (a,b). Lids formed by two rings attached to the cylinders (c,d). $\beta = 0$ (a,c), $\beta = 6$ (b,d), Ha = 10, Re = 1775. $\mu_\Omega = 0.27$, $\Gamma = 10$, Pm = 0, insulating lids. From Szklarski (2007).

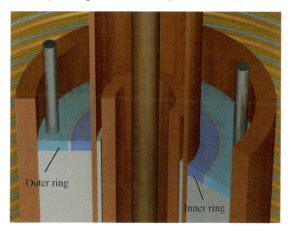

Figure 5.56 The endplates are split into two rings attached to the inner and outer cylinders of the container, where the inner ring is 16 mm wide and the outer ring 24 mm. From Stefani et al. (2009). Copyright 2009 by The American Physical Society.

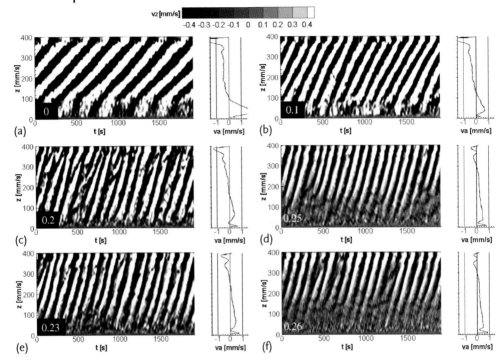

Figure 5.57 Axial velocity measurements with PROMISE 2 in dependence on the rotation ratio $\mu_\Omega = 0$ (a), 0.1 (b), 0.2 (c), 0.23 (d), 0.25 (e) and 0.26 (f). The frequency of the inner cylinder is 0.1 Hz ($u_{in} = 2.51$ cm/s) and the magnetic field ratio is $\beta = 4.5$. The maximum frequency of the traveling wave of 0.01 Hz is at $\mu_\Omega = 0.26$. From Stefani et al. (2009). Copyright 2009 by The American Physical Society.

stable. In all cases a traveling wave has been observed from the lower to the upper endplate (Figure 5.57). The fields are so strong that the pattern of the centrifugally induced Taylor vortices are suppressed in all experiments. There are three important observations:

- the Ekman vortices at the endplates are no longer visible, as they are strongly suppressed,
- the waves also exist beyond the hydrodynamic Rayleigh limit,
- the axial drift becomes faster the flatter the rotation law.

The frequency of the traveling wave slightly beyond the Rayleigh limit is about 10% of the frequency of the inner cylinder. For $\mu_\Omega = 0.26$ it is $u_z \simeq 1.8$ mm/s, that is $u_z/u_{in} \simeq 0.075$ in very good agreement to the values for marginal instability given in Table 5.3. For $\mu_\Omega > 0.26$ the wavelengths again become too long, so that a comparison with the values of Table 5.3 becomes uncertain. The agreement of the observed and predicted values of the frequencies and the axial travel velocities is clear evidence that the HMRI has indeed been realized in the laboratory by the PROMISE experiments.

6
The Tayler Instability (TI)

Thus far the stability of the current-free field with $a_B = 0$ in (5.39) has been discussed, which becomes unstable against nonaxisymmetric perturbations in connection with (stable) differential rotation. The other extreme is $b_B = 0$, which means that finite electric currents flow through the fluid. The currents may have only an axial component which is parallel to the rotation axis for positive a_B and antiparallel for negative a_B. According to the criterion (5.37) such fields are unstable against nonaxisymmetric perturbations even without rotation (Tayler, 1957, 1973; Vandakurov, 1972). This Tayler instability (TI) will be studied in this chapter, in connection also with rotation and with experimental data. The pure form of the TI was unknown in experimental physics for a long time (see Meynet and Maeder, 2005).

The existence of the magnetic Tayler instability has immense astrophysical consequences. Ott *et al.* (2006) concluded from their SN core-collapse simulations (with the rotation of the initial iron core as the free parameter) that the rotation period of a newly born neutron star should not exceed 1 ms, in contrast to the observations that peak with periods around 10–100 ms. Berger *et al.* (2005) argue for an upper limit of 10 km/s rotation velocity for white dwarfs using their spectroscopy of the rotational broadening of the Ca II line. Neutron stars as well as white dwarfs are compact remnants of stellar cores that exhibit a specific angular momentum of 10^{13}–10^{14} cm^2/s. However, for their progenitors, the simulations of Berger *et al.* provide values of more than 10^{16} cm^2/s, indicating more than two missing orders of magnitude between the hydrodynamic theory and the observations. Obviously, the hot stars do produce during their short lifetime rather slowly rotating cores (see Figure 6.1). Suijs *et al.* (2008) indeed show that an evolution scenario for stars with 1.3 M$_\odot$ that includes small-scale Maxwell stresses can explain the extreme spin-down of the stellar core by typically two orders of magnitude.

That the viscosity in the radiative zones of stars must be larger than the molecular value becomes more and more obvious. It is possible to explain the present-day solar rotation law (differential rotation of the convection zone plus solid body rotation of the radiative core) by (i) the standard model of wind-powered stellar spin-down and (ii) a coupling mechanism between convection zone and radiative interior. If the coupling is purely viscous, then viscosities are required of order 10^5 cm^2/s which exceeds the microscopic viscosity by more than four orders of magnitude.

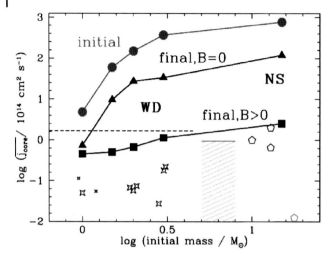

Figure 6.1 The specific angular momentum in stellar evolution calculations for the cores of hot main-sequence stars. The initial models are represented by the bulleted line. The observations (marked by symbols) show much smaller values for the angular momentum than the theoretical final models if magnetic fields are not included (triangled line). Star symbols represent the values for ZZ Ceti stars, the open pentagons correspond to the youngest neutron stars. The dashed line gives the spectroscopic upper limit of the angular momentum of the white dwarfs. From Suijs et al. (2008). Reproduced with permission © ESO.

Alternatively, Rüdiger and Kitchatinov (1996) probed the idea that internal differential rotation interacts with an internal (fossil) poloidal magnetic field. Together with the fossil field the induced toroidal field transports angular momentum outwards so that the solar core becomes decelerated. The rotation becomes uniform along the field lines. If the fossil field only exists in the radiative core (not penetrating the convection zone) then a slender tachocline appears between the core and the bottom of the convection zone. A problem, however, is formed by the remaining 'islands' of fast rotation as due to the nonuniformity of the field lines (Mestel and Weiss, 1987). An enhanced viscosity of 10^4–10^5 cm^2/s is needed to smooth out the isolated peaks of angular velocity to overcome this problem.

The very similar value of 3×10^4 cm^2/s is reported by Eggenberger, Montalbán, and Miglio (2012) to produce an internal rotation law of the red giant KIC 8366239 as consistent with the inversions of the oscillation frequencies observed by means of the KEPLER mission (Beck et al., 2012). Deheuvels et al. (2012) derive from the empirical data an internal rotation profile of the early red giant KIC 7341231 (of lower mass) with a core spinning five times faster than the surface. This is *much* less than one obtains from the stellar evolution codes without extra angular momentum transport from the core to the envelope (Ceillier et al., 2012).

The lithium at the surface of cool main-sequence stars depletes with a timescale of about 1 Gyr. Lithium is burned at temperatures in excess of 2.6×10^6 K that exist about 40 000 km below the base of the convection zone. Consequently, there must be a diffusion process between the base of the convection zone and the regions of the burning temperature. Its timescale must be one or two orders of magnitude shorter than the molecular diffusion in order to explain the lithium decay time. On the other hand, any enhanced chemical mixing coincides with an enhanced mixing of the angular momentum. Considering transport processes in the radiative interior of massive main-sequence stars, Spruit (2002) and Maeder and Meynet (2003, 2005) computed viscosities up to 10^{13} cm^2/s for an equatorial rotational velocity of 300 km/s with which the internal stellar rotation becomes rigid after a few thousand years. Heger, Woosley, and Spruit (2005) and Woosley and Heger (2006) followed on this basis the rotational history of massive main-sequence stars until their collapse.

Rotationally induced mixing is also included in the stellar evolution codes by Heger and Langer (2000). Yoon, Langer, and Norman (2006) presented evolutionary models of rotating stars with different metallicity. As even the lifetime and thus the evolutionary path in the H–R diagram is influenced by the mixing, the action of magnetic instabilities must also influence the ages of young stellar clusters. This may alter the rotation-activity-age relation for low-mass stars, which very much depends on the age determination of young clusters (Barnes, 2010). Brott *et al.* (2008) demonstrate that the magnetic-induced chemical mixing in massive stars must even be reduced to avoid conflicts with the observations.

A dynamo action on the basis of TI, as proposed by Spruit (1999, 2002) and Braithwaite (2006), could not be confirmed thus far (Zahn, Brun, and Mathis, 2007; Gellert, Rüdiger, and Fournier, 2007). The suggestion is that differential rotation and a magnetic kink-type instability could jointly drive a dynamo in stellar radiative zones. If existent, such a dynamo would also be effective for the (magnetic) angular momentum transport and the chemical mixing in stellar interiors.

6.1
Stationary Fluids

An important attribute of the Tayler instability of stationary fluids is the independence of its critical Hartmann number of the magnetic Prandtl number for both axisymmetric (Shalybkov, 2006) and nonaxisymmetric modes (Rüdiger and Schultz, 2010). It is thus possible to compute the critical field amplitudes for fluids with very low Pm with codes which only work for Pm $\simeq 1$. For a proof of this statement replace in the equation system (5.43) and (5.44) for $\Omega = 0$ the frequency Re $\cdot \omega$ by ω and the variables u_ϕ, u_z, b_R by $-iu_\phi, -iu_z, ib_R$. In the resulting system the Pm only survives as a factor of $i\omega$ which for marginal stability is purely imaginary. Hence, the real part of the equation system does not contain the Pm and also not the drift frequency. The only parameter in this equation system is the Hartmann number which, therefore, cannot depend on the magnetic Prandtl number.

The technical possibilities are now discussed to realize such an instability in the laboratory, working with liquid metals in a stationary cylindrical Taylor–Couette container. The basic question is how strong an axial electric current can be to remain stable. Let I_axis be the axial current inside the inner cylinder and I_fluid the axial current through the fluid (i.e., between inner and outer cylinder). The toroidal field amplitudes at the inner and outer cylinders are then

$$B_{in} = \frac{I_{axis}}{5 R_{in}}, \quad B_{out} = \frac{I_{axis} + I_{fluid}}{5 R_{out}}, \quad (6.1)$$

where R, B and I are measured in centimeter, gauss and ampere. Expressing I_{axis} and I_{fluid} in terms of our usual dimensionless parameters one finds

$$I_{axis} = 5\mathrm{Ha}\sqrt{\frac{r_{in}}{1 - r_{in}}}\sqrt{\mu_0 \rho \nu \eta}, \quad I_{fluid} = \frac{\mu_B - r_{in}}{r_{in}} I_{axis} \quad (6.2)$$

in ampere. For $\mu_B = r_{in}$ it follows $I_{fluid} = 0$ which describes the 'vacuum' solution with $a_B = 0$. The opposite approach is to neglect the vacuum solution in (5.39), that is $b_B = 0$ (Roberts, 1956; Tayler, 1957; Pitts and Tayler, 1985; Spies, 1988), which leads to $\mu_B = 1/r_{in}$ and to $I_{axis}(1 - r_{in}^2) = I_{fluid} r_{in}^2$ which means that for very small r_{in} the 'inner' current becomes negligibly small. For our standard case with $r_{in} = 0.5$ the outer current must exceed the inner current by a factor of three (see Table 6.1).

Note also how the required currents depend on the radius ratio r_{in}, but not on the actual physical dimensions R_{in} and R_{out}. Making the entire device bigger thus reduces the current density, inversely proportional to the square of the size. By making the device sufficiently large one can thereby prevent Ohmic heating within the fluid from becoming excessive.

For containers with very wide gaps, that is $r_{in} \ll 1$, it makes sense to work with the 'outer' Hartmann number Ha_{out} defined with the outer magnetic field and the radius of the outer cylinder, that is

$$\mathrm{Ha}_{out} = \frac{B_{out} R_{out}}{\sqrt{\mu_0 \rho \nu \eta}}, \quad (6.3)$$

which can easily be related to Ha by (5.42). One then obtains that

$$I_{axis} = 5 \frac{r_{in}}{\mu_B} \mathrm{Ha}_{out} \sqrt{\mu_0 \rho \nu \eta}, \quad I_{fluid} = 5\left(1 - \frac{r_{in}}{\mu_B}\right) \mathrm{Ha}_{out} \sqrt{\mu_0 \rho \nu \eta}. \quad (6.4)$$

For $\mu_B = 2 r_{in}$ both currents are equal, which for $r_{in} = 0.5$ is true for the almost uniform field with $\mu_B = 1$. This special constellation is marked in bold in Table 6.1.

We are in particular interested in the critical Hartmann numbers for various magnetic profiles produced by the combination of the two currents I_{axis} and I_{fluid}. To this end the (5.43) and (5.44) must be solved numerically with the method described in Chapter 5. Table 6.1 gives the critical Hartmann numbers for the modes with $m = 0$ and $m = 1$ for a container with conducting or insulating walls. The critical currents which follow from (6.2) from these numbers refer to the smaller of the Hartmann numbers for $m = 0$ and $m = 1$. The interval $0 < \mu_B \leq 2$ is stable for $m = 0$, while perturbations with $m = 1$ are only stable in the vicinity of the current-free profile $\mu_B = 0.5$[1]. The stability criteria (5.35) and (5.37) lead to

---

1) Even this current-free field profile becomes unstable under the influence of differential rotation.

Table 6.1 Characteristic Hartmann numbers and electric currents for a container ($r_{in} = 0.5$) with conducting (upper line) and insulating (lower line) walls, using either sodium or galinstan (in brackets). More numbers by Rüdiger et al. (2007b).

μ_B	Ha$^{(0)}$	Ha$^{(1)}$	I_{axis} [kA]	I_{fluid} [kA]
−10	3.96	5.02	0.161 (0.509)	−3.39 (−10.7)
	6.09	4.66	0.190 (0.599)	−3.99 (−12.6)
−4	9.61	12	0.392 (1.24)	−3.53 (−11.1)
	14.6	11.7	0.477 (1.50)	−4.29 (−13.5)
−2	19.8	24.8	0.807 (2.55)	−4.04 (−12.7)
	29.3	25.2	1.03 (3.24)	−5.15 (−16.2)
−1	59.3	63.7	2.42 (7.63)	−7.25 (−22.9)
	73.6	64.6	2.63 (8.31)	−7.89 (−24.9)
1	∞	151	6.16 (19.4)	6.16 (19.4)
	∞	109	4.44 (14.0)	4.44 (14.0)
2	∞	35.3	1.44 (4.54)	4.32 (13.6)
	∞	28.1	1.15 (3.61)	3.45 (10.8)
4	13.2	14.6	0.538 (1.70)	3.77 (11.9)
	20.5	12.5	0.510 (1.61)	3.57 (11.3)
10	4.44	5.4	0.181 (0.571)	3.44 (10.8)
	6.86	4.78	0.195 (0.615)	3.71 (11.7)

the condition $0 < \mu_B \leq 1/r_{in}$ for axisymmetric perturbations, so that for wide gaps even steep magnetic profiles are stable. One finds that all fields with $b_B = 0$ are stable with respect to axisymmetric perturbations.

The area of stability against nonaxisymmetric perturbations is much smaller. It fulfills the relation $0 < \mu_B < \mu_{max}$ with $\mu_{max} = 4r_{in}/(3 + r_{in}^2)$. For $r_{in} = 0.5$ one has $\mu_{max} = 0.62$, and for very thin gaps $\mu_{max} = 1$.

For technical applications we have also considered containers with insulating cylinders. For large $|\mu_B|$ the electrical current I_{fluid} approaches a constant value. This value is larger than a minimum value which in the table is about 3.48 kA. Currents smaller than 3.45 kA for sodium or 10.8 kA for galinstan are thus always stable, but currents exceeding this value can be stable or unstable depending on the value of I_{axis}. Hence, currents I_{fluid} can be stabilized by a proper choice of I_{axis}. The stabilization is easiest with $I_{axis} \cdot I_{fluid} < 0$, that is with antiparallel currents (Stefani et al., 2012). As a demonstration, we take from Table 6.1 for sodium that a I_{fluid} of 5 kA can be stabilized by $I_{axis} \simeq -1$ kA.

The next step is to consider the solutions for uniform current with $b_B = 0$ for various gaps between the cylinders, which for the special case $r_{in} = 0.5$ in Table 6.1 appear in the lines for $\mu_B = 2$. For free choice of r_{in}, Figure 6.2 gives the resulting Hartmann numbers Ha$_{out}$ in dependence on the position of the inner cylinder. For the extremely wide containers with $r_{in} \to 0$ the vacuum condition for the outer boundary has been applied while the inner cylinder has been considered both as

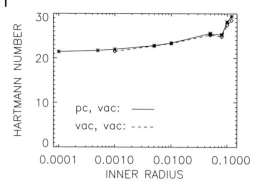

Figure 6.2 The critical Hartmann number Ha_{out} for very wide gaps ($r_{\text{in}} \to 0$) and for all Pm. The outer boundary condition is vacuum, within the inner cylinder is either vacuum or a perfectly conducting material what only makes very little differences. Reproduced by permission of AAS.

perfect conductor or as insulator. The differences of both solutions are very small. In both cases the outer Hartmann number Ha_{out} for $r_{\text{in}} \to 0$ approaches the value 22 for all Pm. With $\sqrt{\mu_0 \rho \nu \eta} \simeq 25.8$ in cgs for the GaInSn alloy the resulting minimum electric current to excite TI according to (6.4) (right) is (only) 2.8 kA. This value does not depend on the size of the container.

Next the *growth rates* of the instability are of interest. In a stationary container only two frequencies exist which may form the growth rate $\omega_{\text{gr}} = -\Im(\omega)$, that is the diffusion frequency $\omega_\eta = \eta/R_0^2$ and the Alfvén frequency $\Omega_A = B_{\text{out}}/\sqrt{\mu_0 \rho} \cdot R_0$ with R_0 as a characteristic length scale. The ratio of both quantities defines the limits of low conductivity ($\Omega_A < \omega_\eta$) and high conductivity ($\Omega_A > \omega_\eta$) which, therefore, depend on the strength of the background field. In both limits the resulting growth rates are different.

The numerical factors may strongly depend on the magnetic Prandtl number. By a numerical solution of (5.43) and (5.44) a formulation can be found in which the influence of Pm is rather compact. We take from Figure 6.3a for the high-conductivity limit of a wide-gap container the linear relation $\bar{\omega}_{\text{gr}} \simeq \gamma \text{Ha}_{\text{out}}$ which yields $\omega_{\text{gr}} \simeq \gamma \Omega_A$ for the physical growth rate value (Goossens, Biront, and Tayler, 1981). The normalized growth rate $\bar{\omega}_{\text{gr}}$

$$\bar{\omega}_{\text{gr}} = \frac{\omega_{\text{gr}} R_{\text{out}}^2}{\bar{\eta}} \tag{6.5}$$

has been used in Figure 6.3a with $\bar{\eta} = \sqrt{\nu \eta}$. For liquid metals this averaged diffusivity is almost equal (see Table 5.1). One finds $\gamma \lesssim 4$ for $\text{Pm} = 1$. For $\text{Pm} = 10^{-6} \gamma$ is still 0.05, so that six orders of magnitudes in Pm are compressed to a factor of 100. Goossens and Veugelen (1978) find 50–100 days for a typical A star.

From Figure 6.3a the growth rates for the low-conductivity limit are also given for $r_{\text{in}} = 0.05$ in their quadratic dependence on the Hartmann number, that is $\bar{\omega}_{\text{gr}} =$

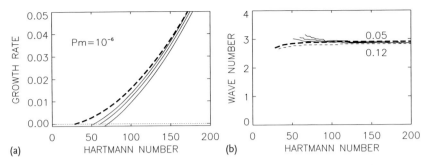

Figure 6.3 The growth rates $\bar{\omega}_{gr}$ (a) according to the relation (6.5) for $Pm = 10^{-6}$ vs. the Hartmann number (6.3). The dashed line gives the growth rates for the stationary container. The solid lines show the influence of rigid rotation with $Re = 500$, 1000 and 2000. Note the stabilizing action of global rotation. $r_{in} = 0.05$, $\mu_B = 20$. The wave number k in the relation $\delta z = (R_{out} - R_{in})(\pi/k)$ for the axial structure of the magnetic pattern (b). From Rüdiger et al. (2012). Reproduced by permission of AAS.

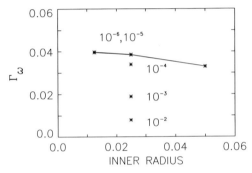

Figure 6.4 The factor Γ_ω from (6.6) in its weak dependence on the (marked) magnetic Prandtl number of the fluid and the value r_{in} of the inner radius of the container.

$\gamma_\omega (\mathrm{Ha}_{out}^2 - \mathrm{Ha}_{crit}^2)$. Hence, in the low-conductivity limit a quadratic dependence

$$\omega_{gr} \simeq \Gamma_\omega \frac{B_{out}^2}{\mu_0 \rho \eta} \tag{6.6}$$

has been realized with $\Gamma_\omega = \gamma_\omega/\sqrt{Pm}$. Note that the size of the container does not appear in this expression. The factor Γ_ω for small Pm has a very weak dependence on Pm (i.e., on the viscosity), but not on the value of r_{in} (Figure 6.4). It can be measured in experiments, while the abovementioned γ for high conductivity can only be determined numerically. In test calculations with small Pm the transition from the low-conductivity to the high-conductivity limit happens for (outer) Hartmann numbers of about 100. Equation (6.6) provides a growth time of 5 s for a toroidal field of 1 kG and for galinstan as the fluid.

6.2 Experiment GATE

Without rotation the current-driven instability leads to a nondrifting steady-state solution. Isosurfaces of the pattern resulting from a 3D nonlinear simulation are shown in Figure 6.5 for all three field components. The boundary conditions at the walls are pseudovacuum. The container with $r_{in} = 0.05$ is periodic in the z-direction with a domain height of $2\pi(R_{out} - R_{in})$. The Prandtl number Pm $= 0.02$ is the minimum value with which the MHD code can still cope. The Hartmann number is slightly supercritical. The two dominating modes with $m = \pm 1$ lead to a kidney-shaped pattern with nearly circular (in the meridian) cells ($\delta z \simeq R_{out} - R_{in}$) in accordance with the linear results in Figure 6.3b for the wave numbers. The field amplitudes are given in units of the outer value of the toroidal background field and are surprisingly high.

Summarizing the previous findings, the realization of the TI in a laboratory should be possible in the form of a stability experiment of a homogeneous current flowing through a cylindrical container filled with galinstan. We know that the easiest excitation of nonaxisymmetric $m = \pm 1$ modes happens for stationary containers and that

- the cylinder should be made from insulating material,
- the critical Hartmann number is $Ha_{out} \simeq 22$, and does not depend on the magnetic Prandtl number Pm,
- the critical current is 2.8 kA, and does not depend on the size of the container,

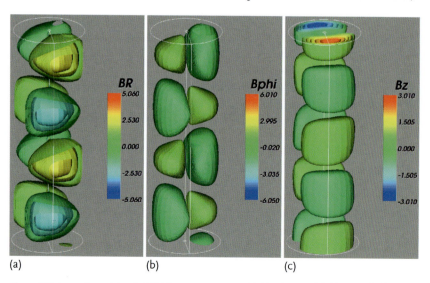

Figure 6.5 Isosurface plots of all field components (a–c). Ha is twice the critical value, the cylinders are considered as pseudovacuum. The fields are given in units of the outer toroidal background field. Note the high values of the saturated magnetic field components. Re $= 0$, $r_{in} = 0.05$, $\mu_B = 20$, Pm $= 0.02$.

- the exponential growth time for $I_{\text{fluid}} = 6\,\text{kA}$ is about 100 s,
- the cells are almost circular, that is the axial size of the 'kidneys' in Figure 6.5 approaches the outer radius of the container.

The experimental apparatus constructed according to these principles consists of an insulating cylinder with a height of 75 cm and a radius R_{out} of 5 cm which is filled with galinstan (Figure 6.6). An inner cylinder with radius R_{in} of 0.6 cm can be inserted. At the top and bottom, the liquid column is in contact with two massive copper electrodes which are connected to an electric power supply which provides up to 8 kA. With 14 fluxgate sensors the fluctuations due to TI are observed. Three of the sensors are positioned along the azimuth in the upper part. Such measurements give the geometry of the field, thus its shape in azimuthal and axial direction as well as the minimum current to excite the TI (Seilmayer et al., 2012).

In all cases of instability the observed pattern of the magnetic perturbations is a nonaxisymmetric one with $m = 1$. For the growth rates Figure 6.7 shows the experimental results in comparison with the theoretical calculations for containers with very wide gaps between the cylinders. The theoretical result does not depend on r_{in} for $r_{\text{in}} \ll 1$ (see Figure 6.4), so that these results may serve as a good proxy for the experiment without any inner cylinder (crosses in Figure 6.7).

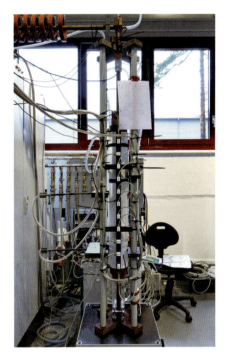

Figure 6.6 The experiment GATE as operated by the HZDR (Seilmayer et al., 2012). A uniform electric current flows through a vertical container filled with galinstan and becomes unstable against nonaxisymmetric modes.

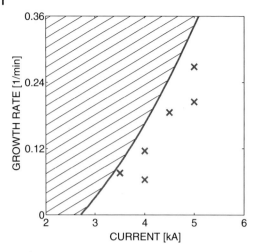

Figure 6.7 Observed and calculated growth rates for GATE without inner cylinder. For very small $r_{\rm in}$ the growth rates are independent of $r_{\rm in}$ (solid curve: $r_{\rm in} = 0.001$, $r_{\rm in} = 0.0001$). The predicted threshold value for the electric current is 2.8 kA. As the eigenvalues are optimized no growth rate measurement should lie in the hatched area.

For low growth rates the experimental data fit well the theoretical curves. The agreement is almost perfect for the container with $r_{\rm in} = 0.12$. For this case one finds a relation $\omega_{\rm gr} \simeq \gamma(I^2 - I_{\rm crit}^2)$ with $\gamma = 2.7 \times 10^{-10}$ so that a value of $\Gamma_\omega \simeq 0.038$ results (see Figure 6.4, Rüdiger et al. (2012)). The theory, however, always provides the maximum growth rates, optimized over the wave number. The observed growth rates should thus never lie above the theoretical values which indeed is the case. For stronger currents the measured growth rates are systematically smaller than the calculated values. One possible explanation is that the fluid becomes warmer as a result of the Ohmic heating.

6.3
Rotating Fluids

Rigid rotation stabilizes the TI (Acheson, 1978; Pitts and Tayler, 1985). The main influence of a uniform rotation is the increase of the critical Hartmann number. This rotational quenching depends on the magnetic Prandtl number of the considered fluid. A rather compact formulation for this situation can be found by use of the modified Reynolds number $\overline{\rm Rm}$ and the Hartmann number Ha,

$$\overline{\rm Rm} = \frac{\Omega R_0^2}{\bar{\eta}}, \quad {\rm Ha} = \frac{B_{\rm in} R_0}{\sqrt{\mu_0 \rho \bar{\eta}}} \tag{6.7}$$

and the growth rate (6.5) which are all inverse in $\bar{\eta}$. First the influence of rigid rotation is considered.

6.3.1
Rigid Rotation

To study the rotational stabilization of TI we focus on magnetic field profiles with $B_\phi \propto R$, that is $\mu_B = 1/r_{in}$. Figure 6.8 gives for a wide-gap container the rotational quenching of the TI vs. the magnetic Prandtl number. For highly supercritical Ha the lines for marginal instability in the Ha–$\overline{\mathrm{Rm}}$ plane are rather straight. In this formulation the maximum stabilization of TI by rigid rotation happens for Pm = 1. Note that the rotational stabilization is much weaker for very small and very large Pm.

One finds that for fast rotation the slope Ω_A/Ω of the lines of marginal stability is almost constant for given Pm. Its dependence on Pm is also weak (Figure 6.8b). One obtains $\Omega_A/\Omega \gtrsim 1$ as the excitation condition for TI under the presence of rigid rotation (Chandrasekhar, 1956; Tataronis and Mond, 1987). Here, the Alfvén frequency $\Omega_A = B_{out}/\sqrt{\mu_0\rho}\,R_{out}$ of the *outer* magnetic field amplitude and the outer radius has been used. Obviously, the TI is a strong-field instability. For Pm = 1 fields with $\Omega_A/\Omega > 0.8$ become unstable while for much smaller or much larger Pm the critical magnetic fields can be weaker by about one order of magnitude.

The growth rates are generally reduced by rotation. In the low-conductivity limit ($\Omega_A < \omega_\eta$) the growth rate (6.6) must be written as

$$\omega_{gr} \propto \frac{\Omega_A^2}{\Omega} \tag{6.8}$$

to model the influence of the basic rotation. On the other hand, for the high-conductivity limit ($\Omega_A > \omega_\eta$) the only possible frequencies are Ω_A and Ω. The diffusion frequency cannot contribute as it does not remain finite for high conductivity. For $\Omega_A \gg \Omega$ the slope of the growth rate in its linear dependence of Ω_A cannot be changed by the rotation. Its only influence is a downward shift ('quenching') of the profile towards lower values of the growth rate (see Figure 6.3a,c).

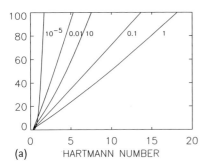

(a)

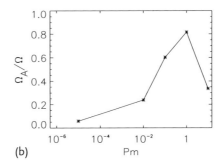

(b)

Figure 6.8 The suppression of TI by rigid rotation for various Pm. $\overline{\mathrm{Rm}}$ vs. Ha. The curves are marked with their magnetic Prandtl numbers. Fluids with Pm = 1 undergo the strongest rotational stabilization. The critical Hartmann number for resting container is 0.31 for all Pm (a). The critical Ω_A/Ω plotted over six orders of magnitude does not vary too strongly (b). $r_{in} = 0.05$. $\mu_B = 20$.

6.3.2
Differential Rotation

The influence of differential rotation is rather complex, mainly depending on the radial profiles of the background field and the rotation law. Figure 6.9 shows for the standard container with $r_{\rm in} = 0.5$ how the stability curves depend on Ha and Re, for Pm $= 10^{-5}$ with $\mu_\Omega = 0$ (stationary outer cylinder) and for various μ_B. For Ha $= 0$ we find that $m = 0$ becomes unstable before $m = 1$ at the critical Reynolds number $\text{Re}_{\rm crit} = 68$, the familiar value for the onset of nonmagnetic Taylor vortices. At the other limiting case, Re $= 0$ (no rotation), we find that $m = 1$ goes unstable at critical Hartmann numbers which increase with decreasing μ_B. These Ha-values are independent of Pm.

How are these two limiting cases Ha $= 0$ and Re $= 0$ connected, and how do the two types of instabilities interact when neither parameter is zero? The two instabilities are certainly connected; the $m = 1$ modes in the two limiting cases are smoothly joined to one another for all Prandtl numbers.

The dashed lines in Figure 6.9 describe the stability behavior of the axisymmetric disturbances, which in the absence of magnetic field form Taylor vortices. They are

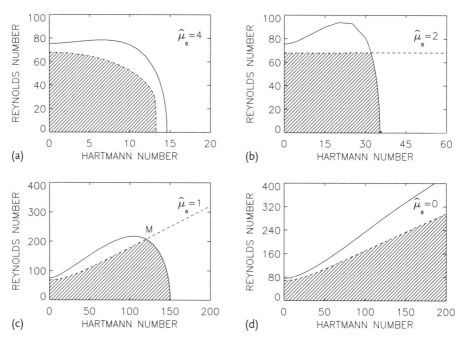

Figure 6.9 Stability areas ('hatched') for various radial profiles of the toroidal magnetic field and for stationary outer cylinder (a–d). $m = 0$ (dashed), $m = 1$ (solid). Radial boundaries are no-slip and of conducting material. Note that for $\mu_B = 2$ the classical axisymmetric solution with Re $= 68$ is not influenced by the axial current. $\mu_\Omega = 0$, $r_{\rm in} = 0.5$, Pm $= 10^{-5}$. From Rüdiger et al. (2007b). Copyright 2007 by The American Physical Society.

stabilized by toroidal fields with flat radial profiles. They coexist, however, with the fields of the profile described by $\mu_B = 2$ which is due to a uniform axial electric current in the fluid conductor. Hence, for stationary outer cylinder and Re > 68 the axisymmetric vortices will always exist even for strong magnetic amplitudes; they cannot be suppressed by the azimuthal magnetic field. For steeper radial profiles ($\mu_B > 2$) the axisymmetric Taylor vortices are even the preferred mode (see also Table 6.1 for details).

One can easily show that for fields which are due to homogeneous currents (i.e., $\mu_B = 2$) the axisymmetric mode ($m = 0$) is not influenced by the magnetic field. To this end the magnetic equations (5.44) can be reduced to

$$\frac{d^2 b_R}{dR^2} + \frac{1}{R}\frac{db_R}{dR} - \frac{b_R}{R^2} - k^2 b_R - i\,\text{Pm}\,\text{Re}\omega\, b_R = 0 \tag{6.9}$$

and

$$\frac{d^2 b_\phi}{dR^2} + \frac{1}{R}\frac{db_\phi}{dR} - \frac{b_\phi}{R^2} - k^2 b_\phi - i\,\text{Pm}\,\text{Re}\omega\, b_\phi$$
$$+ \text{PmRe}\,R\frac{d\Omega}{dR} b_R - R\frac{d}{dR}\left(\frac{B_\phi}{R}\right) u_R = 0. \tag{6.10}$$

It becomes immediately clear that for $B_\phi \propto R$ the system of the magnetic perturbations decouples from the hydrodynamic equations so that they must decay because of missing energy input. The fundamental axisymmetric Taylor vortex is thus *not* destabilized by the magnetism. One can also show that the last term in (6.10) stabilizes the axisymmetric mode for $\mu_B < 2$ and it destabilizes the axisymmetric mode for $\mu_B > 2$. This is only true in the linear regime. An example for the nonlinear regime without and with rotation is given by Figure 6.10. It is visible how the unstable mode $m = 1$ grows and how the neighboring modes $m = 0$ and $m = 2$ start to decay. Only if the unstable mode is strong enough energy also flows in the neighboring modes. Without rotation only a few percent of the energy exists in the neighboring modes but with rotation this ratio grows up to unity. For Re = 350

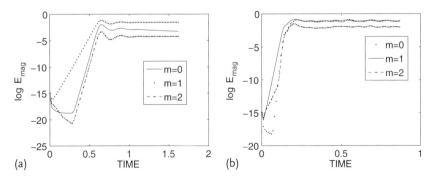

Figure 6.10 Stabilization and destabilization of the TI modes for homogeneous current without (a, Re = 0) and with rotation (b, Re = 350). For $B_\phi \propto R$ the axisymmetric mode is always linearly stable at Re = 0. Ha = 80, $\mu_\Omega = 0$, $\mu_B = 2$, Pm = 1.

also the axisymmetric mode is unstable so that it can develop the same amplitude as $m = 1$ (Figure 6.10b).

The magnetic profile in Figure 6.9 that is closest to being current-free is $\mu_B = 0$, and we find that for Ha $\lesssim 200$ there is no destabilizing influence of the field for either axisymmetric or nonaxisymmetric perturbations. For flat magnetic profiles with $0 < \mu_B < 1$ the magnetic field even stabilizes the flow for both $m = 0$ and $m = 1$. The results are unknown so far for stronger magnetic fields, the given Hartmann number only yields a rather small Lundquist number of order unity.

For $\mu_B = 1$ and $\mu_B = 2$ the nonaxisymmetric mode proves to be the most unstable mode for fields stronger than the crossover point M at which the type of the most unstable mode changes from $m = 0$ to $m = 1$. Note also how for $\mu_B = 1$ for increasing fields the critical Reynolds number increases for the $m = 0$ mode and the $m = 2$ mode before suddenly the $m = 1$ mode becomes unstable even without rotation. We have the interesting situation therefore that weak fields initially stabilize the TC flow, before stronger fields eventually are destabilized by a nonaxisymmetric mode.

The rotation law used in Figure 6.9 is steep enough to be unstable without any magnetic field. Flatter rotation laws should be more interesting. Again the field is due to a homogeneous current ($b_B = 0$). Figure 6.11 presents the instability curves for a container with a very narrow gap for three flat rotation laws and, for comparison, for a steep profile (dotted). The latter shows the behavior known from Figure 6.9. The curves are marked with their ratio μ_Ω. The curve for rigid rotation ($\mu_\Omega = 1$) simply demonstrates its stabilizing influence. Superrotation ($\mu_\Omega > 1$) is even more stabilizing. In this case the energy of the differential rotation cannot be transferred into growing modes. This, however, is possible for subrotation laws ($\mu_\Omega = 0.92$). Here the excitation of TI even becomes subcritical. The critical Hartmann number is reduced by almost one order of magnitude, but the vertical axis (Ha $= 0$) cannot be reached. For sufficiently high Reynolds number the criti-

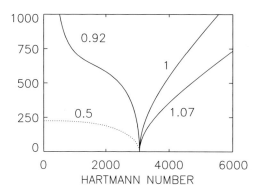

Figure 6.11 Critical $\overline{Rm}$ vs. Ha for homogeneous electric current in a small gap subject to different rotation laws. Rigid rotation ($\mu_\Omega = 1$) and positive shear ($\mu_\Omega > 1$) stabilize the instability while negative shear has a destabilizing effect. Conducting cylinders with $r_{in} = 0.95$, $m = 1$, $\mu_B = 1.05$. The critical Hartmann number for $\Omega = 0$ is 3061, for all Pm. The Rayleigh limit is at $\mu_\Omega = 0.90$. Pm $= 1$. From Rüdiger and Schultz (2010).

cal Hartmann number reaches its minimum and grows again as too strong shear stabilizes the flow.

For the subcritical excitation a minimum Reynolds number and a maximum Reynolds number exist; between them the flow is unstable. For supercritical excitation, on the other hand, only one critical Reynolds number exists which limits the domain of instability for too strong shear. This limit exceeds the limit for rigid rotation by more than two orders of magnitude. While for rigid rotation the instability only appears for $\Omega_A/\Omega > 1$, the instability condition for weak differential rotation is $\Omega_A/\Omega > 0.1$. The latter condition should be fulfilled in stellar objects so that we expect the pure kink-type instability as existing only in differentially rotating stellar layers. In summary, one finds from Figure 6.11 that for superrotation the instability is suppressed while for subrotation the instability is supported unless the rotation is too fast.

For one and the same flat rotation law, Figure 6.12 presents the bifurcation lines for various magnetic profiles μ_B. The vacuum-free field with $b_B = 0$ is given by $\mu_B = 2$ while the curve with $\mu_B = 0.5$ gives the current-free (AMRI) solution with $a_B = 0$. Between both the (black) curves the marginal instability curve for the maximally uniform magnetic profile ($\mu_B = 1$) is plotted in red. It combines the basic properties of the curves for the extrema $a_B = 0$ and $b_B = 0$. The AMRI does not exist without rotation, but it shows the characteristic tilted cone form which also exists in the curves. The locations of the minimum Hartmann number for excitation lie along a straight line almost parallel to the dotted line of Re = Ha, that is Mm = 1. This line obviously separates the domain of TI ($\Omega_A/\Omega > 1$) from the domain of AMRI ($\Omega_A/\Omega < 1$). It is the line of maximum transfer from the rotation energy to the magnetic energy.

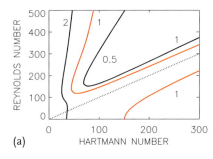 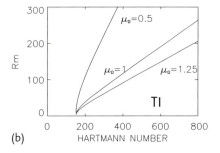

Figure 6.12 The stability map for toroidal fields with different radial profiles under the influence of weak differential rotation ($\mu_\Omega = 0.5$). Black curves: $\mu_B = 0.5$ (AMRI), $\mu_B = 2$ (TI); red: $\mu_B = 1$ (nearly uniform). Dotted line: $\Omega_A = \Omega$ (a). The stabilization of the red curve in the lower right corner ($\mu_B = 1$) by rigid rotation ($\mu_\Omega = 1$) and superrotation ($\mu_\Omega = 1.25$) (b). $r_{in} = 0.5$, Pm = 1.

6.3.3
Eddy Viscosity and Turbulent Diffusivity

6.3.3.1 Eddy Viscosity

The solutions of the linear equations allow an undetermined multiplicative constant. We thus cannot know the sign of the flow and/or the field. However, for quadratic expressions such as the correlation tensor or the electromotive force we can compute the sign, as all parts of the solution are multiplied by the same constant. Let us apply this idea to the angular momentum transport

$$T_{R\phi} = \left\langle u_R u_\phi - \frac{1}{\mu_0 \rho} b_R b_\phi \right\rangle . \tag{6.11}$$

The nonaxisymmetric components of both flow and field may be used in the following as the 'fluctuations' while the axisymmetric components are considered as the background quantities. The averaging procedure is then simply the integration over the azimuth ϕ. If the angular momentum flows towards the minimum of the angular velocity one can introduce an eddy viscosity ν_T via

$$T_{R\phi} = -\nu_T R \frac{d\Omega}{dR} \tag{6.12}$$

with positive eddy viscosity ν_T. Within the linear theory it can be shown that the angular momentum (kinetic plus magnetic) always flows opposite to the gradient of the angular velocity, that is $\nu_T > 0$. With nonlinear simulations the amplitude of this eddy viscosity is also computed. The eddy viscosity is calculated as a function of the Reynolds number of rotation for the parameters used in Figure 6.13a. All these models belong to the rotation-dominated domain AMRI in the instability map. For ν_T/ν a linear growth with Rm is found. Due to the rotational quenching of the instability a maximum exists characteristic for any given magnetic field. The eddy viscosity is normalized with the molecular viscosity so that the results are fully general. The resulting viscosities ν_T/ν do not exceed the value of (say) 30. Such values comply with those of Spruit (2002) despite the different nature of the instability model.

Figure 6.13b shows the results of calculations of maximum values for ν_T/ν for increasing magnetic amplitudes. The Reynolds numbers belonging to the maximum viscosity are also given. We find a saturation of the ratio ν_T/ν for strong magnetic fields. The eddy viscosity exceeds the microscopic viscosity by not more than a factor of 30. The existence of the maximum of ν_T/ν for fixed magnetic amplitude seems to be evident. As a function of the magnetic Reynolds number the relation $\nu_T/\nu \simeq 0.026$ Rm describes the results even with models with magnetic Prandtl numbers down to 0.1 included.

In the TI domain, that is for slow rotation (Re $=$ 30) and for Hartmann numbers exceeding Ha $=$ 180, the eddy viscosity becomes very small. Again the values saturate for increasing magnetic field amplitudes. The normalized eddy viscosities, however, prove to be very small. The small values of the eddy viscosity seem to be

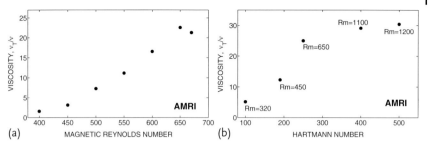

Figure 6.13 The viscosity ν_T/ν for $\Omega > \Omega_A$ (AMRI). For fixed Hartmann number (Ha = 250) a maximum of ν_T/ν exists (a). The curve of the maxima of ν_T/ν vs. Hartmann number (b). The data are marked with the Reynolds number where the eddy viscosity is largest. $\mu_B = 1$, $r_{in} = 0.5$, $\mu_\Omega = 0.5$, Pm = 1. From Rüdiger, Gellert and Schultz (2009). Copyright © 2009 RAS.

characteristic for the considered magnetic-dominated TI regime. It is thus doubtful whether the Tayler instability of strong fields proves to be important for the possible decay of a differential rotation.

6.3.3.2 Turbulent Diffusivity

In the quasi-linear theory the turbulent magnetic diffusivity is a simpler quantity than the eddy viscosity. The total eddy viscosity is formed by both a Reynolds stress part and a Maxwell stress part. This is not the case for the turbulent diffusivity. For a turbulence field with a correlation time τ_{corr} one obtains for the eddy viscosity and the turbulent diffusivity

$$\nu_T \simeq \left(\frac{2}{15}\langle u^2 \rangle + \frac{1}{3}\frac{\langle b^2 \rangle}{\mu_0 \rho}\right)\tau_{corr}, \quad \eta_T \simeq \frac{\tau_{corr}}{3}\langle u^2 \rangle \tag{6.13}$$

(Vainshtein and Kichatinov, 1983). The magnetic fluctuations do not contribute to the magnetic diffusivity. Equation (6.13) suggest that in (nonrotating) turbulent magnetized fluids the effective magnetic Prandtl number always exceeds the minimum value 0.4.

It is standard to express the turbulence-induced electromotive force (EMF) as

$$\mathcal{E} = \langle u \times b \rangle = -\eta_T \operatorname{curl} \bar{B} . \tag{6.14}$$

Here the eddy diffusivity η_T is considered as a positive scalar. In cylindrical geometry the mean current curl $\bar{B}$ may have only a z-component. Hence, along the axis $\mathcal{E}_z = -\eta_T \operatorname{curl}_z \bar{B}$. For $\mu_B = 1$ the electric current through the fluid is positive. Then the expected $\mathcal{E}_z$ must be negative in order to produce positive η_T.

The mode $m = 1$ is the only linearly unstable mode. Higher m only appear due to nonlinear interactions. The resulting spectrum is rather steep. While the energy of $m = 1$ is around 1% of the energy of the external field, the mode $m = 6$ is already four orders of magnitude weaker than $m = 1$. For Re = 500 only around 50 rotations are required before a steady state is reached. Figure 6.14 shows the resulting pattern of the axial EMF for a simulation of the instability with Ha = 200

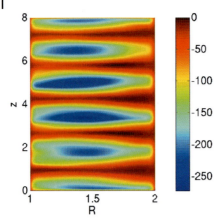

Figure 6.14 The axial EMF (normalized with $\nu B_{in}/R_0$) is always negative. $\mu_B = 1$, $r_{in} = 0.5$, Ha = 200, Pm = 0.01. From Gellert and Rüdiger (2009a). Copyright 2009 by The American Physical Society.

for the small magnetic Prandtl number Pm = 0.01. The EMF is everywhere negative. The cells prove to be rather flat for Pm around unity and their size increases with decreasing Pm.

We take from Figure 6.15a that for different Pm the averaged EMF scales as E/Pm, with the factor $E \simeq 1.2$. The resulting EMF in physical units is

$$\mathcal{E}_z = -E\eta \frac{B_{in}}{R_0}, \qquad (6.15)$$

independent of the molecular viscosity ν. The negative EMFs lead to positive η_T. From (6.14) and (6.15) one finds $\eta_T/\eta \simeq 1.5 E \simeq 1.8$ for all Pm and for stationary cylinders.

Figure 6.15b gives the resulting values of η_T/η from simulations with various magnetic Prandtl numbers Pm without rotation but also with (differential) rotation. They are of the expected order. If, however, the very weak trend to smaller values for smaller Pm is real, the expected value of η_T for sodium or gallium should be only 10% of the molecular value. Such small values also result from the few experiments in laboratories which have been done so far (Reighard and Brown, 2001; Monchaux et al., 2007; Frick et al., 2008; Noskov et al., 2012).

Both the eddy viscosity and the turbulent diffusivity grow with the magnetic Reynolds number Rm, but the Rm-dependence of η_T is much weaker than that of ν_T. Hence, for slow rotation the effective magnetic Prandtl number ν_T/η_T depends strongly on the rotation. It is smaller than unity for slow rotation and it is larger than unity for fast rotation. The reason is that the magnetic fluctuations strongly grow with increasing Rm, and from (6.13) is known that the magnetic fluctuations form an extra part of the eddy viscosity.

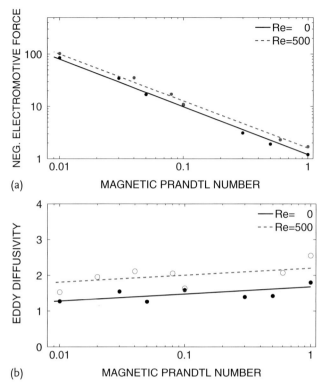

Figure 6.15 The negative axial EMF vs. the magnetic Prandtl number Pm averaged over the entire container (a). The lower curve is for $\Omega = 0$, the upper curve is for Re = 500 and $\mu_\Omega = 0.35$. The ratio η_T/η vs. Pm for Ha = 200 and $\mu_B = 1$ (b). The lower curve is without rotation, the upper curve is with differential rotation included (Re = 500, $r_{in} = 0.5, \mu_\Omega = 0.35$).

6.3.3.3 Mixing of Chemicals

The lithium at the surface of cool main-sequence stars decays with a timescale of 1 Gyr. It is burned at temperatures in excess of 2.6×10^6 K which exists below the base of the convection zone. There must be a slow diffusion process to the location of the burning temperature which is only one or two orders of magnitude faster than the molecular diffusion. The molecular diffusion at the bottom of the solar convection zone results as $D \simeq 30 \, \text{cm}^2/\text{s}$ (Barnes, Charbonneau, and MacGregor, 1999), which must be increased to about $10^3 \, \text{cm}^2/\text{s}$ as the consequence of an instability. The plasma velocities for such a diffusion coefficient are by many orders of magnitude smaller than the convective velocities. There is obviously a diffusion mechanism with a diffusion coefficient $D_T > D$ forming a Schmidt number Sc with

$$\text{Sc} = \frac{\nu_T}{D_T} > 1 \tag{6.16}$$

(see Zahn, 1989, 1992). Eggenberger, Maeder, and Meynet (2005), on the basis of the dynamo scenario with a toroidal magnetic field of about 100 G derived Sc $\simeq$ 100.

Considering the transport processes in the radiation interior of massive (15 $M_\odot$) main-sequence stars Maeder and Meynet (2003, 2005) computed viscosities up to 10^{13} cm^2/s for a rotational velocity of 300 km/s. An extreme chemical mixing is the immediate consequence of Sc = $O(1)$. Brott et al. (2008) demonstrate that the corresponding chemical mixing in massive stars must even be neglected to understand the observations. The real Schmidt number must thus be larger than unity. Often the turbulent magnetic diffusivity has been used as the value for D_T so that the Schmidt number equals the magnetic Prandtl number. Note that the material mixing and also the dissipation of the magnetic field only results by the action of the kinetic part of the momentum transport tensor rather than by its magnetic part so that

$$D_T = \frac{1}{6} \int_0^\infty \int_{-\infty}^\infty \frac{Dk^2 E(k,\omega)}{\omega^2 + D^2 k^4} dk d\omega . \qquad (6.17)$$

There is no magnetic influence in the diffusion equation except the magnetic suppression of the correlation tensor of the fluctuations. If the correlation time of the instability scales with its growth time then the coefficient of the diffusion of chemicals must be $\langle u_R^2 \rangle / \omega_{gr}$ for small molecular diffusion, that is $D \to 0$. For strong fields one can estimate $\omega_{gr} \simeq \Omega_A$. Inserting numbers ($D_T \simeq 10^3$ cm^2/s, $\Omega \simeq 10^{-6}$ s^{-1}) leads to radial velocities of order 1 mm/s. To produce such a very weak turbulence intensity in the radial direction the stable density stratification below the convection zone may suppress the vertical turbulence component so that a highly anisotropic turbulence field evolves (Vincent, Michaud, and Meneguzzi, 1996; Toqué, Lignières, and Vincent, 2006). It has been shown, however, that due to the stellar rotation even a strictly horizontal flow pattern obtains radial components which are able to transport chemicals through the layer below the convection zone. On the other hand, the mixing coefficient (6.17) within the convection zone is rotationally quenched by the basic rotation. The combination of both effects lead to the result that the Li abundance at the stellar surface of young stellar clusters should be correlated with the rotation rate of the star or – what is almost the same – that the depletion becomes anticorrelated with Ω (Rüdiger and Pipin, 2001). Xing, Shi, and Wei (2007) report such a relation for young and low-mass main-sequence stars. With a large but much more diverse sample Strassmeier et al. (2012, their figure 16) derives a relation $n(\text{Li}) \propto \Omega^{0.6}$ (also for single stars) which seems to match the reported scenario.

6.4
The Tayler Generator

Figure 6.16 shows the instability pattern for TI with $\Omega = 0$. The azimuthal wave number of the modes is $m = \pm 1$. For a purely toroidal field, modes with $\pm m$, corresponding to left and right-handed spirals, are degenerate, and necessarily have exactly the same growth rate curves. This is not a surprise as the underlying background field also does not have a helical structure.

Even though $m = \pm 1$ are degenerate, the numerical solutions do not consist of equal mixtures of both modes. Instead, either the left or the right mode wins out, and completely suppresses the other. Which mode one obtains depends on the initial conditions. If these already favor one mode, then that one wins, but if the initial condition is evenly balanced between the two modes, it is ultimately just numerical noise that determines which mode wins.

Spontaneous parity-breaking bifurcations of this type are known in nonmagnetic Taylor–Couette flow (e.g., Hoffmann et al., 2009; Altmeyer et al., 2010), but are almost unknown in magnetohydrodynamic problems. Other examples are given by Chatterjee et al. (2011) and Bonanno et al. (2012). We shall discuss several possibilities to produce both kinetic and current helicity in magnetized TC flows.

In the majority of the models the radial profile of the toroidal field was prescribed. The simplest way to obtain a natural radial profile is to consider the result of an axial shear $d\Omega/dz$ acting on a given uniform axial field B_0 (Braithwaite, 2006). If the induced toroidal field B_ϕ becomes strong enough a bifurcation can be observed leading to a growing nonaxisymmetric field. The largest nonaxisymmetric mode is $m = 1$. Not only the magnitude of B_ϕ must be strong enough but also

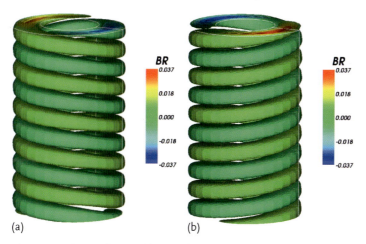

Figure 6.16 The spirals excited by a purely toroidal background field. The two modes are equivalent; their kinetic helicity is $\pm 6.0 \times 10^{-4}$ and their current helicity is $\pm 3.5 \times 10^{-3}$ (both in units of $\Omega_A^2 R_0$), where one spiral is positive (a) and the other spiral negative (b). The modes do not drift in the azimuthal direction. $\mu_B = 1$, Re $= 0$, Ha $= 200$, Pm $= 1$. From Gellert, Rüdiger, and Hollerbach (2011). Copyright © 2011 RAS.

6 The Tayler Instability (TI)

a certain limit B_ϕ/B_0 must be exceeded as an additional poloidal field component suppresses the instability. As an illustration we mention that for Pm = 15 the instability sets in at a Lundquist number of the axial field $S = B_0 R_{out}/\sqrt{\mu_0\rho}\eta = 20$, where the maximum of the induced toroidal field reaches the value of B_0 (Gellert, Rüdiger, and Elstner, 2008). The excitation condition for lower magnetic Prandtl numbers are given in Figure 6.17. In this representation the critical Lundquist numbers of the axial background field for given Rm of the rotation are shown which for small Pm seem to converge. For the smallest possible rotation rate, that is for the minimum of Rm an axial field with $S \simeq 10$ is required to excite the instability.[2]

As before the nonaxisymmetric parts of the flow and the field are considered as the fluctuations. Boundary effects at the top and bottom of the container induce additional flow disturbances and change the structure of the magnetic field. Only the central region around the midplane is thus considered. In Figure 6.18 the resulting energies of both the kinetic and magnetic fluctuations in dependence on the external field B_0 are shown at $R = 0.5 R_{out}$. Note how strongly the magnetic energy dominates the kinetic energy. Their ratio varies between 15 and 30. The instability is clearly dominated by the magnetic field components. The energy q of the magnetic fluctuations to the energy of the external field B_0 results to $q \simeq 16$ which is very close to the magnetic Reynolds number of the fluctuations, $Rm' = u_{rms} R_{out}/\eta \simeq 22$ (see Section 3.1).

The question is whether also the current helicity dominates the kinetic helicity according to their definitions

$$\mathcal{H}_{kin} = \frac{1}{2\pi} \oint \mathbf{u} \cdot \text{curl } \mathbf{u} \, d\phi , \quad \mathcal{H}_{curr} = \frac{1}{2\pi\mu_0\rho} \oint \mathbf{b} \cdot \text{curl } \mathbf{b} \, d\phi , \quad (6.18)$$

which after a transition time takes rather stationary values (Figure 6.19a). Indeed, the current helicity in the center of the cylinder is about 40 times the kinetic helicity.

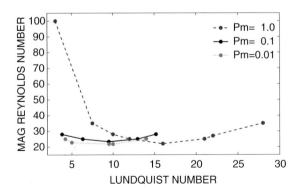

Figure 6.17 The critical magnetic Reynolds number of the rotation vs. the Lundquist number of the axial field for a magnetized container with axial shear and axial field. $r_{in} = 0$.

2) The model has no inner cylinder.

For positive axial shear $\mathcal{H}_{\text{kin}}$ is negative and $\mathcal{H}_{\text{curr}}$ is positive. For the opposite shear $d\Omega/dz$ the signs of $\mathcal{H}_{\text{kin}}$ and $\mathcal{H}_{\text{curr}}$ also change, whereas the final result does not depend on the sign of B_0. Obviously, a possible relation between helicity and both the external field and differential rotation is

$$\mathcal{H} \propto \bar{B}_i \bar{B}_j \Omega_{i,j}, \qquad (6.19)$$

where $\bar{B}_i$ means the axial external field and $\Omega_{i,j}$ the axial shear of the basic rotation. The sign of the pseudoscalar (6.19) does *not* depend on the sign of the magnetic field but does depend on the sign of the shear. The relation (6.19) requires a quadratic helicity law, $\mathcal{H} \propto B_0^2$ which is indeed confirmed by the simulations.

The numerical values given in Figures 6.18 and 6.19 also allow to estimate the correlation coefficients of the kinetic and the current helicity, that is

$$c_{\text{kin}} = \frac{\mathcal{H}_{\text{kin}} R_{\text{out}}}{u_{\text{rms}}^2}, \qquad c_{\text{curr}} = \frac{\mathcal{H}_{\text{curr}} R_{\text{out}}}{b_{\text{rms}}^2}. \qquad (6.20)$$

In both cases the result is about unity. As can be expected for the considered quasi-laminar flows, the correlations are rather strict. Hence, in the container with axial differential rotation helicities are produced according to the simple relations $\mathcal{H}_{\text{kin}} \simeq u_{\text{rms}}^2/R_{\text{out}}$ and $\mathcal{H}_{\text{curr}} \simeq b_{\text{rms}}^2/R_{\text{out}}$, where the latter clearly dominates.

Nonvanishing helicity indicates the existence of an α effect on the basis of TI. It is known that in the presence of an external field $\bar{B}$ and with large magnetic Reynolds number, a dominating contribution from magnetic field fluctuations to the α effect may arise (Pouquet, Frisch, and Léorat, 1976; Brandenburg and Subramanian, 2007). It should have the same sign as the current helicity and/or the

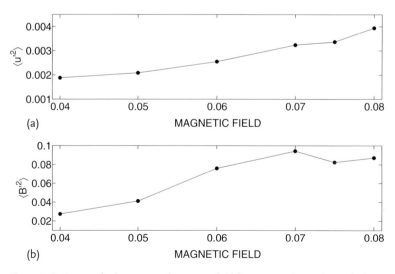

Figure 6.18 Energy of velocity (a) and magnetic field fluctuations (b) vs. the applied magnetic field B_0. $r_{\text{in}} = 0$, $\text{Pm} = 15$. From Gellert, Rüdiger, and Elstner (2008). Reproduced with permission © ESO.

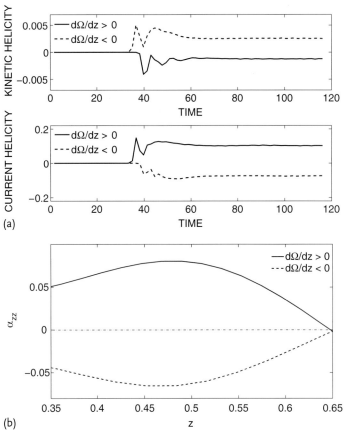

Figure 6.19 $\mathcal{H}_{\text{kin}}$ and $\mathcal{H}_{\text{curr}}$ (a) in the center of the generator ($R = R_{\text{out}}/2$). The external field values are $B_0 = 0.05$ (in units of $\sqrt{\mu_0 \rho} U_\phi$) for negative and $B_0 = 0.06$ for positive shear. The α_{zz} has the same sign as $\mathcal{H}_{\text{curr}}$ and $d\Omega/dz$ (b). Pm = 15. From Gellert, Rüdiger, and Elstner (2008). Reproduced with permission © ESO.

opposite sign as the kinetic helicity. For Figure 6.19b the axial component α_{zz} is directly computed for positive and for negative shear via the z-component of the electromotive force $\mathcal{E} = \langle \mathbf{u} \times \mathbf{b} \rangle$. One finds the same signs for α_{zz} and $d\Omega/dz$. The α_{zz} always has the same sign as the current helicity and the opposite sign as the kinetic helicity. For the correlation coefficient one finds $|\langle \mathbf{u} \times \mathbf{b} \rangle_z|/(u_{\text{rms}} b_{\text{rms}}) \simeq 0.25$, also indicating that an almost laminar flow pattern is considered. In the sense of an order-of-magnitude estimation the normalized α becomes

$$C_\alpha = \frac{\alpha_{zz} R_{\text{out}}}{\eta_{\text{T}}} \propto \frac{\alpha_{zz}}{u_{\text{rms}}} \simeq 1, \tag{6.21}$$

which is surprisingly high compared with the results from other models. Considering the Alfvén frequency for the *toroidal* field one finds this to be in the strong-field regime, $\Omega_{\text{A}} > \Omega$.

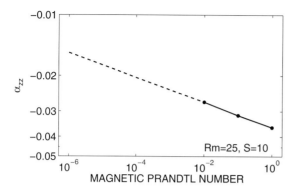

Figure 6.20 The axial component of the α-tensor normalized with the maximal linear velocity $R_{out}\Omega$ of the rotating endplate. The Pm-dependence is rather weak. Courtesy of M. Gellert.

The relation (6.21) has interesting consequences. The potential difference between the endplates in the axial direction is

$$\Delta\Phi = 10^{-8}\alpha_{zz} B_0 R_{out} \qquad (6.22)$$

if the height of the container simply equals its radius. The quantities in (6.22) are to be measured in volt, gauss and centimeter. With (6.21) it follows

$$\Delta\Phi = 10^{-8} C_\alpha \eta B_0 . \qquad (6.23)$$

Hence, a resistivity for liquid metals of order 10^3 cm^2/s (sodium) and an amplitude of 100 G for the axial field as used in PROMISE leads to a potential difference of $\Delta\Phi \simeq C_\alpha$ (in millivolt) which with C_α of order unity can easily be measured. The container filled with gallium or sodium acts as a generator of potential differences between its endplates (via the axial α effect) if an axial field of order 100 G exists and (only!) one of its endplates must be forced to rotate. This is why this (theoretical) constellation may be called a 'Tayler generator' which may serve as the presently best known α experiment without fixed helicity. No helicity is mechanically prescribed by the technical construction of the container.

Numerical results for the dimensionless quantity $\alpha_{zz}/R_{out}\Omega$ are given in Figure 6.20. The simulations only cover the values Pm $= 1$ down to Pm $= 0.01$. The resulting Pm-dependence is very weak. The dashed line suggests an extrapolation to much smaller magnetic Prandtl numbers. If this would be allowed then $\alpha_{zz} \simeq 0.02 R_{out}\Omega$ which yields a potential difference of $\Delta\Phi = 2 \times 10^{-10}\eta^2 \sqrt{\mu_0\rho}\mathrm{Rm}\ S/R_{out}$ between the endplates if again the height of the container equals its radius. For $R_{out} = 10$ cm this expression provides for sodium about 11 mV. For longer containers the potential difference grows linearly.

6.5
Helical Background Fields and Alpha Effect

The two modes with $m = \pm 1$ in Figure 6.16 resulting from the instability of toroidal fields possess the same critical eigenvalue and also the same growth rates for supercritical Ha. The modes with $\pm m$ form kinetic (and current helicity) with opposite signs so that the total helicity should vanish. The reason is that $\mathcal{H}_{\text{kin}}$ and $\mathcal{H}_{\text{curr}}$ are pseudoscalars which require the existence of a finite pseudoscalar in the undisturbed constellation. The well-known pseudoscalar product $\boldsymbol{g} \cdot \boldsymbol{\Omega}$ plays this role in rotating convection zones but in the containers considered here it does not exist.

6.5.1
Helical Fields with Weak Axial Current

If the background field has mixed poloidal and toroidal fields (Tayler, 1980) then there is another pseudoscalar on which helicity (and α effect) may be based. If the toroidal field is not current-free and the background field has an axial component then there is the pseudoscalar $\boldsymbol{B} \cdot \text{curl}\, \boldsymbol{B}$ which is positive, negative or even zero. For finite axial component B_0 it makes sense to work with the normalization (5.49) of the toroidal field. Then one finds for the large-scale current helicity of the background field the expression

$$\boldsymbol{B} \cdot \text{curl}\, \boldsymbol{B} = 2 a_B B_0 = \frac{4\beta}{3} \frac{B_0^2}{R_{\text{out}}}, \tag{6.24}$$

the latter for $r_{\text{in}} = 0.5$ and $\mu_B = 1$. The sign of β determines the sign of the imposed global current helicity. Positive (negative) β represents right-hand (left-hand) spirals. It is also clear from (5.45) that perturbations with negative m describe right-hand spirals (marked by R) and negative m describe left-hand spirals (marked by L). Of course, all results are invariant against the simultaneous transformation $m \to -m$ and $\beta \to -\beta$. Both signs of (6.24), therefore, yield identical instability maps but for different m corresponding to different spirals. If the background field has a helical structure the system no longer exhibits the $\pm z$ symmetry. For axisymmetric modes, the consequence of this is that previously stationary modes become oscillatory, that is traveling wave modes as realized in HMRI. For nonaxisymmetric modes, the breaking of the $\pm z$ symmetry of the basic state breaks the $\pm m$ symmetry of the instability pattern. This corresponds to the fact that modes spiraling either in the same or the opposite sense to the spiral structure of the background field are indeed different. The resulting flow pattern will exhibit a net helicity of a well-defined sign. This $\pm m$ symmetry breaking is also a convenient distinguishing feature between the AMRI and TI; for the AMRI the most unstable magnetic mode spirals in the opposite sense to the imposed field, whereas for the Tayler instability it spirals in the same sense. We shall see, however, that for both cases the kinetic helicity of the unstable fluctuations averaged over ϕ has the same sign.

It is also clear from this consideration that for purely toroidal fields the $\pm m$ symmetry of the instabilities is not broken, so that their net kinetic helicity must always vanish. One would be tempted to suggest that such a simple magnetic configuration cannot work as a dynamo. This finding will also hold for magnetic field configurations where the toroidal component strongly dominates the poloidal one.

A helical background field may have positive polarity, so that it has also a positive current helicity, that is it spirals to the right. If the axial field is weak, for example with $\beta = 100$, then the marginal instability curves for a *steep* rotation law shown in Figure 6.21a strongly resemble the map for $\beta \to \infty$ (see Figure 6.9). The main differences are the slightly smaller Hartmann number of the toroidal field, and the splitting of the spiral modes $m = 1$ and $m = -1$ into two curves with different handedness (L and R). The left-hand modes require a faster rotation than the right-hand modes. For background fields with positive current helicity, we thus find that the right spirals are preferred, whereas for background fields with negative current helicity, R and L would be exchanged, and the left spirals would be preferred.

For $\beta = 10$ the differences between the L and R modes for given m increase, so that the pure Tayler instability exists only as the 1R mode. The 1L mode no longer connects to Re $= 0$, and is not the most unstable mode anywhere in the given domain (Figure 6.21b).

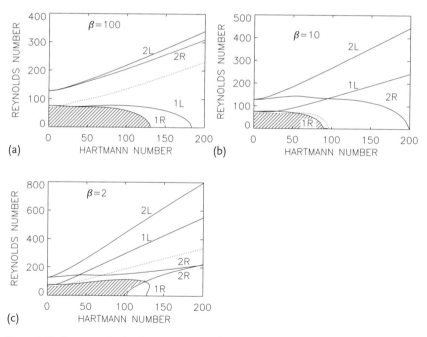

Figure 6.21 Steep rotation law ($\mu_\Omega = 0$); instability maps for $\beta = 100$ (a), $\beta = 10$ (b) and $\beta = 2$ (c). The curves are marked with their m. The notation R (right spiral) stands for negative m and the notation L (left spiral) for positive m. The shaded areas are the stable regions. Without magnetic fields the $m = 0$-curves (dotted) always start at Re $= 68$. Pm $= 1$, $\mu_B = 1$. Copyright 2010 by The American Physical Society.

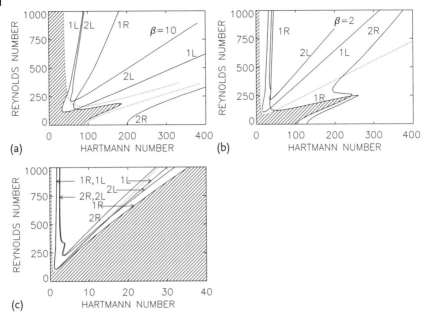

Figure 6.22 Flat rotation law ($\mu_\Omega = 0.5$): $\beta = 10$ (a), $\beta = 2$ (b) and $\beta = 0.1$ (c). Notations as in Figure 6.21. From Rüdiger et al. (2010). Copyright 2010 by The American Physical Society.

The 1R mode ($m = -1$) also dominates for β of order unity. There is, however, an interesting particularity in this case. For very slow rotation, a 2R mode reduces the stability domain. For Re $\simeq 0$, and in a limited range of Ha (Ha $\simeq$ 100–130), this mode forms the first instability (see Bonanno and Urpin, 2008). Even rather weak differential rotation, however, brings the system back to the 1R instability.

For models with helical field and steep rotation law we find the expected splitting between right and left spiral instabilities also for the preferred modes. If the axisymmetric background field is right-handed, then the first unstable mode is also right-handed, that is it possesses positive current helicity. Note how in Figure 6.21 an increasing β brings the characteristic modes with $m = -1$ and $m = 1$ more and more together until they finally merge for $\beta \to \infty$.

For a sufficiently *flat rotation law* the nonmagnetic Taylor vortices disappear. For $\mu_\Omega = 0.5$ the instability curves for purely toroidal fields are given by Figure 6.22. Both the instabilities appear in this case: TI exists in the lower right corner, and AMRI exists in the upper left corner. The two instabilities are separated by a stable branch with Re $\approx$ Ha, where the differential rotation stabilizes the TI.

For $\beta \to \infty$ the modes with positive and negative m are again degenerate (not shown). The AMRI solution with the lowest Reynolds number is a nonaxisymmetric L-mode with positive m. This remains true for helical background fields with large β, but for $\beta \lesssim 1$ the $m = 0$ mode dominates the instability with the lowest Reynolds number (Figure 6.22a,c) – as is true for the standard MRI and HMRI. The transition from nonaxisymmetry to axisymmetry can be accomplished simply

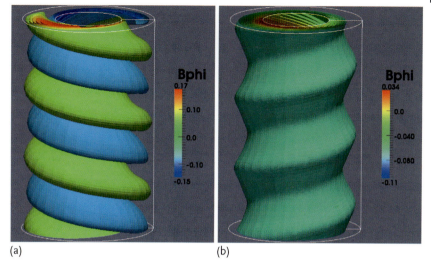

Figure 6.23 The azimuthal components of the magnetic pattern in the AMRI domain (a, Re = 200, Ha = 80) and in the TI domain (b, Re = 30, Ha = 130). The fields are normalized with B_{in}. $\mu_B = 1$, $\mu_\Omega = 0.5$, $\beta = 10$, Pm = 1. Copyright 2010 by The American Physical Society.

by increasing the axial component of the background field. It is thus clear that there is a smooth transition from one form of the MRI to the next.

For purely toroidal fields and no rotation we have Ha = 150 as the critical Hartmann number (Figure 6.12), whose value does not depend on the magnetic Prandtl number. An additional weak axial field component (which does not dominate the toroidal component) reduces the critical Hartmann number to about 100. The most unstable mode is here 1R for $\beta \geq 10$, but is 2R for β of order unity. This result also holds for very weak differential rotation; but for faster rotation the modes higher than $m = 1$ lose their importance for the transition from stability to instability.

The different mode pattern is the characteristic difference to the preferred modes in the TI domain and in the AMRI domain. For background fields with positive current helicity the TI favors instability patterns with right spirals (Figure 6.23). The instability curves of the AMRI limit are more complex. For large β it is formed by the nonaxisymmetric modes with left-handed spirals, while for small β the axisymmetric mode prevails. Consequently, the slopes of the lines change from positive for the nonaxisymmetric modes to negative for the axisymmetric modes (Figure 6.22c). One finds that even the transition from AMRI to standard MRI can be achieved by variation of β.

6.5.2
Uniform Electric Current

For $b_B = 0$ in the definition (5.40) the field in the fluid is due to a uniform current (while it is current-free in the fluid for $a_B = 0$). We are interested in the linear

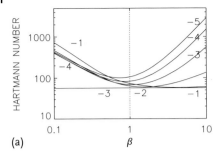
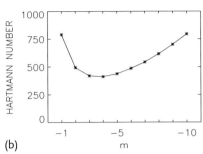
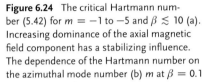

Figure 6.24 The critical Hartmann number (5.42) for $m = -1$ to -5 and $\beta \lesssim 10$ (a). Increasing dominance of the axial magnetic field component has a stabilizing influence. The dependence of the Hartmann number on the azimuthal mode number (b) m at $\beta = 0.1$ (the left vertical axis of the left plot). The thin horizontal line marks the absolute minimum value 29 for Ha. $r_{\rm in} = 0.5$, $\mu_B = 2$, all Pm. From Rüdiger, Schultz, and Elstner (2011). Reproduced with permission © ESO.

stability of such a background field $\bar{B} = (0, \beta R/R_{\rm in}, 1) B_0$, with $B_0 = $ const under the influence of the flow $\bar{U} = (0, R\Omega(R), 0)$. The cylinders are considered as perfect conductors. For $r_{\rm in} = 0.5$ it follows that $\mu_B = 2$. For a nonrotating container the critical Hartmann number (5.42) does not depend on Pm. Without rotation an instability pattern never drifts in the azimuthal direction; the azimuthal drift of the nonaxisymmetric instability pattern vanishes.

The critical Hartmann number for vanishing B_z has the value of 35 for $m = 1$. Figure 6.24 shows that this value is (slightly) reduced if a small and uniform axial component of the magnetic field is added to the system. Hence, a uniform axial field supports the instability of the toroidal field. This effect, however, is rather weak: the critical Hartmann number sinks from about 35 only to 29. For $m > 1$ the destabilization of the toroidal field by axial fields is much stronger so that, for β close to one, all the modes with different m have more or less the same critical Ha.

For $\beta = 4$ we find Ha $= 29$ as the absolute minimum of the stability curve for $m = -1$. For stronger axial fields, the critical Hartmann number basically grows to reach values of about 500 for $\beta \simeq 0.05$. For strong axial fields, modes with $|m| > 1$ possess lower critical Hartmann numbers than the mode with $m = -1$. The differences in the curves with various m are much smaller than those for weak B_0, but the Fourier component with $m = -4$ possesses the lowest critical Hartmann number for $\beta = 0.05$. By a dominating axial field the toroidal field is strongly stabilized in comparison with the situation for purely toroidal fields (see Figure 6.24a). In the limit $\beta \to 0$, of course, the critical magnetic field amplitude goes to ∞.

The maximal stabilization for strong axial field component appears for $m = -1$ so that the most unstable modes have azimuthal mode numbers $|m| > 1$. If B_ϕ and B_z are of the same order, then B_ϕ is more unstable than for very weak or very strong axial fields B_z. Obviously, rather strong toroidal fields can be stored in the container with strong enough uniform axial magnetic fields.

There is a new situation under the presence of global rotation and, in particular, of nonuniform global rotation. The results for the rather flat rotation law with

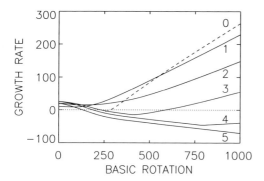

Figure 6.25 Growth rates (in unit of the diffusion frequency) vs. Rm for the modes marked with $|m|$ for differential rotation. The axisymmetric mode (dashed) dominates for fast rotation. $\mu_\Omega = 0.5$, $\mu_B = 2$, $\beta = 1$, Ha = 80, Pm = 1. Reproduced with permission © ESO.

$\mu_\Omega = 0.5$ is here applied to the helical magnetic field with almost equal toroidal field and axial field components. The growth rates for a fixed Hartmann number (Ha = 80) are given in Figure 6.25. For slow rotation all modes are rotationally stabilized. The stabilization is stronger for higher values of the mode number m. For $\Omega \gtrsim \Omega_A$, however, the magnetic instability is re-animated at most for the mode $m = -1$. Finally, this mode becomes dominant, because its growth rate (here in diffusion units) becomes increasingly large, scaling with the rotation frequency itself. The growth rate grows with increasing Ω rather than with Ω_A when $\Omega > \Omega_A$. For differential rotation the higher modes dominate only for small Reynolds numbers. They do not contribute to the instability for high Reynolds numbers because they are dampened by too strong differential rotation.

Figure 6.25 also gives the growth rate of the mode $m = 0$ which for fast rotation possesses the highest growth rate. In this case the growth rate scales with Ω itself. For fast rotation ($\Omega \gg \Omega_A$) the most interesting case of $|B_\phi| \simeq |B_z|$ leads to a dominance of axisymmetric modes quite similar to the result for SMRI. There is always an intersection between the growth rates of $m = 0$ and $m = -1$. Left of this point, the nonaxisymmetric mode dominates the axisymmetric one, and vice versa to the right.

The dependence of the growth rates on the magnetic Prandtl number is not trivial. In Figure 6.26a the growth rates without rotation, and with differential rotation (Figure 6.26b) are given for a fixed Hartmann number. The growth rates of the mode $m = -1$ and the global rotation rate are normalized with $\omega_{\bar{\eta}} = \bar{\eta}/R_0^2$ ($\bar{\eta} = \sqrt{\nu\eta}$). By this formulation one finds that Pm = 1 always leads to maximum growth rates independent of the rotation rate. Either small or large magnetic Prandtl numbers lead to slower growth of the instability. That means that an instability found with numerical simulations for Pm = 1 does not automatically also exist for much smaller or much larger Pm. This is a restriction on the validity of the numerical simulations of the magnetic instabilities. For both stationary or rotating fluids the fields are most unstable (measured via growth rates) for Pm = 1.

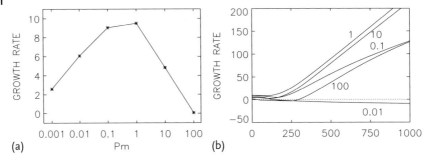

Figure 6.26 Growth rates $\bar{\omega}_{\rm gr}$ for the modes $m = -1$ for fixed Hartmann number and for various Pm. No rotation (a). Growth rates vs. $\overline{\rm Rm}$ for differential rotation (b). The curves are marked with their magnetic Prandtl number. Conducting cylinders, $\mu_\Omega = 0.5$, Ha $= 80$, $\mu_B = 2$, $\beta = 1$. From Rüdiger, Schultz, and Elstner (2011). Reproduced with permission © ESO.

For any given Hartmann number (here Ha $= 80$), one finds two regimes for the rotational influence on the growth rates $\bar{\omega}_{\rm gr}$ in Figure 6.26. There is almost no influence for slow rotation. For a stationary container Figure 6.26 shows a *maximum* growth rate for Pm $= 1$ of about 10. One finds, however, fast differential rotation as basically supporting the instability. From Figure 6.26b the physical growth rate results in $\omega_{\rm gr} \simeq 0.2\Omega$ independent of the diffusivities so that the growth time is rotationally reduced to about one rotation time.

6.5.3
Alpha Effect

A new question is how the different handednesses of the TI and AMRI perturbation patterns transform into the resulting kinetic helicity as well as current helicity. The surprising answer will be that the sign and amplitudes of both the helicities hardly differ for models with slow and fast rotation, or equivalently, for strong and weak magnetic fields. For positive β both helicities are negative and the $\alpha_{\phi\phi}$ becomes positive while the α_{zz} becomes negative.

The profiles of the basic state field and flow may be fixed at our standard values $r_{\rm in} = 0.5$, $\mu_B = 1$ and $\mu_\Omega = 0.5$ but the geometry parameter β will be varied, mapping the whole instability domain from MRI to AMRI/TI. The characteristic amplitudes of the two models are Ha $= 100$ and Re $= 200$ for the first series for fast rotation, and Ha $= 200$, Re $= 20$ for the second one for slow rotation. The first series is thus rotationally dominated ($\Omega > \Omega_{\rm A}$), whereas the second is magnetically dominated ($\Omega_{\rm A} > \Omega$), where the Alfvén frequency is $\Omega_{\rm A} = B_{\rm in}/\sqrt{\mu_0\rho}\,R_0$.

6.5.3.1 The Helicities

For both runs, Figure 6.27 shows the kinetic and magnetic rms values of the flow fluctuations $\sqrt{\langle u^2 \rangle}$ (Figure 6.27a) and the magnetic fluctuations $\sqrt{\langle b/B_{\rm in}\rangle^2}$ (Figure 6.27b, similar to the value $\sqrt{q}$ in Section 3.1). For sufficiently large β its influence is rather weak; the axial component B_0 is then so weak that it has no further

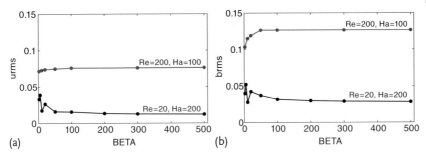

Figure 6.27 The rms values for flow (a) and field (b) fluctuations in units of the Alfvén velocity $R_0\Omega_A$. Note that once β exceeds ~ 100 it has no further influence. $\mu_B = 1$, $r_{in} = 0.5$, $\mu_\Omega = 0.5$, Pm = 1. Copyright © 2011 RAS.

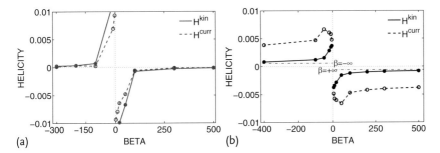

Figure 6.28 The kinetic and current helicities of the nonaxisymmetric perturbations as functions of β for $\Omega_A < \Omega$ (a, Ha = 100, Re = 200) and for $\Omega_A > \Omega$ (b, Ha = 200, Re = 20). The dash-dotted lines indicate the limits $\pm 6 \times 10^{-4}$ of the kinetic helicity of the left-hand and right-hand modes in Figure 6.16. Pm = 1. From Gellert, Rüdiger, and Hollerbach (2011). Copyright © 2011 RAS.

influence. This is not true for small β, where the axial field dominates. For $\beta < 1$ the instability is strongly stabilized and the resulting energies of the perturbations are reduced. In all cases the magnetic energy dominates the kinetic energy by a factor of about 2.5. On the other hand, the ratio q of the energy of the normalized magnetic fluctuations (see Figure 6.27b) complies with the small values for slow and fast rotation which can be read from Figure 6.23. For the two examples given in Figure 6.27 the values q of the energy ratio differ by a factor of ten which is just the ratio of the two Reynolds numbers of the global rotation. Again the normalized energy in the magnetic fluctuations scales with the magnetic Reynolds number. Note that the quantities given in Figure 6.27 result from an averaging procedure over the whole container. The peak values of the magnetic field components can be much larger (see Figure 6.23).

Figure 6.28 shows the kinetic and current helicities for the two series. Both the helicities have the same sign but opposite to the sign of β. That the spirals of the perturbation patterns for slow and fast rotation have an opposite handedness does *not* play a role for the sign of the helicities. For β of order unity, the basic state has a sufficiently strong handedness to force the appearance of a clear parity in the instability pattern. For increasing β (each time using the previous solution as the new initial condition) this parity is preserved to $\beta \to \infty$. Without an axial field, however, the basic state no longer has its own handedness so that both left and right instabilities could simultaneously exist. The initial conditions determine the resulting parities. If they balance each other then the numerical noise will select the winning mode.

One can also find from the data in Figure 6.28 that for slow rotation (TI) the current helicity (in Alfvén velocity units) dominates the kinetic helicity which is not true for fast rotation (AMRI).

6.5.3.2 The Alpha Effect

Because of the structure of the background field it is possible to determine the α effect in both azimuthal and axial directions. According to the rule that the azimuthal α effect is anticorrelated with the (kinetic) helicity we expect the azimuthal α effect to be positive for $\beta > 0$. There are several theories and simulations leading to $\alpha_{\phi\phi}$ and α_{zz} with opposite signs (see Section 3.6 for more details). One finds a similar behavior in many numerical simulations. It also means that any dynamo problem with weak differential rotation cannot be treated with a simple scalar α effect.

Figure 6.29 gives the results for slow and fast rotation. The dimensionless α effect in the form

$$C'_\alpha = \frac{\alpha R_0}{\eta} \tag{6.25}$$

is plotted for the components $\alpha_{\phi\phi}$ and α_{zz} after averaging over the whole container. In both cases $\alpha_{\phi\phi}\beta > 0$ and $\alpha_{zz}\beta < 0$ results almost everywhere in the meridional plane. This anticorrelation between the two components seems not to be caused by the boundaries. The dependence of α on the rotation rate is evidently rather weak. The fact that $\alpha_{\phi\phi}$ is (slightly) smaller for rapid rotation than for slow rotation illustrates just how different these magnetic-induced helicities are from the familiar results for rotating convection. Finally, Figure 6.30 shows how the amplitudes of $\alpha_{\phi\phi}$ vary with β, being roughly inversely proportional for slow and fast rotation. For higher β the α effect (6.25) decreases like C/β with $C \simeq 0.05$. The dynamo number, therefore, becomes

$$C_\alpha = \frac{\alpha R_0}{\eta + \eta_T} = \frac{C}{c\beta} \tag{6.26}$$

with the factor $c = 1 + \eta_T/\eta$. Hence, if it is allowed to replace B_z in the definition of β by B_R and with $|B_\phi|/|B_R| \simeq \sqrt{C_\Omega/C_\alpha}$ one finds for the dynamo number

$$C_\alpha C_\Omega = \frac{C^2}{c^2}, \tag{6.27}$$

which with $C \simeq 0.05$ and $c > 1$ never exceeds unity and is thus certainly too small for dynamo excitation. It is clear, furthermore, that because of $C_\alpha < 1$ an α^2-dynamo cannot work either. The current-driven instability of helical background

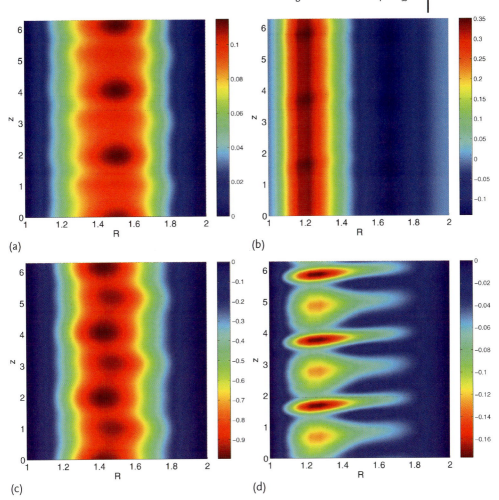

Figure 6.29 The normalized components C'_α of the α-tensor for slow rotation ($\Omega_A/\Omega = 10$; a,c) and fast rotation ($\Omega_A/\Omega = 0.5$; b,d). $\alpha_{\phi\phi}$, positive in both cases (a,b), α_{zz}, negative in both cases (c,d). $\beta = 3$, $\mu_\Omega = 0.5$, $r_{in} = 0.5$, $\mu_B = 1$, Pm = 1. Copyright © 2011 RAS.

fields, therefore, does produce small-scale helicity and α effect, but the resulting numerical values seem to be too small for the operation of large-scale dynamos. According to Figure 6.30 this negative statement holds for both slow and fast rotation.

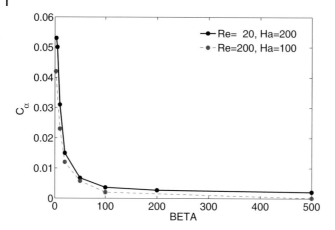

Figure 6.30 The C_α for $\alpha_{\phi\phi}$ for slow and fast rotation. For higher β the α effect decreases like C/β with $C \simeq 0.05$. $\mu_\Omega = 0.5$, $\mu_B = 1$, Pm $= 1$. From Gellert, Rüdiger, and Hollerbach (2011). Copyright © 2011 RAS.

6.6
TI with Hall Effect

If the core of a supergiant rotates rapidly the neutron star will be born after the SN explosion as a fast rotator with an angular velocity near the break-up value, that is 1 kHz. This value exceeds the rotation rate of the fastest young pulsars known by one order of magnitude, so that the question is how a critically rotating protoneutron star (PNS) spins down. One possibility is the angular momentum loss by gravitational wave emission via unstable r-modes (Friedman and Schutz, 1978; Andersson, 1998; Stergioulas and Font, 2001; Lindblom, Tohline, and Vallisnen, 2001). As the viscous damping of the r-modes is smallest at temperatures around 10^9 K, this instability works best as long as the neutron star remains hot.

Another transport of angular momentum from core to envelope can be due to magnetic instabilities. If a toroidal field is induced by the interaction of a differential rotation and a poloidal fossil field then the questions are whether the toroidal field becomes unstable and whether the field and flow fluctuations form an effective Reynolds and/or Maxwell stress. The induced toroidal field also has a dipolar symmetry with respect to the equator, that is the signs of the field belts in the two hemispheres are opposite, which ordinarily has no consequences, unless the instability occurs in the presence of the Hall effect. In this case it might be possible that the onset of the instability, its growth rate and the resulting wave numbers differ in the two hemispheres of the star. If the fields are strong, as known from PNS, and the radiation is influenced by the geometry and amplitude of the magnetic field, then one cannot expect that the star appears equatorially symmetric. Schwope et al. (2005) found an equatorially asymmetric X-ray brightness of the isolated neutron star RBS1223. They assumed the existence of one bright 'spot' in each hemisphere and found two temperature maxima of different strengths. If this asymmetry effect

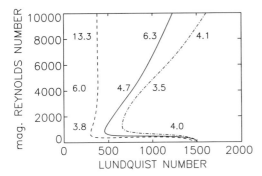

Figure 6.31 Stability limits for $m = 1$ influenced by the Hall effect ($\beta_0 = 0.01$ dashed, $\beta_0 = 0$ solid, $\beta_0 = -0.01$ dot-dashed). The lines are marked with the wave numbers of the marginal instability. Note that the cell structure also strongly depends on the sign of the Hall effect. $r_{in} = 0.5$, $\mu_\Omega = 0.5$, $\mu_B = 1$, Pm = 100. From Rüdiger et al. (2009).

is general for neutron stars then the interior magnetic fields must also be asymmetric with respect to the equator.

The induction equation (2.64) is solved after linearization about the background flow $U_R = U_z = B_R = B_z = 0$ with U_ϕ and B_ϕ as in Section 5.6.1. Again $r_{in} = 0.5$, $\mu_B = 1$ and $\mu_\Omega = 0.5$. The last term in (2.64) represents the Hall effect. As usual the Hall parameter is normalized after (2.67). This parameter can also be considered as the ratio of the magnetic decay time τ_{diff} to the Hall time τ_{Hall}, which for newborn neutron stars only differ by one or two orders of magnitudes, so that the Hall time is rather long and the question is important whether it determines the growth times of the instability or not.

In a very young neutron star with differential rotation the toroidal field results from a poloidal field with dipolar symmetry. The B_ϕ is then also antisymmetric with respect to the equator. One finds that the Hall-TI produces large-scale perturbations with $m = 1$ which, however, include a magnetic flux symmetric with respect to the equator. One can say that the Hall-TI produces an extra quadrupolar component of the toroidal fields from its originally dipolar symmetry. The amplitudes (and the geometry) of the total toroidal field belts are thus different in both hemispheres.

As in (5.34) a nonmagnetic parameter β_0 is introduced via $\beta_0 = R_B/S$ with S as the Lundquist number under use of (5.42). Figures 6.32 show the influence of the Hall effect on the stability of toroidal fields with positive (dashed) and negative (dot-dashed) amplitudes (here represented with the sign of β_0). To mimic the fluids in the crust of neutron stars the calculations have been done for *large* magnetic Prandtl numbers. Without rotation the critical Hartmann number Ha = 150 (for $\mu_B = 1$) is known. For slow differential rotation in both cases the shear leads to a subcritical instability excitation but if it is sufficiently fast, the rotation leads to a positive slope of the curve Rm = Rm(S) representing the rotational suppression. The Hall effect does not change this general picture except each sign of β_0 produces its own stability curve Rm = Rm(S). Positive fields ($\sim$ parallel to the rotation axis) are strongly destabilized by the Hall effect while negative fields ($\sim$ antiparallel to the rotation axis) are strongly stabilized. We are thus confronted with the constellation that the toroidal fields in one hemisphere are destabilized and the toroidal fields in the other hemisphere are stabilized. Whether and how such a magnetic configuration is able to saturate remains an open question. The situation is different from the standard MRI with Hall effect where the orientation of the axial field may vary from star to star but does not vary from hemisphere to hemisphere. If indeed in newborn neutron stars a differential rotation exists which produces strong toroidal fields from a poloidal fossil dipolar field then the two hemispheres cannot be identical with respect to the magnetic geometry. Since for strong magnetic fields the heat transport is strongly influenced by the magnetic geometry, the differences of the two hemispheres should even be observable.

The described differences between the two hemispheres are also reflected by the wave numbers and even the growth rates of the instability. The latter are given in Figure 6.32. They grow with growing Lundquist number. One also finds that the growth times of the instability, despite the Hall effect, scale with the rotation time. The much longer Hall time τ_{Hall} does *not* influence the growth rate of the Hall-TI. Nevertheless, differing signs of the magnetic field lead to different growth rates of the instability where again positive Hall effect acts destabilizing and vice versa. For $S \sim 1000$, Pm ~ 100 and $\beta_0 \sim 10^{-2}$ taken from Figure 6.32 one finds for the Hall parameter $R_B \sim 10$, leading to the typical value $\sim 10^{13}$ G for the neutron star.

We know that strong magnetic fields suppress the heat transport in neutron stars (Schaaf, 1988; Heyl and Hernquist, 2001). The heat transport is blocked in the direction perpendicular to the field lines so that the heat conductivity tensor becomes anisotropic, that is $\chi_{ij} = \chi_1 \delta_{ij} + \chi_2 \bar{B}_i \bar{B}_j$, where χ_1 represents the heat-flux perpendicular to the field which is quenched by strong magnetic fields. With

$$\chi_{ij} = \frac{\chi_0}{1 + R_B^2} \left(\delta_{ij} + R_B^2 \frac{\bar{B}_i \bar{B}_j}{B^2} \right) \tag{6.28}$$

the heat-flux remains finite along the field lines even for $B \to \infty$.

The magnetic-induced anisotropy of the heat-flux tensor is not consistent with a homogeneous surface temperature (Geppert, Küker, and Page, 2006). If the latitu-

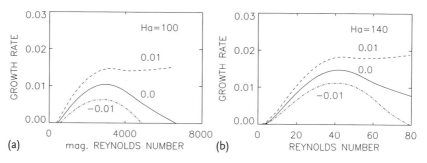

Figure 6.32 The influence of the Hall effect on the growth rates (in units of Ω_{in}) for $m = 1$. The curves are marked by β_0. The values belong to $S = 1000$ (a) and $S = 1400$ (b) in Figure 6.31. $\mu_B = 1$, $r_{\mathrm{in}} = 0.5$, $\mu_\Omega = 0.5$, Pm $= 100$, conducting boundaries.

dinal distribution of the magnetic field is strictly symmetric or antisymmetric with respect to the equator then the surface temperature results as equatorially symmetric. The total magnetic field, however, combines a dipole and a quadrupole under the presence of the Hall-TI. If the differential rotation of the neutron star disappears then the magnetic fields are frozen in, so that the magnetic constellation is conserved for the timescales of the Ohmic decay (also modified by the Hall effect). We would thus expect the two hemispheres of an isolated neutron star to have quite different X-ray activity.

7
Magnetic Spherical Couette Flow

Spherical Couette flow is the flow induced in a spherical shell by imposing a differential rotation between the inner and outer spheres. The simplest configuration consists of a stationary outer sphere and a rotating inner one. In the narrow-gap limit, this yields flow states very similar to the Taylor vortices familiar from cylindrical Taylor–Couette flow (e.g., Mamun and Tuckerman, 1995). For wider shells other flow states having more in common with Kelvin–Helmholtz instabilities arise. Spherical Couette flow is thus an integral part of classical fluid dynamics, and displays a broad variety of solutions even in this simplest possible configuration.

Given its obvious application as an analog of planetary or stellar interiors, it is of interest to extend this 'classical' configuration in two ways. First, rather than having a stationary outer sphere, the geophysically more relevant case is to have both spheres rapidly rotating, with a much smaller differential rotation between them. This leads to the so-called Stewartson layer problem, which we will consider in the following section. Second, and most directly relevant to the focus of this book, is to take the fluid to be electrically conducting, and include magnetohydrodynamic effects. These not only modify the previous nonmagnetic flow states, but result in fundamentally new phenomena having no analog in the nonmagnetic problem. The purpose of this chapter is to review the full range of theoretical, numerical, and experimental work on magnetohydrodynamic spherical Couette flow.

7.1
Stewartson Layers

The classical Stewartson layer problem was first posed by Proudman (1956), and consists of a spherical shell in rapid overall rotation, with additionally an infinitesimal differential rotation between the two spheres. The asymptotic solution was provided by Stewartson (1966), and consists of a shear layer situated on the tangent cylinder C, the cylinder parallel to the axis of rotation and just touching the inner sphere. While it might seem curious that such a cylindrical structure would emerge in a spherical geometry, it arises quite naturally from the Taylor–Proudman theorem, stating that the flow will tend to align itself parallel to the axis of rotation. Outside C the fluid then corotates with the outer boundary, since that satisfies both

Magnetic Processes in Astrophysics, First Edition. G. Rüdiger, L.L. Kitchatinov, and R. Hollerbach.
© 2013 WILEY-VCH Verlag GmbH & Co. KGaA. Published 2013 by WILEY-VCH Verlag GmbH & Co. KGaA.

the Taylor–Proudman theorem as well as the boundary conditions at the two endpoints of a given fluid column. In contrast, inside $\mathcal{C}$ the boundary conditions at the two endpoints are different, making it impossible to satisfy $\partial U/\partial z = 0$ everywhere along the column. The resolution turns out to be that the fluid continues to satisfy $\partial U/\partial z = 0$ in the interior, rotating at a rate intermediate between the inner and outer spheres. The adjustment at the boundaries is accomplished by Ekman layers, of thickness $E^{1/2}$, whereas the jump in angular velocity across $\mathcal{C}$ is accomplished by this Stewartson shear layer, consisting of nested layers of innermost thickness $E^{1/3}$, and outer thicknesses $E^{2/7}$ and $E^{1/4}$. The Ekman number $E = \nu/\Omega\, r_{in}^2$ is an inverse measure of the overall rotation Ω. Figure 7.1, from Hollerbach (1994), shows a numerical solution of the problem, for $E = 10^{-2}$–10^{-5}, and clearly indicates the increasingly narrow shear layer on $\mathcal{C}$. Subsequent numerics by Dormy, Cardin, and Jault (1998) extended this down to $E \approx 10^{-7}$, demonstrating good agreement with many of the expected asymptotic scalings.

Turning next to the magnetic version of this problem, one might intuitively expect the magnetic tension force to suppress the Stewartson layer, at least if the field has a component perpendicular to $\mathcal{C}$. Figure 7.2 illustrates this suppression, for an imposed dipolar field of increasing strength, as measured by the Elsasser number $\Lambda = B_0^2/\Omega\mu\rho\eta$, the ratio of Lorentz to Coriolis forces. The asymptotics of this process were elucidated by Kleeorin et al. (1997), and shown to consist of an intricately nested structure, with the previous three sublayers of thicknesses $E^{1/3}$, $E^{2/7}$ and $E^{1/4}$ all affected at slightly different values of Λ. The primary shear suppression begins once $\Lambda > E^{1/3}$, and once $\Lambda = O(1)$ essentially nothing remains of the original Stewartson layer, as seen also in Figure 7.2. In the Earth's core therefore, where $E = O(10^{-15})$ and $\Lambda = O(1)$, there is almost certainly no Stewartson layer present – despite the fact that there may well be a differential rotation of the inner core (e.g., Deguen, 2012).

Instead of a dipolar field, one could also impose an axial field, which does not have a component perpendicular to $\mathcal{C}$, and therefore might be expected to yield very different results. A similar problem in cylindrical geometry – where the Stewartson layer becomes a *boundary* layer involving thicknesses $E^{1/3}$ and $E^{1/4}$ – was considered asymptotically by Vempaty and Loper (1975, 1978), who showed that for $O(1) <$

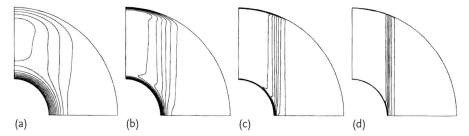

(a)   (b)   (c)   (d)

Figure 7.1 Contours of the angular velocity in the nonmagnetic Stewartson layer problem. The contour interval is 1/15. As shown, $E = 10^{-2}$ (a), 10^{-3} (b), 10^{-4} (c), 10^{-5} (d). From Hollerbach (1994).

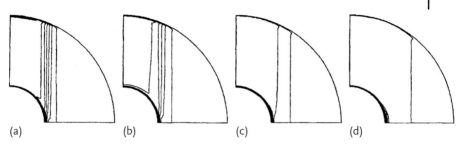

Figure 7.2 As in Figure 7.1, contours of the angular velocity, with a contour interval of 1/15. $E = 10^{-4}$, and $\Lambda = 0.01$ (a), 0.04 (b), 0.16 (c), 0.64 (d). An increasingly strong magnetic field suppresses the Stewartson layer, with essentially all of the jump in angular velocity confined to the inner Ekman–Hartmann layer instead. From Hollerbach (1994).

$\Lambda < E^{-1/3}$ there is an intermediate regime where the $E^{1/3}$ layer is unchanged, the $E^{1/4}$ layer becomes a thinner $(E/\Lambda)^{1/4}$ layer, and there is a new Λ^{-1} layer. For $\Lambda = O(E^{-1/3})$ these three layers merge, and for $\Lambda > E^{-1/3}$ there is a single thinner $(E/\Lambda)^{1/4}$ layer. That is, only very strong fields will have an effect on the Stewartson layer, and even then they will enhance it rather than suppress it. The spherical version of this problem was considered by Hollerbach (1997), with the results in broad agreement with the expected asymptotics.

7.2
Shercliff Layers

This new $(E/\Lambda)^{1/4}$ layer is in fact our first example of a so-called Shercliff layer. These layers, and related jets, are sufficiently important that we will devote an entire section to them. To understand these layers in their simplest possible form, it is helpful to begin by noting that

$$\frac{\Lambda}{E} = \frac{B_0^2/\Omega\mu\rho\eta}{\nu/\Omega r_{\text{in}}^2} = \frac{B_0^2 r_{\text{in}}^2}{\mu\rho\nu\eta} \tag{7.1}$$

does not actually involve the overall rotation Ω at all. It should be possible therefore to study these layers in the classical spherical Couette flow configuration where only the inner sphere is rotating, rather than in the Stewartson configuration where both spheres are rapidly rotating. The advantage of this is that there is then one less parameter to deal with: instead of one parameter (Λ) measuring Lorentz to Coriolis forces, and another (E) measuring viscous to Coriolis forces, the Coriolis term is dropped completely, and there is only one parameter, the Hartmann number

$$\text{Ha} = (\Lambda/E)^{1/2} = \frac{B_0 r_{\text{in}}}{\sqrt{\mu\rho\nu\eta}} \tag{7.2}$$

measuring Lorentz to viscous forces. In terms of Ha, we then expect the Shercliff layers to have thicknesses $\text{Ha}^{-1/2}$. For simplicity and consistency we will present all results in this section in terms of Ha and the 'classical' configuration, even though perhaps half of the relevant literature was done in the 'rapidly rotating' configuration involving E and Λ.

We thus consider a spherical shell with inner and outer radii r_{in} and r_{out} (with r_{in}/r_{out} typically 1/2 or 1/3), where the outer sphere is stationary, and the inner one rotates at rate Ω_{in}. The fluid filling the gap has density ρ, viscosity ν, and magnetic diffusivity η. A magnetic field $\boldsymbol{B}_0$ is externally imposed, where we will consider axial, dipolar, and quadrupolar fields, but all oriented such that their symmetry axis coincides with the axis of rotation of the inner sphere. Let $B_0 = |\boldsymbol{B}_0|$ at $r = r_{out}$, $\theta = 0$ be a measure of the strength of $\boldsymbol{B}_0$ (so the dipolar and quadrupolar fields are actually considerably stronger at the inner boundary).

Scaling length by r_{in}, time by the viscous timescale r_{in}^2/ν, and U by $\Omega_{in}r_{in}$, the nondimensional momentum and induction equations become

$$\frac{\partial U}{\partial t} + \text{Re}\, U \cdot \nabla U = -\nabla p + \nabla^2 U + \text{Ha}^2 (\text{curl}\, b) \times (B_0 + \text{Rm}\, b), \tag{7.3}$$

$$\text{Pm}\frac{\partial b}{\partial t} = \nabla^2 b + \text{curl}(U \times (B_0 + \text{Rm}\, b)). \tag{7.4}$$

$B = B_0 + \text{Rm}\, b$ is the total magnetic field, scaled by B_0, with B_0 the externally imposed field, and $\text{Rm}\, b$ the field induced in the fluid by the action of U.

The four nondimensional parameters appearing in these equations are the Hartmann number Ha as before, the ordinary and magnetic Reynolds numbers

$$\text{Re} = \frac{\Omega_{in} r_{in}^2}{\nu}, \quad \text{Rm} = \frac{\Omega_{in} r_{in}^2}{\eta}, \tag{7.5}$$

and the magnetic Prandtl number $\text{Pm} = \nu/\eta = \text{Rm}/\text{Re}$. Pm is a material property of the fluid, and is very small for all liquid metals, $O(10^{-6})$. Unless Re is very large, Rm will thus also be small. We will therefore simplify the governing equations to

$$\frac{\partial U}{\partial t} + \text{Re}\, U \cdot \nabla U = -\nabla p + \nabla^2 U + \text{Ha}^2\, \text{curl}\, b \times B_0, \tag{7.6}$$

$$0 = \nabla^2 b + \text{curl}(U \times B_0). \tag{7.7}$$

Physically this approximation amounts to stating that (i) the magnetic field is essentially just the imposed field B_0, with $\text{Rm}\, b$ negligible in comparison, (ii) the electric currents though are given by $J = \text{curl}\, b$ (since B_0 has no associated electric currents), and (iii) b is induced by the action of U on B_0, and adjusts effectively instantaneously.

Note incidentally that this $\text{Pm} \to 0$ approximation is identical to the quasi-static approximation considered in Chapter 5 in the context of magnetorotational instabilities. In that context this limit is quite subtle, with the standard MRI disappearing entirely, but the helical and azimuthal MRI surviving. In the Shercliff layer context considered here though, this limit is very robust; none of the results presented here depend fundamentally on either $\text{Pm} \to 0$ or $\text{Pm} \neq 0$.

Having reduced the system down to only two parameters, Ha and Re, in the remainder of this section we will simplify it further by taking Re → 0, corresponding to an infinitesimal rotation of the inner sphere. Finite Re effects will be considered in the following sections. The simplest possible (note that they are linear) equations consistent with Shercliff layer dynamics are thus

$$\frac{\partial U}{\partial t} = -\nabla p + \nabla^2 U + \text{Ha}^2 \, \text{curl} \, b \times B_0 , \tag{7.8}$$

$$0 = \nabla^2 b + \text{curl}(U \times B_0) , \tag{7.9}$$

involving only the spatial structure of B_0, as well as the single parameter Ha measuring its strength.

The boundary conditions associated with (7.8) are

$$U = r \sin \theta \, \hat{e}_\phi \quad \text{at} \quad r = r_{\text{in}} , \quad U = 0 \quad \text{at} \quad r = r_{\text{out}} . \tag{7.10}$$

For b we can consider four possible combinations, according to whether the inner and outer regions $r < r_{\text{in}}$ and $r > r_{\text{out}}$ are independently either insulating or conducting. The most geophysically relevant of these cases is of course to have the inner boundary conducting and the outer boundary insulating, but considered as general MHD the other combinations are also worthy of study, particularly as some of the results will be rather unexpected. (Note that whether 'conducting' means perfectly conducting or merely finitely conducting is relatively unimportant, provided that the total conductance of an exterior region is comparable to that of the fluid.)

Figure 7.3 shows results for axial, dipolar, and quadrupolar imposed fields B_0. For the axial field we see precisely the Shercliff layer on the tangent cylinder C alluded to earlier. Comparing it with the previous nonmagnetic Stewartson layer (Figure 7.1), there are a few subtle differences though; for example, in the Stewartson layer the contours lines are virtually parallel, whereas in the Shercliff layer they spread outward slightly away from the point where C touches the inner sphere (that is, $\partial U / \partial z = 0$ is more closely satisfied in the Stewartson layer than in the Shercliff layer). These differences are reflected in the formal asymptotics; whereas the Stewartson layer involves thicknesses $E^{1/3}$, $E^{2/7}$ and $E^{1/4}$, the Shercliff layer only has the single thickness $\text{Ha}^{-1/2}$ (Roberts, 1967).

Turning next to the dipolar and quadrupolar fields, these too exhibit Shercliff layers, but no longer located on C, which has no special significance for these choices of B_0. As Starchenko (1997) pointed out, the existence and location of the Shercliff layers depends on the topology of B_0, with shear layers occurring on the particular field lines separating regions coupled only to one boundary or the other from regions coupled to both. The reason why these field lines take on a special role is very similar to the reasoning that singled out C in the Stewartson layer problem, except with the Taylor–Proudman theorem replaced by the Ferraro isorotation law. According to this law – a consequence of the magnetic tension force – the angular velocity will tend to arrange itself to be constant along magnetic field lines. On field lines that couple only to the outer boundary, such as those outside C for an axial field, the fluid will therefore be at rest, whereas on field lines that couple only to the inner boundary, such as occur for dipolar and quadrupolar fields, the fluid will corotate with the inner sphere. And finally, on any field lines that link both boundaries the fluid will rotate at some intermediate rate (with so-called Hartmann layers taking over the role previously played by the Ekman layers at the boundaries). We see therefore how a jump in angular velocity, and hence a Shercliff layer, will inevitably exist on all those field lines that separate different regions according to these three types of field line topology.

The results in Figure 7.3 are for the case where both the inner and outer boundaries are insulating. Figure 7.4 shows what happens when the inner boundary is taken to be conducting instead. For the axial field, the inner Hartmann layer has been completely eliminated, and correspondingly the jump in angular velocity across the Shercliff layer has been doubled. That is, instead of rotating at a rate $\sim \Omega_{in}/2$, intermediate between the rotation of the inner sphere and the non-rotation of the outer sphere, the entire region inside the tangent cylinder is now locked to the inner sphere. The reason for this is easy to understand: torques can be transmitted across a conducting boundary not just viscously but magnetically as well, so the fluid inside $\mathcal{C}$ is simply coupled far more strongly to the inner boundary than to the outer one.

This suppression of the inner Hartmann layer also occurs for the dipolar and quadrupolar fields, even if it is not quite so obvious there. More intriguing though –

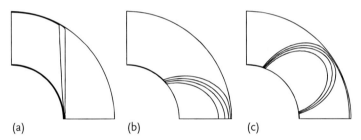

(a)   (b)   (c)

Figure 7.3 Contours of the angular velocity, with a contour interval of 1/5. As shown, the imposed magnetic fields are axial (a), dipolar (b), and quadrupolar (c). Ha = $10^{2.5}$, Re = 0, and insulating inner and outer boundaries. From Hollerbach (2000b).

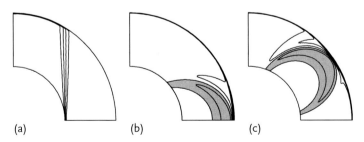

(a)   (b)   (c)

Figure 7.4 As in Figure 7.3, axial (a), dipolar (b) and quadrupolar (c) fields, but with a conducting inner boundary. The gray shading in the dipolar and quadrupolar cases indicates the superrotating regions, with maximum values of $\sim$ 30%.

and completely unexpected – are regions of superrotation, in which the fluid rotates at a rate *greater* than Ω_{in}. This superrotation phenomenon was discovered independently (and essentially simultaneously) by Dormy, Cardin, and Jault (1998) and Starchenko (1997). For increasingly large Ha the degree of superrotation levels off at $\sim$ 30% in both cases; that is, the maximum angular velocity within the fluid is $\sim 1.3\Omega_{in}$.

The existence of such superrotating regions clearly requires an explanation. However, before turning to that we will consider the case where both boundaries are conducting, which was first investigated by Hollerbach (2000b) as part of a systematic study of the three different choices of B_0, and the four possible choices of boundary conditions. One reason why having both boundaries conducting might be an interesting case to consider is this suppression of the Hartmann layer at a conducting boundary: if both boundaries are conducting it is then not clear where the adjustment in angular velocity will come about. And indeed, as indicated in Figure 7.5, there are no Hartmann layers at either boundary, with the adjustment from angular velocity 1 at the inner boundary to 0 at the outer boundary occurring throughout the interior instead (this is most pronounced in the axial field case, inside the tangent cylinder).

Far more dramatic, however, is what happens inside the Shercliff layers, or rather jets in this case. What was previously a $\sim$ 30% superrotation in the dipolar and quadrupolar cases is increased by an order of magnitude, and indeed increases indefinitely with increasing Ha rather than leveling off. Hollerbach (2000b) suggested, based on numerical calculations, that the strength of these jets increases roughly as $Ha^{0.6}$; according to the asymptotics discussed below the correct exponent is actually 1/2. The axial field case is also interesting, and different still. Instead of a superrotating jet, there is a counterrotating jet, with fluid rotating in the opposite direction to the inner sphere. This jet also increases indefinitely with increasing Ha. (Just to complete the picture, we note that the final case, with an axial field, inner boundary insulating and outer boundary conducting, also yields a counterrotating jet, but one that levels off at maximum strength around −0.25, evidently the analog of the dipole/quadrupole, inner boundary conducting, outer boundary insulating $\sim$ 30% superrotation.)

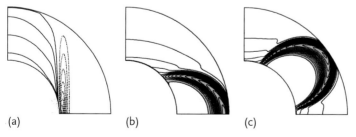

(a) (b) (c)

Figure 7.5 As in Figures 7.3 and 7.4, axial (a), dipolar (b) and quadrupolar (c) fields, but with both boundaries conducting. The gray shading in the dipolar and quadrupolar cases again indicates the superrotating regions, with the maximum values far greater than in Figure 7.4. The dashed contour lines in the axial case indicate the counterrotating jet.

To summarize, we see that the *location* of the Shercliff layer or jet structures is controlled entirely by the topology of the imposed field B_0, always occurring on the particular field lines separating regions linked only to one boundary from those linked to both. What type of structure will arise on these special field lines, a simple shear layer, or a slightly super- or counterrotating jet, or a jet that increases indefinitely with Ha, is determined entirely by the boundary conditions. Having both boundaries insulating yields a simple shear layer – the only one of these structures that is 'intuitively' understandable, in terms of the Ferraro law referred to above. Having both boundaries conducting yields the most dramatic results, namely these $O(Ha^{1/2})$ jets in Figure 7.5.

To understand the origin of these jets, we need to consider the associated electric currents J, shown in Figure 7.6 for the two cases of both boundaries insulating and both boundaries conducting. Focusing attention on the axial field first, we note that the overall pattern is very similar for both choices of boundary conditions, consisting of a broad downward flow inside C, and a narrow upward flow just outside C. The most obvious difference is at the boundaries, where the current must flow within the Hartmann layers in the insulating case, but can pass through the exterior regions in the conducting case. This turns out to have a dramatic effect on the current's amplitude; because it is so much easier to recirculate current through the entire exterior regions rather than the narrow boundary layers, the current is much greater in the conducting case, scaling differently with Ha even: in the conducting case $J = O(1)$, whereas in the insulating case $J = O(Ha^{-1})$.

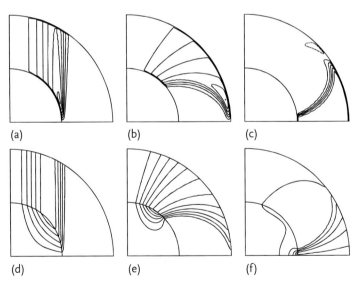

Figure 7.6 Streamlines of the electric currents associated with Figures 7.3 (a–c) and 7.5 (d–f). The sense of circulation is counterclockwise for the axial and dipolar fields (which are also antisymmetric about the equator), and clockwise for the quadrupolar field (which is symmetric about the equator). In the bottom row the recirculation through the outer boundary is not shown.

A further detail needed to understand the simple shear layer in the insulating case versus the counterrotating jet in the conducting case is to notice that in both cases the current right at the equator of the inner sphere is radially outward, so in both cases the Lorentz force $\boldsymbol{J} \times \boldsymbol{B}_0$ is oriented in the $-\hat{\boldsymbol{e}}_\phi$ direction. In the insulating case this force is just sufficient to bring about the adjustment from rotating at rate Ω_{in} at the inner sphere to no rotation in the region outside $\mathcal{C}$; in the conducting case the current, and hence the Lorentz force, is so much greater that the system essentially overcompensates, resulting in the counterrotating jet.

One might also ask, just because the system is able to circulate much greater currents in the conducting rather than the insulating case, why is this also necessary? This is related to the point noted above, that Hartmann layers are suppressed at conducting boundaries, and that far more of the adjustment in angular velocity must therefore take place in the interior instead. The resulting violation of the Ferraro isorotation law then induces a much larger azimuthal field b_ϕ, and hence current $\boldsymbol{J}$ (with the stream function of the current in the meridional plane being given directly by $b_\phi R$, where R is the cylindrical distance from the axis).

Turning next to the dipolar and quadrupolar fields, the picture there is much the same as for the axial field. The overall flow of current is again similar for both choices of boundary conditions, with the most obvious difference being the flow within the Hartmann layers in the insulating case, resulting in $\boldsymbol{J} = O(\text{Ha}^{-1})$, versus the flow through the boundaries in the conducting case, resulting in $\boldsymbol{J} = O(1)$. By working out which way $\boldsymbol{J}$ and $\boldsymbol{B}_0$ are oriented – and specifically at the point where the critical field lines touch the outer sphere – one finds that the Lorentz force in these cases is in the $+\hat{\boldsymbol{e}}_\phi$ direction. In the insulating case this force is again just sufficient to bring about the adjustment from no rotation at the outer sphere to rotating at rate Ω_{in} in the region inside the critical field line; in the conducting case the much larger current again causes the system to overcompensate, resulting in a superrotating jet in these cases. We see therefore that we have not only explained the origin of these jets, but also why the axial field has a counterrotating jet whereas the dipolar and quadrupolar fields have superrotating jets.

Returning also to the case originally considered by Dormy, Cardin, and Jault (1998), with the 30% superrotation in the dipolar field, conducting inner boundary and insulating outer boundary, this comes about because the smaller and thinner (because $\boldsymbol{B}_0$ is much stronger there) inner Hartmann layer restricts the flow of current far more than the larger and thicker outer Hartmann layer, so switching only the inner boundary from insulating to conducting already increases the current by close to an order of magnitude. Because the outer boundary layer ultimately still limits the current to $O(\text{Ha}^{-1})$ though, this superrotation does not increase indefinitely with Ha.

The final items to discuss regarding Shercliff layers are the asymptotic analyses that have been done on this problem, particularly once it was realized how rich the dynamics are, and how crucially they depend on the boundary conditions on $\boldsymbol{b}$. The first partial solution (but enough already to suggest a superrotation phenomenon) was by Starchenko (1997), who considered both insulating and conducting inner boundaries, in each case with an insulating outer boundary. See also Starchenko

(1998a,b) for further developments of this work. In contrast, the work in Dormy, Cardin, and Jault (1998) was entirely numerical, but correspondingly also more complete in many ways. The first essentially complete and quantitatively accurate asymptotic analysis of the 30% superrotation was by Dormy, Jault, and Soward (2002).

Dormy, Jault, and Soward (2002) further conjectured how their analysis would have to be altered to explain the enhanced superrotation in the case with both boundaries conducting. This was far from a complete solution, but they correctly suggested already that asymptotically the exponent ought to be $Ha^{1/2}$ rather than the $Ha^{0.6}$ suggested by the numerical results (Hollerbach, 2000b). That $Ha^{1/2}$ is indeed the correct scaling was proved by Bühler (2009), who also pointed out that the flux in the jets tends to a limit independent of Ha, thereby nicely linking their $Ha^{1/2}$ maximum amplitudes with their $Ha^{-1/2}$ thicknesses.

Finally, the gradual transition from an insulating to a conducting outer boundary (for a conducting inner boundary in all cases) was first explored by Mizerski and Bajer (2007), who suggested that even a relatively weakly conducting outer boundary could significantly enhance the superrotation beyond the original 30% value. The full solution to this problem was provided in an asymptotic *tour de force* by Soward and Dormy (2010), who mapped out the entire parameter regime as it depends on the outer boundary's conductance ϵ, finding that the degree of superrotation is $O(1)$ for $1 \ll Ha^{3/4} \ll \epsilon^{-1}$, $O(\epsilon^{2/3} Ha^{1/2})$ for $1 \ll \epsilon^{-1} \ll Ha^{3/4}$, and $O(Ha^{1/2})$ for $\epsilon^{-1} \ll 1 \ll Ha^{3/4}$.

7.3
Finite Re in an Axial Field

The results in the preceding section were all in the linear, Re → 0 regime, corresponding to an infinitesimal rotation of the inner sphere. In the next few sections we wish to consider finite Re effects, that is, (7.6), (7.7) or even (7.3), (7.4) instead of the linear system (7.8), (7.9). Probably not surprisingly, including nonlinear effects introduces an enormous variety of additional features, including substantial modifications to the axisymmetric basic states, nonaxisymmetric but still large-scale instabilities, and small-scale wave-like phenomena. There is also considerable interplay between experimental and numerical work, with each stimulating further developments in the other. To simplify the discussion we will present these two strands as more separated than they really are, but where appropriate we will also at least attempt to indicate where developments in one area lead directly to further work in another.

7.3.1
Numerics

Beginning with the axial field case, the first results were by Hollerbach and Skinner (2001), who numerically studied the nonaxisymmetric, Kelvin–Helmholtz type instabilities of the Shercliff layer in Figure 7.3, and the counterrotating jet in Figure 7.5. Further numerics were also done by Hollerbach (2009) and Gissinger, Ji, and Goodman (2011). These latter papers were largely motivated by the experiment of Sisan et al. (2004), but we will first discuss all of the numerical results before turning to the experiment.

Figure 7.7, from Hollerbach (2009), presents numerical solutions of (7.6), (7.7) in a shell with radius ratio 1/3, and insulating inner and outer boundary conditions. The various curves denote the linear onset of nonaxisymmetric instabilities, showing the critical Reynolds numbers as they depend on the Hartmann number. The range Ha > 30, to the right of the dotted line, was the case originally considered by Hollerbach and Skinner (2001). In this regime Ha is sufficiently large that a clearly defined Shercliff layer exists, as in Figure 7.3. For sufficiently large Re one would expect such a shear layer to become unstable, yielding a Kelvin–Helmholtz type of instability. As seen in Figure 7.7, one even obtains a reasonably well-defined asymptotic scaling $Re_c \propto Ha^{0.66}$ for the critical onset values.

Recall also the strong similarity between the Stewartson layer in Figure 7.1 and the Shercliff layer in Figure 7.3. If the Shercliff layer becomes unstable for sufficiently large Re, it seems likely that the Stewartson layer should yield very similar instabilities, as was indeed found to be the case by Hollerbach (2003) and Schaeffer and Cardin (2005). There is however also an important distinction between the Stewartson and Shercliff cases: for the Shercliff layer it makes no difference whether the inner sphere rotates one way or the other, but for the Stewartson layer whether the inner sphere rotates faster or slower than the outer sphere has a significant influence on the resulting instabilities (Hollerbach, 2003). Wei and Hollerbach (2008) then considered the combined case, but obtained such a gradual transition

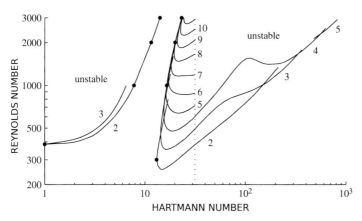

Figure 7.7 The linear stability curves in the (Ha,Re) plane. The numbers beside individual curves denote the azimuthal wave numbers *m*. The two left-most curves are the equatorially antisymmetric radial jet instabilities; the four dots correspond to the four solutions shown in Figure 7.9. The curves between Ha $\approx$ 15 and 30 are the equatorially symmetric return flow instabilities; the four dots correspond to the four solutions shown in Figure 7.10. The curves for Ha > 30 are the Shercliff layer instabilities. From Hollerbach (2009).

from more Schercliff-like to more Stewartson-like behavior that it was difficult to reach any definite conclusions. Finally, Hollerbach and Skinner (2001) additionally considered the instabilities of the counterrotating jet that results if both boundaries are conducting (Figure 7.5), and obtained onset curves scaling as $Re_c \propto Ha^{0.16}$.

As interesting as these various shear layer and jet instabilities may be, the left half of Figure 7.7, Ha < 30, is possibly even more intriguing. This is the regime that was considered by Hollerbach (2009), and yields two types of instabilities that are quite different both from each other and from the previous Kelvin–Helmholtz instabilities. We begin by noting that for such relatively small Hartmann numbers the basic state is substantially different from the previous Schercliff layer structure; as indicated in Figure 7.8, this only begins to emerge in the Ha > 30 regime. In contrast, the associated meridional circulation, which is almost negligible in the Schercliff layer problem, turns out to play an important role in both of these new types of instabilities.

Figure 7.9 shows the first type, the so-called radial jet instability, corresponding to the left-most instability branches in Figure 7.7. We see how the meridional circulation consists of a radial jet in the equatorial plane, with a broader return flow throughout the rest of the shell. The instabilities are always concentrated right on this jet, hence the name. They are smoothly connected to the nonmagnetic limit Ha = 0 (in fact Ha = 1 is already essentially nonmagnetic), where they have been extensively studied before both experimentally (e.g., Munson and Menguturk,

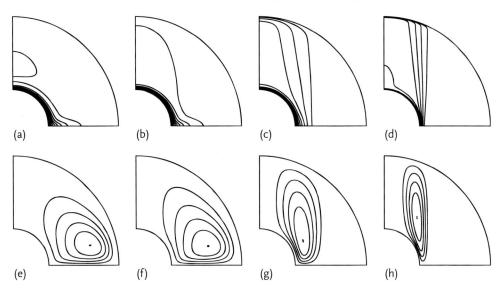

Figure 7.8 The top row (a–d) shows contours of the angular velocity, with a contour interval of 0.1. The bottom row (e–h) shows streamlines of the (counterclockwise) meridional circulation Ψ, with contour interval $\Psi_{max}/5$. From left to right Ha = 1, 4, 16, 64, and Re = 250 for all four. The corresponding Ψ_{max} values are 0.13, 0.093, 0.024, 0.0018, indicating the strong suppression of the meridional circulation for increasingly large Ha. Note also the gradual emergence of the Schercliff layer in the angular velocity. From Hollerbach (2009).

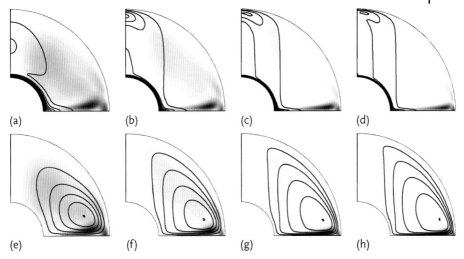

Figure 7.9 The contours indicate the angular velocity (a–d) and the meridional circulation (e–h) at the four dots on the radial jet branch in Figure 7.7. Specifically, from left to right we have (i) Ha = 1, Re = 385, Ψ_{max} = 0.12, (ii) Ha = 7.8, Re = 1000, Ψ_{max} = 0.061, (iii) Ha = 11.5, Re = 2000, Ψ_{max} = 0.044, (iv) Ha = 14.0, Re = 3000, Ψ_{max} = 0.037. The gray shading (the same in both rows) denotes the azimuthally integrated kinetic energy of the nonaxisymmetric ($m = 2$) instabilities. That this gray shading correlates with the equatorial jet in the meridional circulation indicates the 'radial jet' nature of these instabilities.

1975; Egbers and Rath, 1995) and numerically (e.g., Araki, Mizushima, and Yanase, 1997; Hollerbach, Junk, and Egbers, 2006). The initially axisymmetric radial jet develops a wavy nonaxisymmetric structure, alternately slightly above and below the equatorial plane. As seen in Figure 7.7, as Ha is increased the critical Reynolds number also increases, that is, these instabilities are suppressed by the presence of the magnetic field. The reason for this is the suppression of the entire meridional circulation, as indicated in Figures 7.8 and 7.9. The qualitative nature of these modes is essentially unchanged though along the entire instability curves.

Figure 7.10 shows the second type, the so-called return flow instability, corresponding to the middle instability branches in Figure 7.7. Comparing Figures 7.9 and 7.10, we notice how the radial jet no longer extends all the way to the outer boundary, but is forced by the – now stronger – magnetic field to turn around and flow parallel to the field. The instabilities are consistently located at this point where the radial jet starts its return, hence the name. Note in particular how these instabilities have nothing to do with a Shercliff layer, which does not even exist at these (Ha, Re) combinations. Nevertheless, as indicated in Figure 7.7, these instabilities are continuously connected to the Shercliff shear layer instabilities previously studied by Hollerbach and Skinner (2001). In addition to the linear onset curves in Figure 7.7, Hollerbach (2009) further explored the supercritical, fully three-dimensional regime, and in particular the gradual transition from return flow to Shercliff layer instabilities, as Ha is increased at fixed Re = 1200. Very briefly,

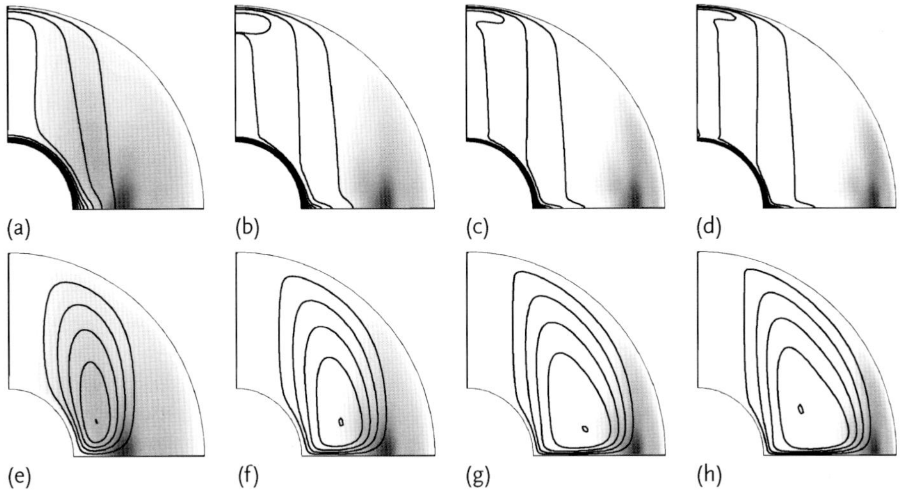

(a) (b) (c) (d)

(e) (f) (g) (h)

Figure 7.10 The contours indicate the angular velocity (a–d) and the meridional circulation (e–h) at the four dots on the return flow branches in Figure 7.7. Specifically, from left to right we have (i) Ha = 13.1, Re = 300, Ψ_{max} = 0.034, (ii) Ha = 16.5, Re = 1000, Ψ_{max} = 0.033, (iii) Ha = 20.0, Re = 2000, Ψ_{max} = 0.027, (iv) Ha = 23.2, Re = 3000, Ψ_{max} = 0.023. The gray shading denotes the azimuthally integrated kinetic energy of the nonaxisymmetric (m = 2, 5, 7, and 8, respectively) instabilities. That this gray shading correlates with the outer edge of the meridional circulation indicates the 'return flow' nature of these instabilities.

the increasingly strong magnetic field forces the radial jet to turn around ever sooner, until the return flow instability is located essentially on the tangent cylinder, at which point it gradually becomes more associated with the angular velocity rather than the meridional circulation.

Summarizing Figure 7.7, for small Ha we have the radial jet instabilities, for moderate Ha we have the return flow instabilities, and for large Ha we have the Shercliff layer instabilities. The return flow and Shercliff layer instabilities are continuously connected, but are nevertheless qualitatively different. Both are also very different from the radial jet instabilities, which are equatorially antisymmetric, whereas the return flow and Shercliff layer instabilities are symmetric. Noteworthy also is the region of stability separating the radial jet and return flow instabilities. This was explored further by Travnikov, Eckert, and Odenbach (2011), who demonstrated that it is a robust feature, existing over a broad range of radius ratios. It would be of considerable interest to know whether this stability strip exists for indefinitely large Re, or whether the radial jet and return flow instability curves eventually cross (or some entirely new instability sets in).

Turning next to Gissinger, Ji, and Goodman (2011), they explored a broad variety of cases, including both axial and dipolar magnetic fields, as well as insulating and conducting inner spheres (the outer boundary was always insulating). They also considered scenarios where the outer sphere was rotating as well, thereby tending to generate a Stewartson layer. Figure 7.11 shows one of their stability diagrams.

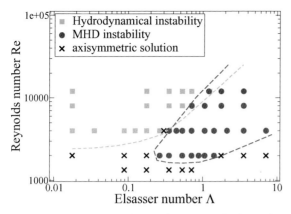

Figure 7.11 The instability diagram for an imposed dipolar field, with insulating inner and outer boundaries, rotating with the ratio $\Omega_{in}/\Omega_{out} = 8$. The squares are the radial jet instabilities, the circles the return flow and/or Shercliff layer instabilities. Note in particular the similarity to the axial field results in Figure 7.7. Their Reynolds number equals 3.45 times our Re, and their Elsasser number equals $Ha/Re^{1/2}$ in our notation. From Gissinger, Ji, and Goodman (2011). Copyright 2011 by The American Physical Society.

This is actually one of their dipole field cases, but we nevertheless present it here, as the similarity with Figure 7.7 is striking. That both axial and dipolar fields suppress the radial jet instability is hardly surprising; any field that has a component perpendicular to the jet (that is, $B_z \neq 0$ in the equatorial plane) will almost certainly do so. What is surprising is that their 'MHD instability' branch should be so similar to the return flow/Shercliff layer branch in Figure 7.7. The stability strip separating the two branches is absent in this case, but there is no reason why it should necessarily also exist in this case.

Figure 7.12 shows some of their results for an axial field and an insulating inner sphere, so directly comparable to Hollerbach (2009) and Hollerbach and Skinner (2001). And indeed, we see precisely the return flow and Shercliff layer instabilities (with azimuthal wave number $m = 2$ in this case, so relatively low down on what would be their equivalent of Figure 7.7). One important difference though is that they did not make the Pm $\to 0$ approximation (7.6), (7.7), but instead solved the full system (7.3), (7.4), with Pm $= 0.01$. That the results are nevertheless qualitatively the same as those in the Pm $\to 0$ approximation validates the earlier assertion that Pm $\neq 0$ versus Pm $\to 0$ is far less significant in this context than it is in many others.

Figure 7.13 shows the effect of switching the inner boundary from insulating to conducting. The change from Figure 7.12 is quite striking: not only has the wave number switched from $m = 2$ to $m = 1$, but more fundamentally, the equatorial symmetry has switched from symmetric to antisymmetric. The complete instability diagram, as in Figure 7.7 or 7.11, was unfortunately not computed for this case. It would be very interesting to see how this mode relates to some of the previously known ones. The only previous modes that are equatorially antisymmetric are the radial jet instabilities, but this mode is clearly very different. Even if it did turn

302 | *7 Magnetic Spherical Couette Flow*

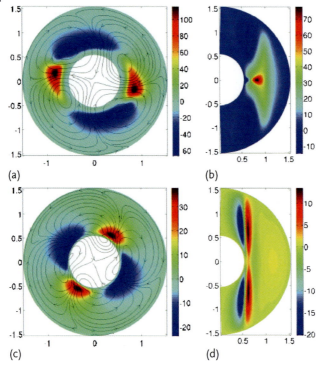

Figure 7.12 The instabilities obtained for Re = 1450, and Ha = 23 in the first two panels (a,b) and Ha = 76 in the second two panels (c,d), where for ease of comparison their parameters have been translated to our Re and Ha notation. The first panel in each pair shows the nonaxisymmetric flow in the equatorial plane, with the contours representing U_r. Note the $m = 2$ dominant wave number in both cases. The second panel in each pair shows the nonaxisymmetric U_r in a meridional section. Ha = 23 corresponds to a return flow instability, Ha = 76 to a Shercliff layer instability. From Gissinger, Ji, and Goodman (2011). Copyright 2011 by The American Physical Society.

out to be continuously connected to the radial jet instabilities, it would still be an essentially new type of instability therefore.

7.3.2
The Maryland Experiment

Turning next to the Sisan *et al.* (2004) experiment that motivated both Hollerbach (2009) and Gissinger, Ji, and Goodman (2011), this consists of a spherical shell with $r_{in} = 5$ cm and $r_{out} = 15$ cm. The rotating inner sphere is made of high conductivity copper, the stationary outer sphere of nonmagnetic stainless steel. Compared with the conductivity of the fluid (liquid sodium), the inner boundary is strongly conducting, whereas the outer boundary is effectively insulating. The inner sphere rotates between 2.5 and 50 Hz, which translates to Reynolds numbers between 6×10^4 and 1.2×10^6. An axial magnetic field up to 0.2 T (which translates

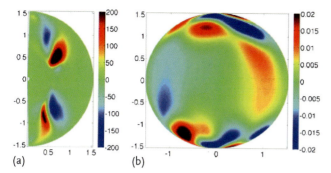

Figure 7.13 The $m = 1$ instability obtained for Re = 1450, Ha = 19, and a conducting inner sphere. The nonaxisymmetric U_r in a meridional section (a) and the nonaxisymmetric B_r at the surface of the outer sphere (b) are shown. From Gissinger, Ji, and Goodman (2011). Copyright 2011 by The American Physical Society.

to Ha ≈ 1200) is applied by a pair of electromagnets. Measurements are done by ultrasound Doppler velocimetry for the flow, and an array of external Hall probes for the induced field.

Figure 7.14 summarizes the different observed states, as they depend on the Lundquist number S and the magnetic Reynolds number Rm, where we recall that

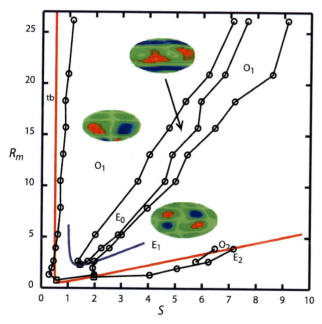

Figure 7.14 The experimentally observed phase diagram in the (S,Rm) plane. E_m and O_m indicate equatorially even/odd modes, with dominant azimuthal wave number m. The color insets show the radial component of the induced field. The red and purple curves are theoretical stability boundaries for the first and second MRI modes. From Sisan et al. (2004). Copyright 2004 by The American Physical Society.

$S = \mathrm{Ha} \cdot \mathrm{Pm}^{1/2}$ and $\mathrm{Rm} = \mathrm{Re} \cdot \mathrm{Pm}$, with $\mathrm{Pm} \approx 10^{-5}$ for liquid sodium. The notation E and O refers to even/odd, that is, equatorially symmetric or antisymmetric. The subscripts indicate the dominant azimuthal wave numbers of the observed modes. The color insets show the spatial structure of the radial component of the induced magnetic field. Note also that all of these large-scale structures are superimposed on a fully turbulent background flow; even the minimum achievable Reynolds number is far beyond the laminar regime.

The red and purple colored lines are theoretical results from the dispersion relation for the magnetorotational instability. Based on the excellent fit, Sisan et al. (2004) interpreted their results as the MRI, albeit a turbulent analog of the classical MRI discussed in Chapter 5. Note, however, that the critical Hartmann and Reynolds numbers which are typical for MRI excitation in cylinders are more than one order of magnitude larger than the minimum values given in Figure 7.14. The differentially rotating spheres considered by Kitchatinov and Rüdiger (1997) require even higher rotation rates and field amplitudes to excite MRI modes than the Taylor–Couette flows do. It is also not clear why nonaxisymmetric modes would be preferred, when the classical MRI is mainly axisymmetric. Also, having an MRI emerge from a turbulent background state somewhat contradicts the special interest in the MRI as a magnetically induced instability of an otherwise hydrodynamically stable flow. Nevertheless, if the Sisan et al. interpretation is correct, it would be very significant, and would be the first experimental observation of the MRI.

Both Hollerbach (2009) and Gissinger, Ji, and Goodman (2011), however, suggested that the Sisan et al. results may instead be turbulent analogs of these Shercliff layer, return flow, or other instabilities. At least some of the qualitative features are similar, after all, such as the absence of instabilities if the field is either too weak or too strong. These other instabilities could also more naturally explain both the preference for nonaxisymmetric modes, as well as the intermingling of symmetric and antisymmetric structures. The Gissinger et al. observation that switching the inner sphere from conducting to insulating changes the instabilities in a very specific way is a further point in favor of this interpretation, as the same effect was also observed in the experiment (Sisan, 2004).

So, how should one decide which interpretation is correct? More data from the experiment would obviously help. Detailed profiles of angular velocity would be of considerable interest, to check whether there are any indications of a Shercliff layer. Figure 7.15 shows an example of such a measured profile, but for zero field, where no Shercliff layer would be expected. Also, while the observation of a Shercliff layer would make it quite plausible that any subsequent instabilities are related to it, the absence of a Shercliff layer would still not necessarily prove that instabilities are the MRI. Note in particular how neither the radial jet instabilities in Figure 7.9 nor the return flow instabilities in Figure 7.10 have anything like a Shercliff layer.

One interesting possibility might be to redo the experiment, but with a different fluid, such as gallium. In particular, if the Sisan et al. MRI interpretation is correct, the relevant parameter should be Rm, whereas if the Hollerbach–Gissinger 'other instabilities' interpretation is correct, it should be Re. Comparing results with two fluids having different magnetic Prandtl numbers could thus help to distin-

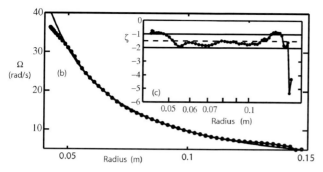

Figure 7.15 The experimentally measured angular velocity profile for zero field and an inner sphere rotation rate of 30 Hz. For comparison, the smooth curve indicates a Kepler profile. The inset gives the velocity exponent $\zeta = d(\ln \Omega)/d(\ln r)$; the Keplerian value is $-3/2$. From Sisan et al. (2004). Copyright 2004 by The American Physical Society.

guish between the two possibilities. (It is possible of course that the background turbulence would be so strong that the effective, turbulent magnetic Prandtl number would be much the same in both fluids after all. This merely illustrates the dangers in any interpretations that rely on the phrase 'turbulent analog of.') We conclude this section by noting that further experiments in this area are indeed planned (Stefani et al., 2012), and will hopefully help to resolve these outstanding questions, as well as yield further fundamentally new results.

7.3.3
The Princeton Experiment

The Princeton MRI experiment (Ji et al., 2006; Nornberg et al., 2010; Roach et al., 2012) has already been discussed in Chapter 5, in the context of Taylor–Couette flows. As a cylindrical experiment, it does not quite fit with the rest of this chapter, on spherical Couette flows. Nevertheless, some of the results on Shercliff layers and their instabilities are so closely related to the corresponding spherical results that it is appropriate to include them here.

Very briefly, the apparatus consists of two nested cylinders with inner radius 7 cm, outer radius 20 cm, and height 28 cm. The top and bottom endplates are split into several independently controlled rings. To study aspects such as whether Rayleigh-stable flows are also nonlinearly stable (e.g., Ji et al., 2006; Schartman et al., 2012), these rings are set so as to make the angular velocity profile in the interior as smooth and 'Keplerian' as possible. However, it is also possible to adjust them to deliberately create an abrupt jump in angular velocity in the middle of the gap. The imposed axial field will then induce a Shercliff layer throughout the entire apparatus, and for a sufficiently large jump in angular velocity one would expect Kelvin–Helmholtz instabilities very similar to those in spherical geometry.

This version of the experiment was done by Roach et al. (2012) and Spence et al. (2012), with accompanying numerics by Gissinger, Goodman, and Ji (2012). Figure 7.16 shows the numerically computed basic state; note how the shear is indeed

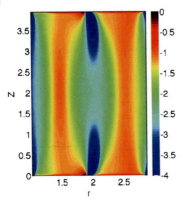

Figure 7.16 A Shercliff layer induced in cylindrical geometry by adjusting the rotation of the endplates to create a jump in angular velocity at the midpoint in radius, $r = 2$. The contours show the local shear parameter $\zeta = d(\ln \Omega)/d(\ln r)$; note how $|\zeta|$ is largest within the Shercliff layer. Reprinted with permission from Gissinger, Goodman, and Ji (2012). Copyright 2012, American Institute of Physics.

largest at the midpoint of the endplates, and is then extended throughout the depth of the container by the influence of the magnetic field. Figure 7.17 compares the saturated instabilities obtained numerically and experimentally. The agreement is excellent, in both cases consisting of an $m = 1$ spiral vortex pattern rotating in the azimuthal direction. The magnetic Reynolds number is 0.02 in the simulation and 0.002 in the experiment. The smallness of these values indicates the inductionless nature of these Shercliff layer instabilities; that is, in terms of our previous discussion, Pm $\to$ 0 and Pm $\neq$ 0 yield the same results.

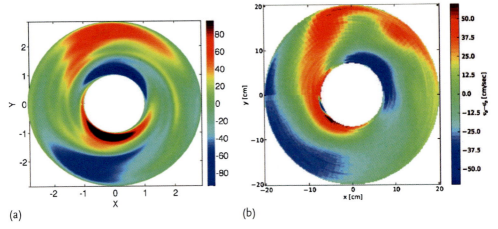

Figure 7.17 The nonaxisymmetric part of U_φ in the midplane of the apparatus, with the numerical simulation (a) and experimental measurements (b). Re = 725 and Ha = 34 (again converted to our definitions). Reprinted with permission from Gissinger, Goodman, and Ji (2012). Copyright 2012, American Institute of Physics.

7.4
The Grenoble DTS Experiment

The DTS ('Derviche Tourneur Sodium') experiment in Grenoble was first planned a decade ago (Cardin et al., 2002), and has since resulted in a substantial number of papers (Nataf et al., 2006, 2008; Nataf and Gagnière, 2008; Schmitt et al., 2008, 2013; Brito et al., 2011; Figueroa et al., 2013; Nataf, 2013). The spherical shell in this case has $r_{in} = 7.4$ cm and $r_{out} = 21$ cm, so slightly larger than the Sisan et al. (2004) setup. The radius ratio of 0.35 was chosen to match that of the Earth's liquid iron outer core. As previously with Sisan et al., the inner sphere is made of copper and the outer sphere of stainless steel, so a conducting inner boundary and an essentially insulating outer one. Unlike Sisan et al., where only the inner sphere rotates, in the DTS setup both spheres can independently rotate (about the same axis), at rates up to ± 30 Hz for the inner sphere and 15 Hz for the outer. In terms of Re, this is comparable to the $\sim 10^6$ Sisan et al. values.

The DTS experiment imposes a dipolar field. This is achieved by stacking highly magnetized rare-earth cobalt bricks into the inner sphere, resulting in a dipole moment of 700 A m$^($, aligned with the rotation axis. The resulting field varies between 345 mT at the poles of the inner sphere and 8 mT at the equator of the outer sphere. (For comparison recall that Sisan et al. had axial fields up to 200 mT.) Because of the way it is imposed, by permanent magnets from the inside rather than electromagnets from the outside, the field strength cannot be adjusted, unfortunately. The very strong variation throughout the shell is also an inevitable consequence of choosing

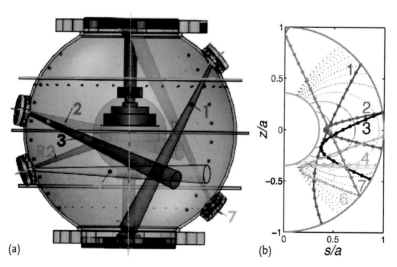

Figure 7.18 A sketch of the DTS experiment, showing details such as the magnets stacked inside the inner sphere, as well as the seven ultrasound beams (a). These seven beams have been projected onto a meridional section (b). The dashed lines indicate the dipolar magnetic field. From Brito et al. (2011). Copyright 2011 by The American Physical Society.

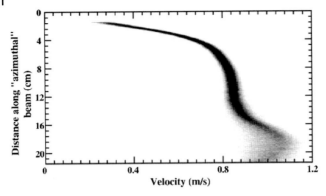

Figure 7.19 An angular velocity profile measured by ultrasound. A 40 s long record was used, with 30 ms between profiles, and 128 shots per profile. Note how the resulting probability density function is remarkably sharp, indicating that the flow is almost steady. From Nataf et al. (2008). Copyright 2008, with permission from Elsevier.

a dipolar field, but does mean that different parts of the shell can potentially be in different dynamical regimes.

As in Sisan et al. measurements are done by both ultrasound Doppler velocimetry and an array of external Hall probes. Figure 7.18 illustrates the basic setup, including the arrangement of magnetized bricks inside the inner sphere, as well as seven different ultrasound beams, giving extensive coverage throughout the shell. Figure 7.19 shows an example of one such ultrasound signal; note the relatively small variations about a well-defined mean state.

Given several such signals, all measuring different linear combinations of (U_r, U_θ, U_ϕ) at different positions throughout the shell, the question then is how to invert such raw data to separate out the angular velocity and meridional circulation? Brito et al. (2011) formulate this as a problem in classical inverse theory: letting the large-scale, axisymmetric part of the flow be given by

$$U = v\hat{e}_\phi + \text{curl}(\psi \hat{e}_\phi), \qquad (7.11)$$

expand the desired quantities v (azimuthal velocity) and ψ (meridional circulation) as

$$v(r,\theta) = \sum_{k=1}^{K}\sum_{l=1}^{L} v_{kl} f_{1k}(r) P^{(1)}_{2l-1}(\cos\theta),$$

$$\psi(r,\theta) = \sum_{k=1}^{K}\sum_{l=1}^{L} \psi_{kl} f_{2k}(r) P^{(1)}_{2l}(\cos\theta). \qquad (7.12)$$

The $P_n^{(1)}$ are associated Legendre functions; taking only odd ones for v and even ones for ψ imposes the equatorial symmetry expected for the large-scale part of the flow. The precise choices of radial expansion functions $f_{1k}(r)$ and $f_{2k}(r)$ are not important here, except to note that the $U_r = 0$ boundary condition is built in (i.e., $f_{2k}(r_\text{in}) = f_{2k}(r_\text{out}) = 0$), but no conditions are imposed on the tangential components, reflecting the fact that this inversion is not intended to include the Hartmann layers in which these flow components adjust to the boundaries.

If the sums in (7.12) are truncated at some finite level ($K = 8$, $L = 4$), there are then only a limited number of expansion coefficients v_{kl} and ψ_{kl}, and standard inverse-theory methods allow one to determine which values best fit the given data. Figure 7.20 shows an example of such a reconstructed large-scale flow. Particularly noteworthy is the angular velocity (which is more than ten times greater than the meridional flow). Near the equator of the inner sphere, where the field is strong, there is a clear tendency for the contours to align with the field lines, in accordance with the Ferraro isorotation law. (Just above the inner sphere's pole, where the

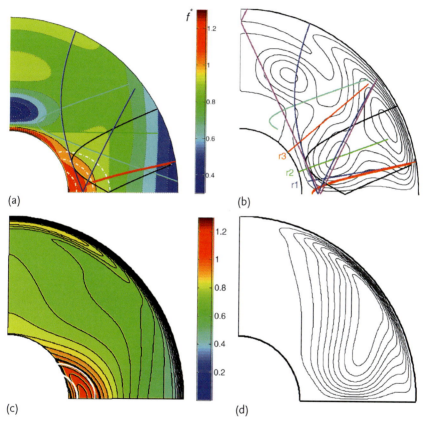

Figure 7.20 Frames illustrating the large-scale flow reconstructed using (7.12) (a,b). The angular velocity is shown (a), normalized by the inner sphere's rotation rate. The values greater than one near the inner sphere's equator thus indicate superrotation. Streamlines of the meridional circulation are shown next (b). In both panels the superimposed curves indicate the meridional projections of the ultrasound beams, as in Figure 7.18. The second two frames (c,d) illustrate a corresponding numerical solution. Note how the angular velocity in particular is very similar to the experimental results. From Brito et al. (2011). Copyright 2011 by The American Physical Society.

field is strongest, the Ferraro law does not seem to hold, but this is also the region where the ultrasound beams provide the least coverage, so the inversion would be most questionable here.) In contrast, near the equator of the outer sphere, where the field is weak, the contours are more vertical, indicating the influence of the Taylor–Proudman theorem. (The outer sphere is in fact stationary in this case, but the entire fluid is rotating sufficiently rapidly that a certain alignment along the vertical is indeed to be expected.)

Figure 7.20 also presents a corresponding numerical solution. Both the Ferraro regime near the inner sphere and the Taylor–Proudman regime near the outer sphere are in surprisingly good agreement with the experimental results. Figure 7.20 in fact has a further interesting feature: the 'Ferraro regions' are rotating faster than the imposed rotation of the inner sphere. We recall that we have seen such superrotating regions before, in Figures 7.4 and 7.5. In those cases, however, the superrotation occurred near the equator of the outer rather than the inner sphere. In the general MHD spherical Couette flow context (i.e., not specifically tuned to the DTS experiment), such a shift in the location of the superrotation with increasing Re was first observed by Hollerbach, Canet, and Fournier (2007), as indicated in Figure 7.21. Comparing the numerics of Hollerbach et al. and Brito et al., it is worth noting also that (as previously in the axial field case) Hollerbach et al. worked in the Pm → 0 limit, whereas Brito et al. considered finite Pm. That

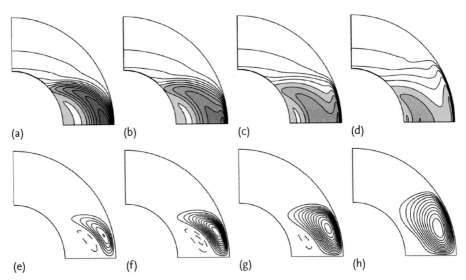

Figure 7.21 The top row (a–d) shows the angular velocity, with a contour interval of 0.2, and gray shading indicating superrotation. The bottom row (e–h) shows streamlines of the meridional circulation, with solid lines corresponding to counterclockwise flow and dashed to clockwise. Both boundaries are conducting, Ha = 100, and from left to right Re = 300, 1000, 3000, 10 000. As Re is increased, the superrotating region gradually shifts from the outer boundary, as in Figure 7.5, toward the inner boundary, as in Figure 7.20. From Hollerbach, Canet, and Fournier (2007). Copyright 2007 Elsevier Masson SAS.

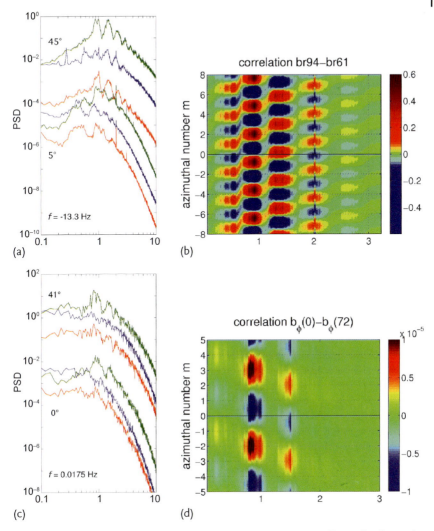

Figure 7.22 The first panel (a) shows frequency spectra of (B_r, B_θ, B_ϕ), color-coordinated as (red, blue, green), measured on the surface of the outer sphere at latitudes 5° and 45° as indicated (offset vertically for clarity). The frequencies on the horizontal axis are normalized by the inner sphere's rotation frequency -13.3 Hz; a time series consisting of 4000 rotations was used. Note how the spectra are dominated by distinct peaks. The second panel (b) shows the cross-correlation between two B_r records separated by 128° in longitude, revealing that distinct peaks in frequency on the horizontal axis correspond to well-defined integer wave numbers m on the vertical axis. The third and fourth panels (c,d) show the same as the first two, but applied to a numerical solution. Note the – at least qualitative – similarity, indicating that the numerical model is capturing much of the experimental physics. Courtesy of Figueroa et al. (2013).

312 | *7 Magnetic Spherical Couette Flow*

features such as this shift in the location of the superrotation occur in both limits indicates once again that the distinction between the two is relatively unimportant in this case.

Having successfully measured and interpreted the large-scale, steady part of the flow, the DTS team were further interested in also understanding the smaller-scale, fluctuating components. These are best studied by the external Hall probes rather than the ultrasound beams; having an array of probes spread over the entire outer surface allows cross-correlations to be measured that could not be obtained by ultrasound. Figure 7.22 shows an example of how the raw data is first Fourier transformed into frequency space, and then cross-correlated to identify peaks at specific frequencies and azimuthal wave numbers.

A corresponding numerical study was carried out by Figueroa *et al.* (2013), and the same type of data analysis was done on the numerical solution. As illustrated in Figure 7.22, the results are at least qualitatively in agreement with the experimental data. Note also that these are turbulent flows, requiring very high spatial resolution to compute. Accurately identifying the relevant frequencies also requires very long time series, corresponding to several hundred inner sphere rotations. These calculations are thus computationally very intensive. (For comparison, several hundred rotations in the experiment takes less than one minute.)

Figure 7.23 shows an example of a large-scale mode obtained by Fourier analyzing the entire numerical data set; see also Schmitt *et al.* (2013) for a linear modal approach that nevertheless yields similar structures. Note how this particular mode extends throughout the entire shell, particularly with regard to the B_r perturba-

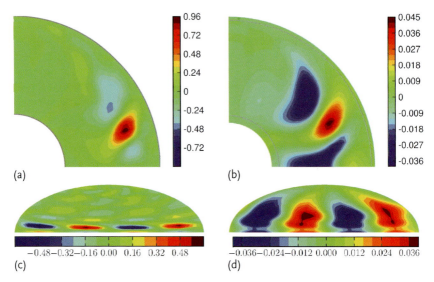

Figure 7.23 A large-scale mode extracted from the numerical solution. This particular example has wave number $m = 2$ and normalized frequency 0.9. The top two panels (a,b) show meridional slices of U_r and B_r, and the bottom two (c,d) show map-views of U_r and B_r near the outer boundary, at normalized radius 0.95. Courtesy of Figueroa *et al.* (2013).

tions. Instabilities concentrated in the outer boundary layer were also obtained; Hollerbach, Canet, and Fournier (2007) found similar boundary layer instabilities in purely axisymmetric calculations.

7.5 Other Waves and Instabilities

7.5.1 Inertial Oscillations

Inertial oscillations are waves in rotating fluids in which the restoring force is provided by the Coriolis force. They exist in completely nonmagnetic situations, and are therefore somewhat different from much of the rest of this chapter. Nevertheless, the experiments of Kelley et al. (2007, 2010), in which magnetic fields were used as tracers of the underlying flow, clearly do fall under the category of magnetic spherical Couette flow, and therefore belong in this review.

The Kelley et al. experiment is in many ways a scaled-up version of the previous Sisan et al. (2004) setup; where Sisan et al. had $r_{in} = 5$ cm and $r_{out} = 15$ cm, Kelley et al. have $r_{in} = 10$ cm and $r_{out} = 30$ cm. The fluid within the shell is again liquid sodium. The two spheres can independently rotate, at rates up to ~ 40 Hz. An axial magnetic field up to 0.04 T is externally applied. For comparison we recall that Sisan et al. achieved up to 0.2 T. It is this considerably weaker field that makes MHD effects dynamically far less important in Kelley et al. than in Sisan et al.

As before in both Sisan et al. and DTS, measurements are done by an array of Hall probes of the induced field; it is this measurement technique rather than the fluid dynamics as such that constitutes the 'magnetic' aspect of this experiment. Figure 7.24 shows examples of two (out of seven) large-scale inertial waves measured by Kelley et al., along with numerically computed eigenmodes. The agreement is astonishingly good, given that the experiment has a fully turbulent background state, whereas the theoretical modes are the linear eigenmodes arising on a $U = 0$ background state.

Kelley et al. (2007) suggested that their modes were triggered by so-called overreflection at the Stewartson layer. The theory of overreflection was originally developed for sound waves in a moving medium (Ribner, 1957), but should indeed be applicable to inertial waves as well. The key feature is that waves incident upon a shear flow (such as the Stewartson layer) are not only partially reflected, but under the right circumstances the reflection coefficient may be greater than one, that is, the wave is amplified (in the process drawing energy out of the shear flow).

These ideas were further developed by Kelley et al. (2010), who postulated that the angle of incidence to be used in the overreflection theory was simply the inertial wave's characteristic angle $\cos^{-1}(\omega/2\Omega)$, and found that this yielded results broadly consistent with the observed seven inertial modes. Compare also Figures 7.23 and 7.24; based on similarities such as these Kelley et al. (2010) suggested that some of the modes observed in DTS may also be overreflected inertial waves (or

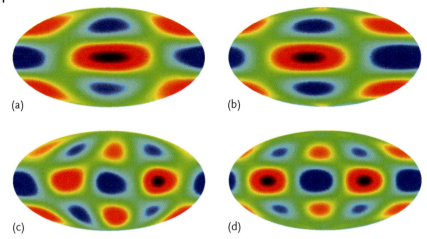

Figure 7.24 Shown are the large-scale modes observed in the experiment (a,c), as well as the corresponding numerical eigenmodes (b,d). Specifically, the induced magnetic field is shown as a Mollweide projection. The top row has $\Omega_{in}/2\pi = -12.2$ Hz and $\Omega_{out}/2\pi = 29.9$ Hz, and the bottom row has $\Omega_{in}/2\pi = 12.0$ Hz and $\Omega_{out}/2\pi = 29.9$ Hz, corresponding in both cases to Re = $O(10^7)$. From Kelley et al. (2007). Reprinted by permission of Taylor and Francis Ltd (http://tandf.co.uk/journals).

perhaps more accurately magneto-inertial waves (e.g., Schmitt, 2010), since magnetic effects are almost certainly dynamically more significant in DTS than in the Kelley et al. 0.04 T field).

Matsui et al. (2011) carried out high-resolution numerical simulations of the Kelley et al. setup, and obtained similar inertial waves. The general picture of Stewartson layer instabilities therefore seems to be that the initial onset in the laminar regime consists of Kelvin–Helmholtz type instabilities that are concentrated largely on the shear layer itself (Hollerbach, 2003; Hollerbach et al., 2004; Schaeffer and Cardin, 2005), but that in the increasingly supercritical regime the instabilities consist more and more of these almost free inertial oscillations that are only very weakly dampened, and hence require only relatively little forcing, via this overreflection mechanism.

Finally, we note that the same Maryland group have recently built a very large ($r_{in} = 0.51$ m, $r_{out} = 1.46$ m) apparatus, and are using it to further study inertial oscillations in spherical Couette flow (Rieutord et al., 2012). Thus far at least these are entirely nonmagnetic studies though, without any imposed field, and hence without the ability to probe the flow through measurements of an induced field.

7.5.2
Torsional Oscillations

All the results discussed thus far have imposed steady rotation rates of the inner and outer spheres; any time-dependent phenomena have been the result of instabilities arising on the stationary background state. Jault (2008) considered the situation where a differential rotation between the two spheres is applied impulsively, and the subsequent transients are studied. That is, the initial condition consists of a spherical shell in rapid solid-body rotation, with an externally applied magnetic field. In this rotating frame, a δ function is applied to the inner sphere's rotation rate, abruptly rotating it by some finite angle with respect to the outer sphere, and thereafter corotating with the outer sphere again.

Figure 7.25 shows the subsequent evolution. The first two panels illustrate the very short-term response, in the induced magnetic field B_ϕ. We see first of all how the shear in the (transient) Ekman–Hartmann boundary layer acts upon the external magnetic field to induce a strong B_ϕ field in the boundary layer. The other feature to note is the gradual development of a certain structure extending along the tangent cylinder – and this despite the fact that the externally imposed field is *not* purely axial (as indicated also in Figure 7.26). The second two panels illustrate the longer-term response, in the angular velocity. We see how the structure that was initially concentrated on the tangent cylinder has split into two waves, so-called torsional oscillations, propagating away from the tangent cylinder in both directions.

Jault's interest in studying this problem was two-fold: first, to demonstrate that the resulting structures will be axially invariant, even though the imposed field is not. That is, the Taylor–Proudman theorem is more important here than the Ferraro isorotation law. Second, what is the most appropriate measure of the magnetic field's strength? In the suppression of the Stewartson layer considered in Section 7.1, for example, it was the Elsasser number $\Lambda = B_0^2/\Omega\mu\rho\eta$, which we note involves the magnetic diffusivity η. Jault argued – and backed it up with his calculations – that on timescales short compared with the magnetic diffusion time, this cannot be the relevant parameter. Instead, the relevant measure is the so-called Lehnert number

$$\lambda = \frac{B_0}{\Omega(\mu\rho)^{1/2}l}, \tag{7.13}$$

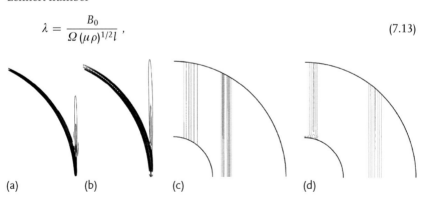

(a) (b) (c) (d)

Figure 7.25 Numerical solutions of the impulsive inner sphere rotation problem, with $\lambda = 1.72 \times 10^{-4}$ and $\Lambda = 0.52$. The first two panels (a,b) show the induced azimuthal magnetic field at times 0.86×10^{-2} and 1.72×10^{-2}, on the Alfvén timescale. The second two panels (c,d) show the angular velocity at the later times 0.52 and 1.03. Reprinted from Jault (2008), Copyright 2008, with permission from Elsevier.

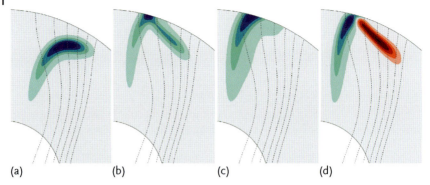

Figure 7.26 The first panel (a) shows the Alfvén wave packet traveling outward along the magnetic field lines, indicated by the dashed lines (Jault's results in Figure 7.25 used the same imposed field). The second panel (b) shows reflection with the same sign for Pm = 0.1, the third panel (c) shows total absorption for Pm = 1, and the fourth panel (d) shows reflection with the opposite sign for Pm = 10. From Schaeffer et al. (2012), by permission of Oxford University Press.

where l is a typical length scale. Physically λ can be thought of as the ratio of the periods of an inertial wave and an Alfvén wave. If l is taken to be the Earth's core radius $\lambda = O(10^{-4})$, indicating the strong dominance of rotational over magnetic effects on these short timescales. In contrast, the Elsasser number $\Lambda = O(1)$ in the core, indicating that on longer timescales Coriolis and Lorentz forces are more comparable. The importance of both Λ and λ was also emphasized by Cardin et al. (2002); the DTS experiment was designed to be in the regime $\Lambda = O(1)$ and $\lambda < O(1)$.

7.5.3
Alfvén Waves

Another variant on the Jault (2008) problem was considered by Schaeffer et al. (2012), who considered the case where the only rotation is the δ function impulse, but otherwise both spheres are at rest. The waves generated in this case are pure Alfvén waves, as indicated in Figure 7.26. One particularly intriguing aspect of this work is what happens once the waves reach the outer boundary and reflect back inward. If Pm < 1 the incident and reflected waves are in phase, if Pm > 1 they are out of phase, and if Pm = 1 there is essentially no reflected wave at all, with all of the incident energy dissipated in the Hartmann boundary layer. In the Earth's core of course Pm $\neq$ 1, so any results that only occur at Pm = 1 are not relevant to the real Earth. Many numerical geodynamo simulations do use Pm = 1 though, so it is well worth recognizing that this case may be special in some ways.

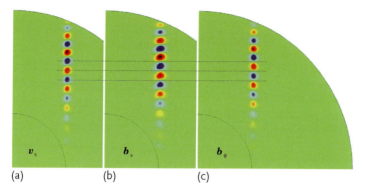

Figure 7.27 Snapshots of U_s (a), B_s (b), and B_ϕ (c), where s is cylindrical radius, in the magnetostrophic MRI. Note the particular phase relationship between the various quantities, exactly as in the standard MRI. From Petitdemange, Dormy, and Balbus (2008). Copyright 2008 by The American Geophysical Union.

7.5.4
The Magnetostrophic MRI

Another interesting wave phenomenon in magnetic spherical Couette flow is due to Petitdemange, Dormy, and Balbus (2008), who showed that it is possible to obtain a variant of the magnetorotational instability in situations that might be relevant to the Earth's core. They considered the Stewartson–Shercliff case of a rapid overall rotation together with an axial magnetic field, essentially the same configuration as in Hollerbach (1997) and Wei and Hollerbach (2008), but now at finite Pm rather than in the Pm → 0 limit. Figure 7.27 shows the resulting small-scale vortices, exactly like the classical MRI (as in Chapter 5). If the differential rotation in the Earth's core has the right profile, it would be possible therefore for the MRI to occur. The MRI is nevertheless unlikely to play as important a role in planetary interiors as it does in accretion disks, for the following reason: the MRI is essentially a mechanism for tapping into the kinetic energy stored in the differential rotation, energy that would otherwise be unavailable if the flow is Rayleigh-stable. In accretion disks this reservoir of energy is enormous, making the MRI correspondingly important. In contrast, in planetary interiors the energy in the differential rotation is far less – and what little differential rotation there is may well have been generated by tapping into the magnetic energy, rather than vice versa. On energetic grounds alone therefore the MRI is likely to be less important. Nevertheless, the possibility that the MRI can occur at all may play a role in limiting how steep the differential rotation profile can become, for example.

7.6
Linear Combinations of Axial and Dipolar Fields

Returning to the basic theme of Section 7.2, we recall that Shercliff layers/jets always occurred on certain special field lines separating regions linked only to one boundary from those linked to both. It is thus of interest to classify all possible field topologies, and see if there are other choices, fundamentally different from the ones considered thus far. This question was addressed by Hollerbach (2001), who argued as follows: there are three different types of field lines, ones linking both boundaries, ones linked only to the inner, and ones linked only to the outer. (Loops closed entirely within the shell are excluded by the requirement that the imposed field be current-free within the fluid.) Given these three types, there are eight topologies one can imagine, corresponding to each type being independently either present or absent. One of these eight, having all field lines linking both boundaries, and none only the inner or only the outer, would require the field to have a monopole component, and is therefore excluded. Another, having all three field line types absent, is of course just nonmagnetic spherical Couette flow, and is thus not relevant here. We are left then with six possible field topologies, which we will denote *bi*, *bo*, *bio*, *i*, *o*, *io*, where the *b/i/o* denotes the presence of field lines linking both boundaries, inner only, and outer only. An axial field corresponds to type *bo*; a dipole (or quadrupole) to type *bi*. The question that Hollerbach (2001) posed is, what about the types *bio*, *i*, *o*, and *io*? Is it possible to construct (reasonably simple) fields having these topologies, and what do the resulting Shercliff layers/jets look like?

Very conveniently, all four of the 'missing' topologies can be constructed simply by taking suitable linear combinations of axial and dipolar fields. Let

$$\mathbf{B}_a = \cos\theta\,\hat{\mathbf{e}}_r - \sin\theta\,\hat{\mathbf{e}}_\theta, \quad \mathbf{B}_d = 8r^{-3}\cos\theta\,\hat{\mathbf{e}}_r + 4r^{-3}\sin\theta\,\hat{\mathbf{e}}_\theta \quad (7.14)$$

denote the two basic ingredients, and consider the linear combinations

$$\mathbf{B}_0 = \mathbf{B}_d + \epsilon\,\mathbf{B}_a. \quad (7.15)$$

That is, start with a dipole, and add an increasingly strong axial field, aligned either parallel or antiparallel. (The entire field can subsequently be renormalized so that the rms value is 1, say.) As indicated in Figure 7.28 – and not surprisingly of course – for both positive and negative ϵ, $\mathbf{B}_0$ starts out as type *bi*, like $\mathbf{B}_d$, and ends up as *bo*, like $\mathbf{B}_a$. However, between those limits negative values of ϵ yield types *i*, *io*, and *o*, and positive values of ϵ yield type *bio*, thereby generating the four previously missing field topologies.

Figures 7.29 and 7.30 present the resulting flow structures, for insulating and conducting boundaries, in the linear limit Re = 0. The results are broadly similar to those in Section 7.2; insulating boundaries yield shear layers, whereas conducting boundaries yield jets. The field with an X-type neutral point at $r = 1.5$ demonstrates that it is possible to have superrotating and counterrotating jets simultaneously. The *i* and *o* fields also yield counterrotating and superrotating (respectively)

7.6 Linear Combinations of Axial and Dipolar Fields | 319

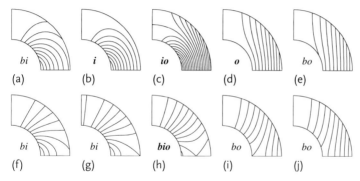

Figure 7.28 Linear combinations of the type $B_d + \epsilon B_a$. The top row (a–e) has $\epsilon = -1/2$, -1, $-8/1.5^3$, -8, -16, respectively. At $\epsilon = -1$ it is of type i, with all field lines linked only to the inner boundary. Between $\epsilon = -1$ and -8 it is of type io, with some field lines linked only to the inner boundary and some only to the outer, but none to both. At $\epsilon = -8$ it is of type o, with all field lines linked only to the outer boundary. The bottom row (f–j) has $\epsilon = 1/4$, $1/2$, $4/1.5^3$, 4, 8, respectively. At $\epsilon = 1/2$ an X-type neutral point appears at the equator of the outer boundary, at $\epsilon = 4/1.5^3$ it is in the middle of the gap, and the field is of type bio, and at $\epsilon = 4$ the X-type neutral point disappears again at the equator of the inner boundary.

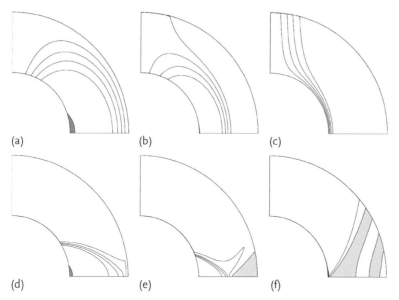

Figure 7.29 Contours of the angular velocity, with insulating inner and outer boundaries. The top row has $\epsilon = -1$ (a), $-8/1.5^3$ (b), -8 (c), the bottom row $1/2$ (d), $4/1.5^3$ (e), 4 (f), corresponding to Figure 7.28b–d,g–i. The contour interval is 0.2, light shading indicates counterrotation, and dark shading superrotation. $Ha = 10^3$ and $Re = 0$.

vortices in the polar regions. Hollerbach (2001) further investigated the scalings with Ha of some of these structures, and found that the jets in the $\epsilon > 0$ cases increase with Ha much as before in Section 7.2, but the vortices in the $\epsilon < 0$ cases

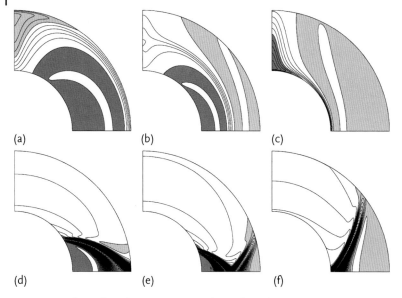

Figure 7.30 All panels (a–f) as in Figure 7.29, but with conducting inner and outer boundaries. In the bottom row $\epsilon = 1/2$ has maximum superrotation 7.7, $\epsilon = 4/1.5^3$ yields values between -5.1 and 9.9, and $\epsilon = 4$ has maximum counterrotation -10.0.

saturate, at $\sim 70\%$ for the counterrotating one, and $\sim 100\%$ for the superrotating one.

The next issue to address is what happens at finite Re, as before in Sections 7.3 and 7.4. Wei and Hollerbach (2010) explored this regime for some of the $\epsilon < 0$ fields. As shown in Figure 7.31, the counterrotating vortex for example is eventually completely eliminated. How the $\epsilon > 0$ fields behave at finite Re has not been studied at all before. Figure 7.32 presents sample results illustrating how the counterrotating jet is also strongly suppressed, and gradually shifts outward, analogous to the inward shift of the superrotating jet in Figure 7.21, for example. Note though that the counterrotating jet in a purely axial field exhibits far less suppression, and no outward shift (Hollerbach and Skinner, 2001), indicating that even comparative-

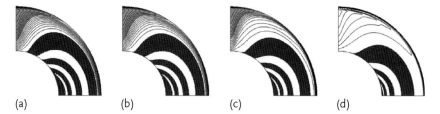

Figure 7.31 Contours of the angular velocity in the conducting boundary, $\epsilon = -1$, Ha $= 10^3$ case, at Re $= 300$ (a), 1000 (b), 3000 (c), 10 000 (d). The contour interval is 0.2, light shading indicates counterrotation, and dark shading superrotation. From Wei and Hollerbach (2010), with permission from Springer Science and Business Media.

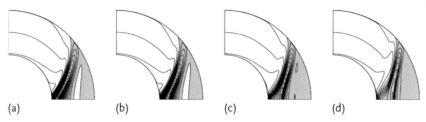

(a)   (b)   (c)   (d)

Figure 7.32 Contours of the angular velocity in the conducting boundary, $\epsilon = 4$, $Ha = 10^{2.5}$ case, at $Re = 100$ (a), 300 (b), 1000 (c), 3000 (d). The contour interval is 0.2, shading indicates counterrotation.

ly minor changes in the structure of B_0 can significantly affect the resulting solutions. Mapping out the entire range of possibilities – including also the possibility of nonaxisymmetric instabilities, which were not considered at all in Figures 7.31 or 7.32 – would undoubtedly require far more work.

Given all these new field configurations that can be achieved simply by taking linear combinations of axial and dipolar fields, and the new flow structures that result, it would be of interest to study some of them experimentally as well. Recalling the previous experiments in Maryland and Grenoble, adding an axial field to the DTS setup would probably be easier than adding a dipolar field to the Sisan et al. setup. An axial field up to 0.345 T would be required to achieve all the $\epsilon < 0$ topologies up to where $B_r = 0$ on the inner sphere; only half that would already be sufficient to achieve all the $\epsilon > 0$ topologies up to where the X-type neutral point is located at the equator of the inner boundary. (For comparison, we recall that Sisan et al. achieved field strengths up to 0.2 T.)

Furthermore, only the i and o topologies, having B_r identically zero on the outer and inner boundaries, respectively, depend crucially on having linear combinations of *exactly* axial and dipolar fields. All the other topologies are qualitatively unchanged if the fields are perturbed slightly. Adding a pair of Helmholtz coils, which does not yield an exactly axial field, would therefore already be enough to yield many interesting new combinations. One might hope that such a field could eventually be added to DTS.

7.7
Dynamo Action

The previous sections have all considered spherical Couette flow in the presence of externally imposed magnetic fields. Far more challenging – but also enormously interesting – would be to consider the possibility of a spherical Couette flow dynamo. This is in fact the objective of the latest 3 m diameter experiments planned in Maryland and Madison (e.g., Lathrop and Forest, 2011), but thus far dynamo action has not been achieved. As noted in Section 7.5.1 though, Rieutord et al. (2012) have already used the Maryland apparatus to study nonmagnetic inertial oscillations.

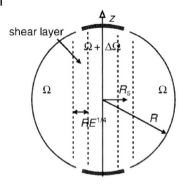

Figure 7.33 A sketch of the 'split sphere' setup, in which the main body of the sphere rotates at rate Ω, but the polar caps rotate at rate $\Omega + \Delta\Omega$. The result is a Stewartson layer involving inner thickness $E^{1/3}$ and outer thickness $E^{1/4}$. The $E^{2/7}$ layer is absent in this case, but otherwise the layers are much the same as in the spherical shell case. Reprinted from Schaeffer and Cardin (2006), Copyright 2006, with permission from Elsevier.

The first to numerically investigate the possibility of a spherical Couette flow dynamo were Schaeffer and Cardin (2006). Their setup was actually slightly different from the classical, differentially rotating spherical shell. Instead, they considered a split sphere, as in Figure 7.33. The flow that this produces is very similar to the Stewartson layer. The advantage of generating it in a split sphere like this is that the height of fluid columns then varies smoothly with cylindrical radius, allowing use of the so-called quasi-geostrophic approximation, in which vertical variations of the flow are dealt with analytically. The resulting flow (analytical in z, numerical in the equatorial plane) is then tested for kinematic dynamo action using a fully 3D code for the induction equation. The simplifications that result from solving for U and B so differently allow one to explore parameter regimes – in particular small magnetic Prandtl numbers – that would not be accessible otherwise.

Figure 7.34 shows an example of one of their dynamos. Not surprisingly, the toroidal field in particular is strongly concentrated near the vicinity of the Stewartson layer. The poloidal field also has a significant external dipole component. This particular solution is oscillatory; for other parameter values steady solutions were obtained.

Guervilly and Cardin (2010) were the first to study dynamo action in the true spherical Couette flow configuration. Because the same, fully 3D numerical approach is used for both U and B, it is possible also to include the Lorentz force whereby B acts back on U, and study the magnetic field's saturation. The price to pay is that many of the parameters, such as the Ekman or magnetic Prandtl numbers, cannot be pushed to quite such extreme values as in the reduced Schaeffer and Cardin (2006) approach. Guervilly and Cardin also considered both the Stewartson layer configuration in which both spheres are rapidly rotating, as well as the radial jet configuration in which only the inner sphere rotates. Figure 7.35 shows an example of a dynamo in the radial jet regime; note how the field is indeed concentrated in the equatorial plane. From the point of view of building an experimen-

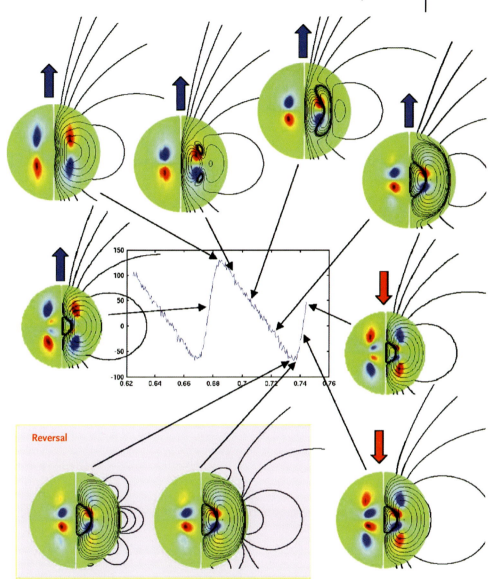

Figure 7.34 The central panel shows the instantaneous growth rate of the magnetic energy as a function of time; that this quantity is on average positive indicates kinematic dynamo action. The surrounding plots show the spatial structure of the axisymmetric part of the field, including the nature of the reversal process. $E = 10^{-6}$, $\Delta\Omega/\Omega = -0.08$ and Pm = 0.1. Reprinted from Schaeffer and Cardin (2006), Copyright 2006, with permission from Elsevier.

tal spherical Couette dynamo, the main conclusion of Guervilly and Cardin's work is that magnetic Reynolds numbers of several hundred are most likely required,

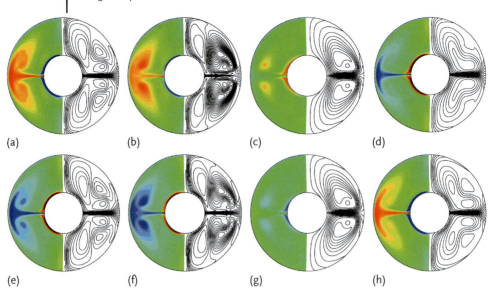

Figure 7.35 Eight snapshots in time of the axisymmetric part of the field (a–h), toroidal on the left and poloidal on the right. The oscillating nature of the solution is particularly clear in the toroidal field. From Guervilly and Cardin (2010). Reprinted by permission of Taylor and Francis Ltd (http://tandf.co.uk/journals).

which is unfortunately rather challenging to achieve even in devices as large as the 3 m Maryland setup.

One reason why spherical Couette flow might not be a very efficient dynamo is that it is difficult to achieve the right balance of azimuthal and meridional flow, at least throughout the interior, away from narrow boundary layers. In the Stewartson layer regime the flow is strongly dominated by the azimuthal velocity, and the meridional circulation is very small. In contrast, the radial jet regime has a much stronger meridional circulation, but the azimuthal velocity is significant only within the boundary layer on the inner sphere. Furthermore, spinning the inner sphere faster and faster, in the hopes of achieving dynamo action eventually, only makes this boundary layer ever thinner, reducing its dynamo efficiency in the process. Finke and Tilgner (2012) asked what would happen if blades were mounted on the inner sphere, thereby improving the coupling between the sphere and the fluid, and preventing the boundary layer from becoming too thin. Their calculations, using a spherical shell code in which the effect of the blades was mimicked by an artificial body force near the inner sphere, yielded a reduction by as much as a factor of 10 for the critical rotation rate. They suggest that even with blades a device such as the one in Maryland would require an inner sphere rotation rate of ~ 14 Hz to achieve dynamo action.

Finally, let us imagine a configuration that one can think of as an extension of the Figure 7.33 split-sphere setup. Instead of imposing $\Omega + \Delta\Omega$ inside the polar caps and Ω elsewhere, impose a completely general zonal flow profile $U_\phi = f(\theta)$ on the boundary. Instead of a single Stewartson layer, this will generate a differ-

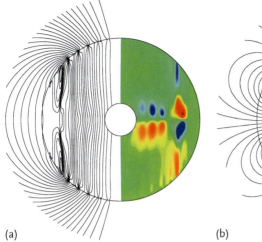

(a) (b)

Figure 7.36 Snapshots of the axisymmetric parts of the magnetic field, with the poloidal field as the left half and the toroidal field as the right half of each panel (a,b). Imposing many narrow jets on the boundary yields the solution (a), a few broad jets yields the solution (b). Reprinted from Guervilly, Cardin, and Schaeffer (2012), Copyright 2012, with permission from Elsevier.

ential rotation profile varying throughout the sphere. Rossby waves similar to the Stewartson layer instabilities can still develop though, leading to complicated three-dimensional flows. Guervilly, Cardin, and Schaeffer (2012) considered two such flows, one a Jupiter-like profile with many narrow jets, and the other a Neptune-like profile with fewer, broader jets, and demonstrated that both work as dynamos. Figure 7.36 illustrates the resulting magnetic field structures. The suggestion then is that convective driving in the conducting regions may not be necessary for dynamo action in giant planets at all; the forcing imposed by the zonal flows in the overlying nonmagnetic regions could already be sufficient. At the very least, these results suggest that the pattern of zonal flows in the overlying regions could have a strong influence on the pattern of the magnetic fields ultimately generated in the conducting regions below.

Dynamos such as these, driven by imposing essentially arbitrary profiles of U_ϕ on the boundaries, are particularly interesting also in light of the Madison Plasma Couette Experiment (Collins *et al.*, 2012; Katz *et al.*, 2012), in which very small-scale magnetic fields are first used to confine an otherwise unmagnetized plasma, and similarly small-scale injected currents at the boundaries are then used to generate Lorentz forces, and thereby drive the plasma at the edges, just as if U_ϕ had been imposed directly. By varying the electron temperature the magnetic Prandtl number can also be adjusted (unlike in liquid metal experiments), making a broad range of ordinary and magnetic Reynolds numbers accessible. The prototype described by Collins *et al.* and Katz *et al.* is cylindrical, with both height and diameter slightly less than 1 m. A 3 m diameter spherical experiment is under development (e.g., Lathrop and Forest, 2011); see also Khalzov *et al.* (2012a,b) for numerics similar in spirit to Guervilly, Cardin, and Schaeffer (2012), but applied to the experiment. Future developments in MHD spherical Couette dynamos from both the Madison and Maryland experiments promise to be very interesting.

References

Abt, H.A. and Morrell, N.I. (1995) *Astrophys. J. Suppl.*, **99**, 135.
Acheson, D.J. and Hide, R. (1973) *Rep. Prog. Phys.*, **36**, 159.
Acheson, D.J. (1978) *Philos. Trans. R. Soc. A*, **289**, 459.
Alonso, R., Auvergne, M., Baglin, A., et al. (2008) *Astron. Astrophys.*, **482**, 21.
Altmeyer, S., Hoffmann, Ch., Heise, M., et al. (2010) *Phys. Rev. E*, **81**, 066313.
Ammler-von Eiff, M. and Reiners, A. (2012), *Astron. Astrophys.*, **542**, A116.
Anantharamaiah, K.R., Radhakrishnan, V., and Shaver, P.A. (1984) *Astron. Astrophys.*, **138**, 131.
Anders, E. and Grevesse, N. (1989) *Geochim. Cosmochim. Acta*, **53**, 197.
Andersson, N. (1998) *Astrophys. J.*, **502**, 708.
Antia, H.M., Basu, S., and Chitre, S.M. (1998) *Mon. Not. R. Astron. Soc.*, **298**, 543.
Araki, K., Mizushima, J., and Yanase, S. (1997) *Phys. Fluids*, **9**, 1197.
Ardeljan, N.V., Bisnovatyi-Kogan, G.S., and Moiseenko, S.G. (2005) *Mon. Not. R. Astron. Soc.*, **359**, 333.
Arlt, R., Sule, A., and Rüdiger, G. (2007) *Astron. Astrophys.*, **441**, 1171.
Arlt, R., Sule, A., and Rüdiger, G. (2007) *Astron. Astrophys.*, **461**, 295.
Arlt, R., Sule, A., and Filter, R. (2007) *Astron. Nachr.*, **328**, 1142.
Arlt, R. (2009) *Sol. Phys.*, **255**, 143.
Arlt, R. and Rüdiger, G. (2011a) *Astron. Nachr.*, **332**, 70.
Arlt, R. and Rüdiger, G. (2011b) *Mon. Not. R. Astron. Soc.*, **412**, 107.
Arlt, R. and Fröhlich, H.E. (2012) *Astron. Astrophys.*, **543**, A7.
Asplund, M., Grevesse, N., Sauval, A.J., et al. (2009) *Annu. Rev. Astron. Astrophys.*, **47**, 481.
Augustson, K.C., Brown, B.P., Brun, A.S., et al. (2012) *Astrophys. J.*, **756**, 169.
Auriére, M., Wade, G.A., Silvester, J., et al. (2007) *Astron. Astrophys.*, **475**, 1053.
Avila, M., Belisle, M.J., Lopez, J.M., et al. (2008) *J. Fluid Mech.*, **601**, 381.
Avila, M. (2012) *Phys. Rev. Lett.*, **108**, 124501.
Bagnulo, S., Landi Degl'Innocenti, M., Landolfi, M., et al. (2002) *Astron. Astrophys.*, **394**, 1023.
Balachandran, S.C. and Bell, R.A. (1998) *Nature*, **392**, 791.
Balbus, S.A. and Terquem, C. (2001) *Astrophys. J.*, **552**, 235.
Balbus, S.A. (2011) *Nature*, **470**, 475.
Balbus, S.A., Latter, H., and Weiss, N. (2012) *Mon. Not. R. Astron. Soc.*, **420**, 2457.
Balthasar, H., Vázquez, M., and Wöhl, H. (1986) *Astron. Astrophys.*, **155**, 87.
Banerjee, R. and Jedamzik, K. (2003) *Phys. Rev. Lett.*, **91**, 251301.
Bao, S. and Zhang, H. (1998) *Astrophys. J.*, **496**, L43.
Barnes, G., Charbonneau, P., and MacGregor, K.B. (1999) *Astrophys. J.*, **511**, 466.
Barnes, J.R., Collier Cameron, A., James, D.J., et al. (2000) *Mon. Not. R. Astron. Soc.*, **314**, 162.
Barnes, J.R., Collier Cameron, A., Donati, J.-F., et al. (2005) *Mon. Not. R. Astron. Soc.*, **357**, L1.
Barnes, S.A. (2003) *Astrophys. J.*, **586**, 464.
Barnes, S.A. (2007) *Astrophys. J.*, **669**, 1167.
Barnes, S.A. (2010) *Astrophys. J.*, **722**, 222.
Basu, S. (1997) *Mon. Not. R. Astron. Soc.*, **288**, 572.

Basu, S. and Antia, H.M. (1997) *Mon. Not. R. Astron. Soc.*, **287**, 189.

Basu, S. and Antia, H.M. (2010) *Astrophys. J.*, **133**, 572.

Beck, P.G., Montalbán, J., Kallinger, T., et al. (2012) *Nature*, **481**, 55.

Beck, R. (1993) in *The Cosmic Dynamo* (eds F. Krause et al.), Kluwer Academic Publishers, p. 283.

Beck, R., Poezd, A.D., Shukurov, A., et al. (1994) *Astron. Astrophys.*, **289**, 94.

Beck, R. (1996) in *High-Sensitivity Radio Astronomy* (ed. D. Jackson), University of Manchester, p. 117.

Beck, R. and Hoernes, P. (1996) *Nature*, **379**, 47.

Beck, R., Brandenburg, A., Moss, D., et al. (1996) *Annu. Rev. Astron. Astrophys.*, **34**, 155.

Beck, R. (2002) in *Highlights of Astronomy* (ed. H. Rickman), ASP, p. 712.

Beck, R. (2007) *Astron. Astrophys.*, **470**, 539.

Becker, W. and Pavlov, G. (2001) in *The Century of Space Science* (eds J. Bleeker et al.), Kluwer, Chapter 8.

Berger, L., Koester, D., Napiwotzki, R., et al. (2005) *Astron. Astrophys.*, **444**, 565.

Biermann, L. (1951) *Z. Astrophys.*, **28**, 304.

Bonanno, A., Elstner, D., Rüdiger, G., et al. (2002) *Astron. Astrophys.*, **390**, 673.

Bonanno, A. and Urpin, V. (2008) *Astron. Astrophys.*, **488**, 1.

Bonanno, A. and Urpin, V. (2011) *Astron. Astrophys.*, **525**, A100.

Bonanno, A. (2012) *Geophys. Astrophys. Fluid Dyn.*, **107**, 11.

Bonanno, A., Brandenburg, A., Del Sordo, F., et al. (2012) *Phys. Rev. E*, **86**, 016313.

Bräuer, H.J. and Krause, F. (1974) *Astron. Nachr.*, **295**, 223.

Braithwaite, J. and Spruit, H.C. (2004) *Nature*, **431**, 819.

Braithwaite, J. (2006) *Astron. Astrophys.*, **453**, 678.

Brandenburg, A., Tuominen, I., Moss, D., et al. (1990) *Sol. Phys.*, **128**, 243.

Brandenburg, A., Moss, D., and Tuominen, I. (1992) *Astron. Astrophys.*, **265**, 328.

Brandenburg, A., Nordlund, Å, Stein, R.F., et al. (1995) *Astrophys. J.*, **446**, 741.

Brandenburg, A., Jennings, R.L., Nordlund, Å, et al. (1996) *J. Fluid Mech.*, **306**, 325.

Brandenburg, A., Saar, S.H., and Turpin, C.R. (1998) *Astrophys. J.*, **498**, L51.

Brandenburg, A., Moss, D., and Soward, A.M. (1998) *Proc. R. Soc. Lond. A*, **454**, 1283.

Brandenburg, A. (2001) *Astrophys. J.*, **550**, 824.

Brandenburg, A. (2005a) *Astrophys. J.*, **625**, 539.

Brandenburg, A. (2005b) *Astron. Nachr.*, **326**, 787.

Brandenburg, A. and Subramanian, K. (2005) *Phys. Rep.*, **417**, 1.

Brandenburg, A. and Subramanian, K. (2007) *Astron. Nachr.*, **328**, 507.

Brandenburg, A., Rädler, K.H., Rheinhardt, M., et al. (2008) *Astrophys. J.*, **686**, 740.

Brandenburg, A., Kleeorin, N., and Rogachevskii, I. (2010) *Astron. Nachr.*, **331**, 5.

Brandenburg, A., Kemel, K., Kleeorin, N., et al. (2011) *Astrophys. J.*, **740**, L50.

Brandenburg, A., Sokoloff, D., and Subramanian, K. (2012) *Space Sci. Rev.*, **169**, 123.

Brandenburg, A., Gressel, O., Käpylä, P., et al. (2013) *Astrophys. J.*, **762**, 127.

Brandt, P.N., Scharmer, G.B., Ferguson, S.H., et al. (1988) *Nature*, **335**, 238.

Breitschwerdt, D., de Avillez, M.A., Feige, J., et al. (2012) *Astron. Nachr.*, **333**, 486.

Brito, D., Alboussière, T., Cardin, P., et al. (2011) *Phys. Rev. E*, **83**, 066310.

Brott, I., Hunter, I., Anders, P., et al. (2008) *AIP Conf. Proc.*, **990**, 273.

Brown, T.M. (1985) *Nature*, **317**, 591.

Brown, B.P., Browning, M.K., Brun, A.S., et al. (2008) *Astrophys. J.*, **689**, 1354.

Brown, B.P., Browning, M.K., Brun, A.S., et al. (2010) *Astrophys. J.*, **711**, 424.

Brown, B.P., Miesch, M.S., Browning, M.K., et al. (2011) *Astrophys. J.*, **731**, 69.

Brun, A.S. and Toomre, J. (2002) *Astrophys. J.*, **570**, 865.

Brun, A.S. and Zahn, J.-P. (2006) *Astron. Astrophys.*, **457**, 665.

Bühler, L. (2009) *Theor. Comput. Fluid Dyn.*, **23**, 491.

Burgess, C., Dzhalilov, N.S., Maltoni, M., et al. (2003) *Astrophys. J.*, **588**, L65.

Burin, M.J., Schartman, E., and Ji, H. (2010) *Exp. Fluids*, **48**, 763.

Burrows, A., Dessart, L., Livne, E., et al. (2007) *Astrophys. J.*, **664**, 416.

Cally, P.S. (2001) *Sol. Phys.*, **199**, 231.

Cally, P.S. (2003) *Mon. Not. R. Astron. Soc.*, **339**, 957.

Cally, P.S., Dikpati, M., and Gilman, P.A. (2003) *Astrophys. J.*, **582**, 1190.

Cameron, R.H. and Schüssler, M. (2012) *Astron. Astrophys.*, **548**, A57.

Cardin, P., Brito, D., Jault, D., et al. (2002) *Magnetohydrodynamics*, **38**, 177.

Carilli, C.L. and Taylor, G.B. (2002) *Annu. Rev. Astron. Astrophys.*, **40**, 319.

Cattaneo, F. and Vainshtein, S.I. (1991) *Astrophys. J.*, **376**, L21.

Ceillier, T., Eggenberger, P., García, R.A., et al. (2012) *Astron. Nachr.*, **333**, 971.

Chan, K.L. (2001) *Astrophys. J.*, **548**, 1102.

Chabrier, G. and Baraffe, I. (1997) *Astron. Astrophys.*, **327**, 1039.

Chabrier, G. and Küker, M. (2006) *Astron. Astrophys.*, **446**, 1027.

Champagne, F.H., Harris, V.G., and Corrsin, S. (1970) *J. Fluid Mech.*, **41**, 81.

Chandrasekhar, S. (1956) *Proc. Natl. Acad. Sci. USA*, **42**, 273.

Chandrasekhar, S. (1960) *Proc. Natl. Acad. Sci. USA*, **46**, 53.

Chandrasekhar, S. (1961) *Hydrodynamic and Hydromagnetic Stability*, Oxford University Press.

Charbonneau, P. and MacGregor, K.B. (1992) *Astrophys. J.*, **397**, L63.

Charbonneau, P. and MacGregor, K.B. (1993) *Astrophys. J.*, **417**, 762.

Charbonneau, P., Christensen-Dalsgaard, J., Henning R., et al. (1999) *Astrophys. J.*, **527**, 445.

Charbonneau, P. and MacGregor, K.B. (2001) *Astrophys. J.*, **559**, 1094.

Charbonneau, P. (2005) *Living Rev. Sol. Phys.*, **2**.

Chatterjee, P., Mitra, D., Brandenburg, A., et al. (2011) *Phys. Rev. E*, **84**, 5403.

Cheung, N.C.M. and Cameron, R.H. (2012) *Astrophys. J.*, **750**, 6.

Choudhuri, A.R., Schüssler, M., and Dikpati, M. (1995) *Astron. Astrophys.*, **303**, L29.

Choudhuri, A.R. (2008) *Adv. Space Res.*, **41**, 868.

Christensen-Dalsgaard, J., Gough, D.O., and Thompson, M.J. (1991) *Astrophys. J.*, **378**, 413.

Chyży, K.T., Beck, R., Kohle, S., et al. (2000) *Astron. Astrophys.*, **355**, 128.

Chyży, K.T. (2008) *Astron. Astrophys.*, **482**, 755.

Chyży, K.T. and Buta, R. (2008) *Astrophys. J.*, **677**, 17.

Collier Cameron, A., Davidson, V.A., Hebb, L., et al. (2009) *Mon. Not. R. Astron. Soc.*, **400**, 451.

Collins, C., Katz, N., Wallace, J., et al. (2012) *Phys. Rev. Lett.*, **101**, 115001.

Couette, M. (1890) *Annu. Chem. Phys.*, **21**, 433.

Couvidat, S., García, R.A., Turck-Chièze, S., et al. (2003) *Astrophys. J.*, **597**, L77.

Cowling, T.G. (1933) *Mon. Not. R. Astron. Soc.*, **94**, 39.

Cowling, T.G. (1945) *Mon. Not. R. Astron. Soc.*, **105**, 166.

Croll, B., Walker, G.A.H., Kuschnig, R., et al. (2006) *Astrophys. J.*, **648**, 607.

Dall'Osso, S., Shore, S.N., and Stella, L. (2009) *Mon. Not. R. Astron. Soc.*, **398**, 1869.

Dame, T.M. and Thaddeus, P. (1994) *Astrophys. J.*, **436**, L173.

de Avillez, M.A. and Breitschwerdt, D. (2007) *Astrophys. J.*, **665**, 35.

Deguen, R. (2012) *Earth Planet. Sci. Lett.*, **333**, 211.

Deheuvels, S., García, R.A., Chaplin, W.J., et al. (2012) *Astrophys. J.*, **756**, 19.

Denissenkov, P.A. (2010) *Astrophys. J.*, **719**, 28.

Deiss, B.M., Reich, W., Lesch, H., et al. (1997) *Astron. Astrophys.*, **321**, 55.

Denissenkov, P.A., Pinsonneault, M., Terndrup, D.M., et al. (2010) *Astrophys. J.*, **716**, 1269.

Deville, M.O., Fischer, P.F., and Mund, E.H. (2002) *High-Order Methods for Incompressible Fluid Flow*, Cambridge University Press.

Dicke, R.H. (1970) *Annu. Rev. Astron. Astrophys.*, **202**, 432.

Dickey, J.M. and Lockman, F.J. (1990) *Annu. Rev. Astron. Astrophys.*, **28**, 215.

Dikpati, M. and Charbonneau, P. (1999) *Astrophys. J.*, **518**, 508.

Dikpati, M. and Gilman, P.A. (2001) *Astrophys. J.*, **559**, 428.

Dikpati, M., Cally, P.S., and Gilman, P.A. (2004) *Astrophys. J.*, **610**, 597.

Dimmelmeier, H., Font, J.A., and Müller, E. (2002) *Astron. Astrophys.*, **393**, 523.

Donahue, R.A., Saar, S.H., and Baliunas, S.L. (1996) *Astrophys. J.*, **466**, 384.

Donati, J.F. and Collier Cameron, A. (1997) *Mon. Not. R. Astron. Soc.*, **291**, 1.

Donati, J.F., Mengel, M., Carter, B.D., et al. (2000) *Mon. Not. R. Astron. Soc.*, **316**, 699.

Donati, J.F., Collier Cameron, A., and Petit, P. (2003) *Mon. Not. R. Astron. Soc.*, **345**, 1187.

Donati, J.F., Collier Cameron, A., Semel, M., et al. (2003) *Mon. Not. R. Astron. Soc.*, **345**, 1145.

Donati, J.F., Forveille, T., Collier Cameron, A., et al. (2006) *Science*, **311**, 633.

Donati, J.F. and Landstreet, J.D. (2009) *Annu. Rev. Astron. Astrophys.*, **47**, 333.

Donnelly, R.J. and Ozima, M. (1960) *Phys. Rev. Lett.*, **4**, 497.

Dormy, E., Cardin, P., and Jault, D. (1998) *Earth Planet. Sci. Lett.*, **160**, 15.

Dormy, E., Jault, D., and Soward, A.M. (2002) *J. Fluid Mech.*, **452**, 263.

Dubrulle, B., Marié, L., Normand, Ch., et al. (2005) *Astron. Astrophys.*, **429**, 1.

Dudley, M.L. and James, R.W. (1989) *Proc. R. Soc. Lond. A.*, **425**, 407.

Durney, B.R. and Latour, J. (1978) *Geophys. Astrophys. Fluid Dyn.*, **9**, 241.

Durney, B.R. (1987) in *The Internal Solar Angular Velocity* (eds B.R. Durney et al.), Reidel, p. 235.

Durney, B.R. (1989) *Astrophys. J.*, **338**, 509.

Durney, B.R. (1996) *Sol. Phys.*, **169**, 1.

Duschl, W.J., Strittmatter, P.A., and Biermann, P.L. (2000) *Astron. Astrophys.*, **357**, 1123.

Duvall, T.L. and Gizon, L. (2000) *Sol. Phys.*, **192**, 177.

Dziembowski, W.A. and Kosovichev, A. (1987) *Acta Astron.*, **37**, 341.

Dziourkevitch, N., Elstner, D., and Rüdiger, G. (2004) *Astron. Astrophys.*, **423**, L29.

Egbers, C. and Rath, H.J. (1995) *Acta Mech.*, **111**, 125.

Eggenberger, P., Maeder, A., and Meynet, G. (2005) *Astron. Astrophys.*, **440**, L9.

Eggenberger, P., Montalbán, J., and Miglio, A. (2012) *Astron. Astrophys.*, **544**, L4.

Egorov, P., Rüdiger, G., and Ziegler, U. (2004) *Astron. Astrophys.*, **425**, 725.

Elperin, T., Golubev, I., Kleeorin, N., et al. (2005) *Phys. Rev. E*, **71**, 036302.

Elsasser, W.M. (1946) *Phys. Rev.*, **69**, 106.

Elstner, D. and Rüdiger, G. (2007) *Astron. Nachr.*, **328**, 1130.

Elstner, D., Bonanno, A., and Rüdiger, G. (2008) *Astron. Nachr.*, **329**, 717.

Elstner, D., Gressel, O., and Rüdiger, G. (2009) in *Cosmic Magnetic Fields: From Planets, to Stars and Galaxies* (eds K.G. Strassmeier et al.), p. 467.

Elstner, D. and Gressel, O. (2012) astro-ph/1206.5097.

Ferrario, L., Pringle, J.E., Tout, C.A., et al. (2009) *Mon. Not. R. Astron. Soc.*, **400**, L71.

Ferraro, V.C.A. (1937) *Mon. Not. R. Astron. Soc.*, **97**, 458.

Ferrière, K. (1992a) *Astrophys. J.*, **391**, 188.

Ferrière, K. (1992b) *Astrophys. J.*, **389**, 286.

Ferrière, K. (1996) *Astron. Astrophys.*, **310**, 438.

Ferrière, K. and Schmitt, D. (2000) *Astron. Astrophys.*, **358**, 125.

Figueroa, A., Schaeffer, N., Nataf, H.-C., et al. (2013) *J. Fluid Mech.*, **716**, 445.

Finke, K. and Tilgner, A. (2012) *Phys. Rev. E*, **86**, 016310.

Fletcher, A., Berkhuijsen, E.M., Beck, R., et al. (2004) *Astron. Astrophys.*, **414**, 53.

Fletcher, A. (2010) in *The dynamic interstellar medium* (eds R. Kothes et al.), p. 197.

Fletcher, A., Beck, R., Shukurov, A., et al. (2011) *Mon. Not. R. Astron. Soc.*, **412**, 2396.

Flores-Soriano, M. and Strassmeier, K.G. (2013) *Astron. Astrophys.*, in prep.

Forgács-Dajka, E. and Petrovay, K. (2002) *Astron. Astrophys.*, **389**, 629.

Fournier, A., Bunge, H.P., Hollerbach, R., et al. (2004) *Geophys. J. Int.*, **156**, 682.

Frick, P., Beck, R., Shukurov, A., et al. (2000) *Mon. Not. R. Astron. Soc.*, **318**, 925.

Frick, P., Denisov, S., Noskov, V., et al. (2008) *Astron. Nachr.*, **329**, 706.

Friedman, J.L. and Schutz, B.F. (1978) *Astrophys. J.*, **199**, L157.

Fröhlich, H.E. and Schultz, M. (1996) *Astron. Astrophys.*, **311**, 451.

Fröhlich, H.E., Küker, M., Hatzes, A.P., et al. (2009) *Astron. Astrophys.*, **506**, 263.

Fröhlich, H.E., Frasca, A., Catanzaro, G., et al. (2012) *Astron. Astrophys.*, **543**, A146.

Fromang, S. and Papaloizou, J. (2007) *Astron. Astrophys.*, **476**, 1113.

Fromang, S., Papaloizou, J., Lesur, G., et al. (2007) *Astron. Astrophys.*, **476**, 1123.

Gailitis, A. (1970) *Magn. Gidrodin.*, **6**, 19.

Gailitis, A. (1994) in *Energy Transfer in Magnetohydrodynamic Flows* (eds A. Alemany et al.), Plenum Press, p. 177.

Gailitis, A., Lielausis, O., Platacis, E., et al. (2001) *Phys. Rev. Lett.*, **86**, 3024.

Garaud, P. (2001) *Mon. Not. R. Astron. Soc.*, **324**, 68.
Garaud, P. (2007) *Astron. Nachr.*, **328**, 1146.
Garaud, P. and Brummell, N.H. (2008) *Astrophys. J.*, **674**, 498.
Garaud, P. and Garaud, J.D. (2008) *Mon. Not. R. Astron. Soc.*, **391**, 1239.
Gellert, M., Rüdiger, G., and Fournier, A. (2007) *Astron. Nachr.*, **328**, 1162.
Gellert, M. and Rüdiger, G. (2008) *Astron. Nachr.*, **329**, 709.
Gellert, M., Rüdiger, G., and Elstner, D. (2008) *Astron. Astrophys.*, **479**, L33.
Gellert, M. and Rüdiger, G. (2009a) *Phys. Rev. E*, **80**, 6314.
Gellert, M. and Rüdiger, G. (2009b) *J. Fluid Mech.*, **623**, 375.
Gellert, M., Rüdiger, G., and Hollerbach, R. (2011) *Mon. Not. R. Astron. Soc.*, **414**, 2696.
Gellert, M., Rüdiger, G., and Schultz, M. (2012) *Astron. Astrophys.*, **541**, A124.
Geppert, U., Küker, M., and Page, D. (2006) *Astron. Astrophys.*, **457**, 937.
Giesecke, A., Ziegler, U., and Rüdiger, G. (2005) *Phys. Earth Planet. Inter.*, **152**, 90.
Gilbert, A.D. and Sulem, P.L. (1990) *Geophys. Astrophys. Fluid Dyn.* 51, 243.
Gilman, P.A. and Benton, E.R. (1968) *Phys. Fluids*, **11**, 2397.
Gilman, P.A. (1986) in *Physics of the Sun*, (ed. P. A. Sturrock), Reidel, p. 95.
Gilman, P.A. and Miller, J. (1986) *Astrophys. J.*, **61**, 585.
Gilman, P.A. and Fox, P.A. (1997) *Astrophys. J.*, **484**, 439.
Gilman, P.A. and Dikpati, M. (2000) *Astrophys. J.*, **528**, 552.
Gilman, P.A. and Miesch, M.S. (2004) *Astrophys. J.*, **611**, 568.
Gilman, P.A., Dikpati, M., and Miesch, M.S. (2007) *Astrophys. J.*, **170**, 203.
van Gils, D.P.M., Huisman, S.G., Bruggert, G.W., et al. (2011), *Phys. Rev. Lett.*, **106**, 024502.
Gissinger, C., Ji, H., and Goodman, J. (2011) *Phys. Rev. E*, **84**, 026308.
Gissinger, C., Goodman, J., and Ji, H. (2012) *Phys. Fluids*, **24**, 074109.
Gizon, L. and Rempel, M. (2008) *Sol. Phys.*, **251**, 241.
Glatzmaier, G.A. (1985) *Astrophys. J.*, **291**, 300.
Goossens, M. and Veugelen, R. (1978) *Astron. Astrophys.*, **70**, 277.
Goossens, M., Biront, D., and Tayler, R.J. (1981) *Astrophys. Space Sci.*, **75**, 521.
Gray, D.F. (1982) *Astrophys. J.*, **261**, 259.
Gray, D.F. (1989a) *Astrophys. J.*, **347**, 1021.
Gray, D.F. (1989b) *Publ. Astron. Soc. Pac.*, **101**, 1126.
Gray, D.F. and Brown, K.I.T. (2006) *Publ. Astron. Soc. Pac.*, **118**, 1112.
Gressel, O., Elstner, D., Ziegler, U., et al. (2008) *Astron. Astrophys.*, **486**, L35.
Gressel, O., Ziegler, U., Elstner, D., et al. (2009) in *Cosmic Magnetic Fields: From Planets, to Stars and Galaxies* (eds K.G. Strassmeier et al.), p. 81.
Gressel, O., Elstner, D., and Rüdiger, G. (2011) in *Advances in Plasma Astrophysics* (eds A. Bonanno et al.), p. 348.
Gressel, O., Bendre, A., and Elstner, D. (2013) *Mon. Not. R. Astron. Soc.*, **429**, 967.
Guervilly, C. and Cardin, P. (2010) *Geophys. Astrophys. Fluid Dyn.*, **104**, 221.
Guervilly, C., Cardin, P., and Schaeffer, N. (2012) *Icarus*, **218**, 100.
Guirado, J.C., Marti-Vidal, I., and Marcaide, J.M. (2008) in *A Giant Step: From Multi- to Micro-Arcsecond Astrometry* (eds J.J. Jin et al.), Kluwer, p. 496.
Hackman, T., Pelt, J., Mantere, M., et al. (2013) astro-ph/1211.0914.
Hall, D.S. (1991) in *The Sun and Cool Stars: Activity, Magnetism, Dynamos* (eds I. Tuominen et al.), p. 353.
Hanasz, M. and Lesch, H. (1997) *Astron. Astrophys.*, **321**, 1007.
Hanasz, M., Kowal, G., Otmianowska-Mazur, K., et al. (2004) *Astrophys. J.*, **605**, 33.
Hanasz, M., Otmianowska-Mazur, K., Kowal, G., et al. (2006) *Astron. Nachr.*, **327**, 469.
Harlander, U., Gellert, M., Egbers, C., et al. (2013) *J. Fluid Mech.*, in prep.
Hartmann, L.W. and Noyes, R.W. (1987) *Annu. Rev. Astron. Astrophys.*, **25**, 271.
Hathaway, D.H., Nandy, D., Wilson, R.M., et al. (2003) *Astrophys. J.*, **589**, 665.
Hathaway, D.H. and Rightmire, L. (2010) *Science*, **327**, 1350.
Hawley, J.F., Gammie, C.F., and Balbus, S.A. (1996) *Astrophys. J.*, **464**, 690.
Heger, A. and Langer, N. (2000) *Astrophys. J.*, **544**, 1016.
Heger, A., Woosley, S.E., and Spruit, H.C. (2005) *Astrophys. J.*, **626**, 350.

Henry, G.W., Eaton, J.A., Hamer, J., et al. (1995) *Astrophys. J. Suppl.*, **97**, 513.

Herzenberg, A. (1958) *Philos. Trans. R. Soc. A*, **250**, 543.

Heyl, J.S. and Hernquist, L. (2001) *Mon. Not. R. Astron. Soc.*, **324**, 292.

Hoffmann Ch., Heise M., Altmeyer S., et al., 2009, *Phys. Rev. E*, **80**, 066308.

Hollerbach, R. (1994) *Proc. R. Soc. Lond. A*, **444**, 333.

Hollerbach, R. (1997) *Acta Astronom. Geophys. Univers. Comen.*, **19**, 263.

Hollerbach, R. (2000a) *Int. J. Numer. Methods Fluids*, **32**, 773.

Hollerbach, R. (2000b) in *Physics of Rotating Fluids* (eds C. Egbers et al.), Springer, p. 295.

Hollerbach, R. (2001) in *Dynamo and Dynamics, a Mathematical Challenge* (eds P. Chossat et al.), Kluwer, p. 189.

Hollerbach, R. and Skinner, S. (2001) *Proc. R. Soc. Lond. A*, **457**, 785.

Hollerbach, R. and Rüdiger, G. (2002) *Mon. Not. R. Astron. Soc.*, **337**, 216.

Hollerbach, R. (2003) *J. Fluid Mech.*, **492**, 289.

Hollerbach, R. and Fournier, A. (2004) in *MHD Couette Flows: Experiments and Models* (eds R. Rosner et al.), p. 114.

Hollerbach, R., Futterer, B., More, T., et al. (2004) *Theor. Comput. Fluid Dyn.*, **18**, 197.

Hollerbach, R. and Rüdiger, G. (2005) *Phys. Rev. Lett.*, **95**, 124501.

Hollerbach, R., Junk, M., and Egbers, C. (2006) *Fluid Dyn. Res.*, **38**, 257.

Hollerbach, R., Canet, E., and Fournier, A. (2007) *Eur. J. Mech. B*, **26**, 729.

Hollerbach, R. (2009) *Proc. R. Soc. Lond. A*, **465**, 2003.

Hollerbach, R., Teeluck, V., and Rüdiger, G. (2010) *Phys. Rev. Lett.*, **104**, 044502.

Howard, R., Adkins, J.M., Boyden, J.E., et al. (1983) *Sol. Phys.*, **83**, 321.

Howard, R.F. and LaBonte, B.J. (1980) *Astrophys. J.*, **239**, L33.

Howard, R.F. and LaBonte, B.J. (1983) in *Solar and Stellar Magnetic Fields: Origins and Coronal Effects*, (ed. J.O. Stenflo), Reidel, p. 101.

Howe, R., Komm, R.W., and Hill, F. (2002) *Astrophys. J.*, **580**, 1172.

Howe, R., Hill, F., Komm, R.W., et al. (2011) *J. Phys. Conf. Ser.*, **271**, 12074.

Hubrig, S., North, P., and Mathys, G. (2000) *Astrophys. J.*, **539**, 352.

Hubrig, S., North, P., and Medici, A. (2000) *Astron. Astrophys.*, **359**, 306.

Hughes, D.W., Rosner, R., and Weiss, N.O. (2007) *The Solar Tachocline*, Cambridge University Press, Cambridge.

Huisman, S.G., van Gils D.P.M., Grossmann, S., et al., 2012, *Phys. Rev. Lett.*, **108**, 024501.

Hupfer, C., Käpylä, P., and Stix, M. (2005) *Astron. Nachr.*, **326**, 223.

Hupfer, C., Käpylä, P., and Stix, M. (2006) *Astron. Astrophys.*, **459**, 935.

Huré, J.M., Richard, D., and Zahn, J.P. (2001) *Astron. Astrophys.*, **367**, 1087.

Janka, H.T. and Mönchmeyer, R. (1989) *Astron. Astrophys.*, **226**, 69.

Jault, D. (2008) *Phys. Earth Planet. Inter.*, **166**, 67.

Jeffers, S.V. and Donati, J.F. (2008) *Mon. Not. R. Astron. Soc.*, **390**, 635.

Ji, H., Goodman, J., and Kageyama, A. (2001) *Mon. Not. R. Astron. Soc.*, **325**, L1.

Ji, H., Burin, M., Schartman, E., et al. (2006) *Nature*, **444**, 343.

Jouve, L., Brown, B.P., and Brun, A.S. (2010) *Astron. Astrophys.*, **509**, 32.

Käpylä, P.J., Korpi, M.J., and Tuominen, I. (2004) *Astron. Astrophys.*, **422**, 739.

Käpylä, P.J. and Brandenburg, A. (2008) *Astron. Astrophys.*, **488**, 9.

Käpylä, P.J., Korpi, M.J., and Brandenburg, A. (2009) *Astron. Astrophys.*, **500**, 633.

Käpylä, P.J., Brandenburg, A., and Korpi, M.J. (2010) *Astrophys. J.*, **719**, 67.

Käpylä, P.J., Mantere, M.J., Guerrero, G., et al. (2011) *Astron. Astrophys.*, **531**, A162.

Käpylä, P.J., Brandenburg, A., Kleeorin, N., et al. (2012) *Mon. Not. R. Astron. Soc.*, **422**, 2465.

Käpylä, P.J., Mantere, M.J., and Brandenburg, A. (2012) *Astrophys. J.*, **755**, 222.

Kaisig, M., Rüdiger, G., and Yorke, H.W. (1993) *Astron. Astrophys.*, **274**, 757.

Katz, N., Collins, C., Wallace, J., et al. (2012) *Rev. Sci. Instruments*, **83**, 063502.

Keinigs, R.K. (1983) *Phys. Fluids*, **76**, 2558.

Kelley, D.H., Triana, S.A., Zimmerman, D.S., et al. (2007) *Geophys. Astrophys. Fluid Dyn.*, **101**, 469.

Kelley, D.H., Triana, S.A., Zimmerman, D.S., et al. (2010) *Phys. Rev. E*, **81**, 026311.

Kemel, K., Brandenburg, A., Kleeorin, N., et al. (2012a) *Astron. Nachr.*, **333**, 95.

Kemel, K., Brandenburg, A., Kleeorin, N., et al. (2012b) *Sol. Phys.*, **280**, 321.

Khalzov, I.V., Brown, B.P., Kaplan, E.J., et al. (2012a) *Phys. Plasmas*, **19**, 104501.

Khalzov, I., Brown, B., Cooper, C.M., et al. (2012b) *Phys. Plasmas*, **19**, 112106.

Khokhlova, V.L. (1975) *Astron. Zh.*, **52**, 950.

Kippenhahn, R. (1963) *Astrophys. J.*, **137**, 664.

Kippenhahn, R. and Weigert, A. (1994) *Stellar Structure and Evolution*, Springer.

Kirillov, O.N. and Stefani, F. (2010) *Astrophys. J.*, **712**, 52.

Kirillov, O.N. and Stefani, F. (2011) *Phys. Rev. E*, **84**, 036304.

Kirillov, O.N. and Stefani, F. (2012) *Astrophys. J.*, **756**, 83.

Kirillov, O.N., Stefani, F., and Fukumoto, Y. (2012) *Astrophys. J.*, **756**, 83.

Kitchatinov, L.L. (1990) *Geophys. Astrophys. Fluid Dyn.*, **54**, 145.

Kitchatinov, L.L. (1991) *Astron. Astrophys.*, **243**, 483.

Kitchatinov, L.L. and Rüdiger, G. (1993) *Astron. Astrophys.*, **276**, 96.

Kitchatinov, L.L., Pipin, V.V., and Rüdiger, G. (1994) *Astron. Nachr.*, **315**, 157.

Kitchatinov, L.L., Rüdiger, G., and Küker, M. (1994) *Astron. Astrophys.*, **292**, 125.

Kitchatinov, L.L. and Mazur, M.V. (1997) *Astron. Astrophys.*, **324**, 821.

Kitchatinov, L.L. and Rüdiger, G. (1997) *Mon. Not. R. Astron. Soc.*, **286**, 757.

Kitchatinov, L.L. and Rüdiger, G. (1999) *Astron. Astrophys.*, **344**, 911.

Kitchatinov, L.L. and Rüdiger, G. (2004) *Astron. Astrophys.*, **424**, 565.

Kitchatinov, L.L. and Rüdiger, G. (2005) *Astron. Nachr.*, **326**, 379.

Kitchatinov, L.L. and Rüdiger, G. (2006) *Astron. Astrophys.*, **453**, 329.

Kitchatinov, L.L. and Rüdiger, G. (2008a) *Astron. Astrophys.*, **478**, 1.

Kitchatinov, L.L. and Rüdiger, G. (2008b) *Astron. Nachr.*, **329**, 372.

Kitchatinov, L.L. and Rüdiger, G. (2009) *Astron. Astrophys.*, **504**, 303.

Kitchatinov, L.L. and Rüdiger, G. (2010) *Astron. Astrophys.*, **513**, L1.

Kitchatinov, L.L. and Olemskoy, S.V. (2011) *Mon. Not. R. Astron. Soc.*, **411**, 1059.

Kitchatinov, L.L. and Brandenburg, A. (2012) *Astron. Nachr.*, **333**, 230.

Kleeorin, N.I., Rogachevskii, I.V., and Ruzmaikin, A.A. (1989) *Pis'ma Astron. Zh.*, **15**, 639.

Kleeorin, N., Rogachevskii, I., Ruzmaikin, A., et al. (1997) *J. Fluid Mech.*, **344**, 213.

Knobloch, E. (1992) *Mon. Not. R. Astron. Soc.*, **255**, 25.

Knobloch, E. (1996) *Phys. Fluids*, **8**, 1446.

Köhler, H. (1969) *Differentielle Rotation als Folge anisotroper turbulenter Viskosität*, Thesis, University of Göttingen.

Köhler, H. (1970) *Sol. Phys.*, **13**, 3.

Köhler, H. (1973) *Astron. Astrophys.*, **25**, 467.

Kolmogorov, A. (1941) *Dokl. Akad. Nauk SSSR*, **30**, 301.

Komm, R.W., Howard, R.F., and Harvey, J.W. (1993) *Sol. Phys.*, **147**, 207.

Komm, R.W., Howe, R., Hill, F., et al. (2005) *Astrophys. J.*, **631**, 636.

Komm, R.W., Hill, F., and Howe, R. (2008) *J. Phys. Conf. Ser.*, **118**, 12035.

Kondić, T., Rüdiger, G., and Hollerbach, R. (2011) *Astron. Astrophys.*, **535**, L2.

Kondić, T., Rüdiger, G., and Arlt, R. (2012) *Astron. Nachr.*, **333**, 202.

Korpi, M.J., Brandenburg, A., Shukurov, A., et al. (1999a) *Astron. Astrophys.*, **350**, 230.

Korpi, M.J., Brandenburg, A., Shukurov, A., et al. (1999b) *Astrophys. J.*, **514**, L99.

Korzennik, S.G. and Eff-Darwich, A. (2011) *J. Phys. Conf. Ser.*, **271**, 012067.

Koschmieder, E.L. (1993) *Benard Cells and Taylor Vortices*, Cambridge University Press.

Kosovichev, A.G. (1996) *Astrophys. J.*, **469**, L61.

Kotake, K., Sawai, H., Yamada, S., et al. (2004) *Astrophys. J.*, **608**, 391.

Kővári, Z., Strassmeier, K.G., Granzer, T., et al. (2004) *Astron. Astrophys.*, **417**, 1047.

Kővári, Z., Korhonen, H., Kriskovics, L., et al. (2012) *Astron. Astrophys.*, **539**, A50.

Kowal, G., Otmianowska-Mazur, K., and Hanasz, M. (2006) *Astron. Astrophys.*, **445**, 915.

Kraft, R.P. (1967) *Astrophys. J.*, **150**, 551.

Krause, F. and Rüdiger, G. (1974) *Astron. Nachr.*, **295**, 185.

Krause, F. and Rädler, K.H. (1980) *Mean-Field Magnetohydrodynamics and Dynamo Theory*, Pergamon Press.

Krause, F. and Beck, R. (1998) *Astron. Astrophys.*, **335**, 789.

Krause, M., Beck, R., and Hummel, E. (1989) *Astron. Astrophys.*, **217**, 17.

Krause, M., Hummel, E., and Beck, R. (1989) *Astron. Astrophys.*, **217**, 4.

Kritsuk, A.G., Norman, M.L., Padoan, P., et al. (2007) *Astrophys. J.*, **665**, 416.

Kronberg, P.P., Dufton, Q.W., Li, H., et al. (2001) *Astrophys. J.*, **560**, 178.

Küker, M. and Rüdiger, G. (1999) *Astron. Astrophys.*, **346**, 922.

Küker, M., Rüdiger, G., and Schultz, M. (2001) *Astron. Astrophys.*, **374**, 301.

Küker, M. and Rüdiger, G. (2007) *Astron. Nachr.*, **328**, 1050.

Küker, M. and Rüdiger, G. (2008) *J. Phys. Conf. Ser.*, **118**, 012029.

Küker, M. and Rüdiger, G. (2011) *Astron. Nachr.*, **332**, 83.

Küker, M., Rüdiger, G., and Kitchatinov, L.L. (2011) *Astron. Astrophys.* 530, A48.

Künstler, A., Carroll, T.A., and Strassmeier, K.G. (2013) *Astron. Astrophys.* in prep.

Kulkarni, S.R. and Fich, M. (1985) *Astrophys. J.*, **289**, 792.

Kulsrud, R.M. and Anderson, S.W. (1992) *Astrophys. J.*, **396**, 606.

Kulsrud, R.M. (1999) *Annu. Rev. Astron. Astrophys.*, **37**, 37.

Kurzweg, U.H. (1963) *J. Fluid Mech.*, **17**, 52.

Lakhin, V.P. and Velikhov, E.P. (2007) *Phys. Lett. A*, **369**, 98.

Landstreet, J.D. and Mathys, G. (2000) *Astron. Astrophys.*, **359**, 213.

Landstreet, J.D., Bagnulo, S., Andretta, V., et al. (2007) *Astron. Astrophys.*, **470**, 685.

Lanza, A.F., Pagano, I., Leto, G., et al. (2009) *Astron. Astrophys.*, **493**, 193.

Lathrop, D.P. and Forest, C.B. (2011) *Phys. Today*, **64**, 40.

Le Bars, M. and Le Gal, P. (2007) *Phys. Rev. Lett.*, **99**, 064502.

Lebedinskii, A.I. (1941) *Astron. Zh.*, **18**, 10.

Lesur, G. and Longaretti, P.Y. (2005) *Astron. Astrophys.*, **444**, 25.

Lindblom, J., Tohline, J.E., and Vallisnen, M.M. (2001) *Phys. Rev. Lett.*, **86**, 1152.

Liu, W., Goodman, J., and Herron, I. (2006) *Phys. Rev. E*, **74**, 6302.

Liverts, E., Shtemler, Yu., and Mond, M. (2012), in *Waves and Instabilities in Space and Astrophysical Plasmas* (eds P.-L. Sulem et al.), San Francisco, p. 136.

Lockman, F.J. (1984) *Astrophys. J.*, **283**, 90.

Lüst, R. (1952) *Z. Naturforsch.*, **7a**, 87.

Lynden-Bell, D. and Pringle, J.E. (1974) *Mon. Not. R. Astron. Soc.*, **168**, 603.

MacGregor, K.B. and Charbonneau, P. (1999) *Astrophys. J.*, **519**, 911.

Mac Low, M.M. and McCray, R. (1988) *Astrophys. J.*, **324**, 776.

Maeder, A. and Meynet, G. (2003) *Astron. Astrophys.*, **411**, 543.

Maeder, A. and Meynet, G. (2005) *Astron. Astrophys.*, **440**, 1041.

Mamun, C.K. and Tuckerman, L. (1995) *Phys. Fluids*, **7**, 80.

Markey, P. and Tayler, R.J. (1973) *Mon. Not. R. Astron. Soc.*, **163**, 77.

Marsden, S.C., Waite, I.A., Carter, B.D., et al. (2005) *Mon. Not. R. Astron. Soc.*, **359**, 711.

Marsden, S.C., Donati, J.F., Semel, M., et al. (2006) *Mon. Not. R. Astron. Soc.*, **370**, 468.

Marsden, S.C., Jardine, M.M., Ramírez Vélez, J.C., et al. (2011) *Mon. Not. R. Astron. Soc.*, **413**, 1939.

Mathis, S., Zahn, J.P., and Brun, A.S. (2008) in *The Art of Modeling Stars in the 21st Century* (eds L. Deng et al.), Cambridge University Press, p. 255.

Mathys, G. (2008) *Contrib. Astron. Obs. Skaln. Pleso*, **38**, 217.

Matsui, H., Adams, M., Kelley, D., et al. (2011) *Phys. Earth Planet. Inter.*, **188**, 194.

Matt, S.P., Do Cao, O., Brown, B.P., et al., (2011) *Astron. Nachr.*, **332**, 897.

Meibom, S., Mathieu, R.D., and Stassun, K.G. (2009) *Astrophys. J.*, **695**, 679.

Merbold, S., Brauckmann, H.J., and Egbers, C. (2013) *Phys. Rev. E.*, **87**, 023014.

Mestel, L. and Weiss, N.O. (1987) *Mon. Not. R. Astron. Soc.*, **226**, 123.

Mestel, L. (1999) *Stellar Magnetism*, Clarendon Press.

Meynet, G. and Maeder, A. (2005) *Astron. Astrophys.*, **429**, 581.

Michael, D. (1954) *Mathematica*, **1**, 54.

Miesch, M.S., Elliott, J.R., Toomre, J., et al. (2000) *Astrophys. J.*, **532**, 593.

Miesch, M.S. (2003) *Astrophys. J.*, **586**, 663.

Miesch, M.S. and Gilman, P.A. (2004) *Sol. Phys.*, **220**, 287.

Miesch, M.S., Brun, A.S., De Rosa, M.L., et al. (2008) *Astrophys. J.*, **673**, 557.

Mitra, D., Tavakol, R., Brandenburg, A., et al. (2009) *Astrophys. J.*, **697**, 923.

Mitra, D. and Brandenburg, A. (2012) *Mon. Not. R. Astron. Soc.*, **420**, 2170.

Mizerski, K.A. and Bajer, K. (2007) *Phys. Earth Planet. Inter.*, **160**, 245.

Mönchmeyer, R. and Müller, E. (1989) in *Timing Neutron Stars* (eds H. Ögelman et al.), p. 549.

Moffatt, K.H. (1972) *J. Fluid Mech.*, **53**, 385.

Molemaker, M.J., McWilliams, J.C., and Yavneh, I. (2001) *Phys. Rev. Lett.*, **86**, 5270.

Monchaux, R., Berhanu, M., Bourgoin, M., et al. (2007) *Phys. Rev. Lett.*, **98**, 044502.

Moss, D. (1987) *Mon. Not. R. Astron. Soc.*, **226**, 297.

Moss, D. (1990) *Mon. Not. R. Astron. Soc.*, **243**, 537.

Moss, D. (2008) *Geophys. Astrophys. Fluid Dyn.*, **102**, 195.

Münch, G. and Zirin, H. (1961) *Astrophys. J.*, **133**, 11.

Munson, B.R. and Menguturk, M. (1975) *J. Fluid Mech.*, **69**, 705.

Nakano, T., Fukushuma, T., Unno, W., et al. (1979) *Publ. Astron. Soc. Japan*, **31**, 713.

Nataf, H.C., Alboussière, T., Brito, D., et al. (2006) *Geophys. Astrophys. Fluid Dyn.*, **100**, 281.

Nataf, H.C. and Gagnière, N. (2008) *C. R. Phys.*, **9**, 702.

Nataf, H.C., Alboussière, T., Brito, D., et al. (2008) *Phys. Earth Planet. Inter.*, **170**, 60.

Nataf, H.C. (2013) *C. R. Phys.*, **14**, 248.

Newton, H.W. and Nunn, M.L. (1951) *Mon. Not. R. Astron. Soc.*, **111**, 413.

Noguchi, K., Pariev, V.I., Colgate, S.A., et al. (2002) *Astrophys. J.*, **575**, 1151.

Nornberg, M.D., Ji, H., Schartman, E., et al. (2010) *Phys. Rev. Lett.*, **104**, 074501.

Noskov, V., Denisov, S., Stepanov, R., et al. (2012) *Phys. Rev. E*, **85**, 016303.

Ogilvie, G.I. and Pringle, J.E. (1996) *Mon. Not. R. Astron. Soc.*, **279**, 152.

Oláh, K., Moór, A., Strassmeier, K.G., et al. (2013) *Astron. Nachr.*, **334**, 614.

Ooyama, K. (1966) *J. Atmos. Sci.*, **23**, 43.

Ortega, V.G., Jilinski, E., de La Reza, R., et al. (2007) *Mon. Not. R. Astron. Soc.*, **377**, 441.

Ossendrijver, M., Stix, M., and Brandenburg, A. (2001) *Astron. Astrophys.*, **376**, 713.

Ott, C.D., Ou, S., Tohline, J.E., et al. (2005) *Astrophys. J.*, **625**, L119.

Ott, C.D., Burrows, A., Thompson, T.A., et al. (2006) *Astrophys. J.*, **164**, 130.

Page, D., Geppert, U., and Weber, F. (2006) *Nucl. Phys. A*, **777**, 497.

Pandey, B.P. and Wardle, M. (2012) *Mon. Not. R. Astron. Soc.*, **426**, 1436.

Paoletti, M.S. and Lathrop, D.P. (2011) *Phys. Rev. Lett.*, **106**, 024501.

Paoletti, M.S., van Gils, D.P.M., Dubrulle, B., et al. (2012) *Astron. Astrophys.*, **547**, A64.

Papaloizou, J. and Pringle, J.E. (1978) *Mon. Not. R. Astron. Soc.*, **182**, 423.

Parker, E.N. (1971) *Astrophys. J.*, **164**, 491.

Parker, E.N. (1992) *Astrophys. J.*, **401**, 137.

Petitdemange, L., Dormy, E., and Balbus, S.A. (2008) *Geophys. Res. Lett.*, **35**, L15305.

Pitts, E. and Tayler, R.J. (1985) *Mon. Not. R. Astron. Soc.*, **216**, 139.

Poezd, A., Shukurov, A., and Sokoloff, D.D. (1993) *Mon. Not. R. Astron. Soc.*, **264**, 285.

Pouquet, A., Frisch, U., and Léorat, J. (1976) *J. Fluid Mech.*, **77**, 321.

Priede, J. and Gerbeth, G. (2009) *Phys. Rev. E*, **79**, 046310.

Priede, J. (2011) *Phys. Rev. E*, **84**, 066314.

Pringle, J.E. (1981) *Annu. Rev. Astron. Astrophys.*, **19**, 137.

Pringle, J.E. (1996) *Mon. Not. R. Astron. Soc.*, **281**, 357.

Proudman, I. (1956) *J. Fluid Mech.*, **1**, 505.

Pulkkinen, P.J., Tuominen, I., Brandenburg, A., et al. (1993) *Astron. Astrophys.*, **267**, 265.

Rädler, K.H. (1969) *Monatsber. Dtsch. Akad. Wiss. Berl.*, **11**, 272.

Rädler, K.H. and Stepanov, R. (2006) *Phys. Rev. E*, **73**, 056311.

Rädler, K.H. and Rheinhardt, M. (2007) *Geophys. Astrophys. Fluid Dyn.*, **101**, 117.

Rashba, T.I., Semikoz, V.B., and Valle, J.W.F. (2006) *Mon. Not. R. Astron. Soc.*, **370**, 845.

Rast, M.P., Ortiz, A., and Meisner, R.W. (2008) *Astrophys. J.*, **673**, 1209.

Reighard, A.B. and Brown, M.R. (2001) *Phys. Rev. Lett.*, **86**, 2794.

Reiners, A. and Schmitt, J.H.M.M. (2003a) *Astron. Astrophys.*, **398**, 647.

Reiners, A. and Schmitt, J.H.M.M. (2003b) *Astron. Astrophys.*, **412**, 813.

Reiners, A. (2006) *Astron. Astrophys.*, **446**, 267.

Rempel, M. (2005) *Astrophys. J.*, **622**, 1320.

Rempel, M. (2007) *Astron. Nachr.*, **328**, 1096.

Rempel, M. (2011) in *The Sun, the Solar Wind, and the Heliosphere* (eds M.P. Miralles et al.), Springer, p. 23.

Rengarajan, T.N. (1984) *Astrophys. J.*, **283**, 63.

Reshetnyak, M. (2006) *Phys. Solid Earth*, **42**, 449.

Reynolds, R.J. (1989) *Astrophys. J.*, **339**, L29.

Ribes, J.C. and Nesme-Ribes, E. (1993) *Astron. Astrophys.*, **276**, 549.

Ribner, H.S. (1957) *J. Acoust. Soc. Am.*, **29**, 435.

Rice, J.B. and Strassmeier, K.G. (1996) *Astron. Astrophys.*, **316**, 164.

Richard, D. and Zahn, J.P. (1999) *Astron. Astrophys.*, **347**, 734.

Rieutord, M., Brandenburg, A., Mangeney, A., et al. (1994) *Astron. Astrophys.*, **286**, 471.

Rieutord, M., Triana, S.A., Zimmerman, D.S., et al. (2012) *Phys. Rev. E*, **86**, 026304.

Roach, A.H., Spence, E.J., Gissinger, C., et al. (2012) *Phys. Rev. Lett.*, **108**, 154502.

Roberts, P.H. (1956) *Astrophys. J.*, **124**, 430.

Roberts, P.H. (1967) *Proc. R. Soc. Lond. A*, **300**, 94.

Roberts, P.H. and Stix, M. (1972) *Astron. Astrophys.*, **18**, 453.

Roberts, P.H. and Soward, A. (1975) *Astron. Nachr.*, **296**, 49.

Roettenbacher, R.M., Harmon, R.O., Vutisalchavakul, N., et al. (2011) *Astron. J.*, **141**, 138.

Rogachevskii, I. and Kleeorin, N. (2003) *Phys. Rev. E*, **68**, 036301.

Rogachevskii, I. and Kleeorin, N. (2007) *Phys. Rev. E*, **76**, 056307.

Rucinski, S.M., Walker, G.A.H., Matthews, J.M., et al. (2004) *Publ. Astron. Soc. Pac.*, **116**, 1093.

Rüdiger, G. (1974) *Astron. Nachr.*, **295**, 275.

Rüdiger, G., Tuominen, I., Krause, F., et al. (1986) *Astron. Astrophys.*, **166**, 306.

Rüdiger, G. (1989) *Differential Rotation and Stellar Convection: Sun and Solar-Type Stars*, Gordon and Breach Science Publishers & Akademieverlag Berlin.

Rüdiger, G. and Kitchatinov, L.L. (1990) *Astron. Astrophys.*, **236**, 503.

Rüdiger, G. and Kitchatinov, L.L. (1993) *Astron. Astrophys.*, **269**, 581.

Rüdiger, G. and Elstner, D. (1994) *Astron. Astrophys.*, **281**, 46.

Rüdiger, G. and Brandenburg, A. (1995) *Astron. Astrophys.*, **296**, 557.

Rüdiger, G. and Kitchatinov, L.L. (1996) *Astrophys. J.*, **466**, 1078.

Rüdiger, G. and Kitchatinov, L.L. (1997) *Astron. Nachr.*, **318**, 273.

Rüdiger, G., Brandenburg, A., and Pipin, V.V. (1999) *Astron. Nachr.*, **320**, 135.

Rüdiger, G. and Pipin, V.V. (2001) *Astron. Astrophys.*, **375**, 149.

Rüdiger, G. and Zhang, Y. (2001) *Astron. Astrophys.*, **378**, 302.

Rüdiger, G. and Küker, M. (2002) *Astron. Astrophys.*, **385**, 308.

Rüdiger, G. and Shalybkov, D.A. (2002) *Phys. Rev. E*, **66**, 016307.

Rüdiger, G., Elstner, D., and Ossendrijver, M. (2003) *Astron. Astrophys.*, **406**, 15.

Rüdiger, G., Schultz, M., and Shalybkov, D. (2003) *Phys. Rev. E*, **67**, 046312.

Rüdiger, G. and Hollerbach, R. (2004) *The Magnetic Universe*, Wiley-VCH Verlag GmbH, Weinheim.

Rüdiger, G. and Shalybkov, D. (2004) *Phys. Rev. E*, **69**, 016303.

Rüdiger, G. (2004) *AIP Conf. Proc.*, **733**, 71.

Rüdiger, G. and Kitchatinov, L.L. (2005) *Astron. Astrophys.*, **434**, 629.

Rüdiger, G., Egorov, P., and Ziegler, U. (2005) *Astron. Nachr.*, **326**, 315.

Rüdiger, G., Egorov, P., Kitchatinov, L.L., et al. (2005a) *Astron. Astrophys.*, **431**, 345.

Rüdiger, G., Hollerbach, R., Schultz, M., et al. (2005b) *Astron. Nachr.*, **326**, 409.

Rüdiger, G. and Kitchatinov, L.L. (2006) *Astron. Nachr.*, **327**, 298.

Rüdiger, G., Hollerbach, R., Stefani, F., et al. (2006) *Astrophys. J.*, **649**, L145.

Rüdiger, G. and Hollerbach, R. (2007) *Phys. Rev. E*, **76**, 068301.

Rüdiger, G., Hollerbach, R., Schultz, M., et al. (2007a) *Mon. Not. R. Astron. Soc.*, **377**, 1481.

Rüdiger, G., Schultz, M., Shalybkov, D., et al. (2007b) *Phys. Rev. E*, **76**, 056309.

Rüdiger, G. and Schultz, M. (2008) *Astron. Nachr.*, **329**, 659.

Rüdiger, G. and Shalybkov, D.A. (2009) *Astron. Astrophys.*, **493**, 375.

Rüdiger, G., Shalybkov, D.A., Schultz, M., et al. (2009) *Astron. Nachr.*, **330**, 12.

Rüdiger, G., Gellert, M., and Schultz, M., (2009) *Mon. Not. R. Astron. Soc.*, **399**, 996.

Rüdiger, G. and Kitchatinov, L.L. (2010) *Geophys. Astrophys. Fluid Dyn.*, **104**, 273.

Rüdiger, G. and Schultz, M. (2010) *Astron. Nachr.*, **331**, 121.

Rüdiger, G., Gellert, M., Schultz, M., et al. (2010) *Phys. Rev. E*, **82**, 016319.

Rüdiger, G., Schultz, M., and Elstner, D. (2011) *Astron. Astrophys.*, **530**, A55.

Rüdiger, G., Schultz, M., and Gellert, M. (2011) *Astron. Nachr.*, **332**, 17.

Rüdiger, G., Kitchatinov, L.L., and Brandenburg, A. (2011) *Sol. Phys.*, **269**, 3.

Rüdiger, G., Kitchatinov, L.L., and Schultz, M. (2012) *Astron. Nachr.*, **333**, 84.

Rüdiger, G., Gellert, M., Schultz, M., et al. (2012) *Astrophys. J.*, **755**, 181.

Rüdiger, G., Küker, M., and Schnerr, R.S. (2012) *Astron. Astrophys.*, **546**, A23.

Rüdiger, G., Kitchatinov, L.L., and Elstner, D. (2012) *Mon. Not. R. Astron. Soc.*, **425**, 2267.

Rüdiger, G. and Brandenburg, A. (2013) *Phys. Rev. E*, subm.

Rust, D.M. and Kumar, A. (1994) *Sol. Phys.*, **155**, 69.

Ruzmaikin, A.A., Shukurov, A.M., and Sokoloff, D.D. (1988) *Magnetic Fields of Galaxies*, Kluwer.

Saar, S.H. and Brandenburg, A. (1999) *Astrophys. J.*, **524**, 295.

Schaaf, M.E. (1988) *Astron. Astrophys.*, **205**, 335.

Schaaf, M.E. (1990) *Astron. Astrophys.*, **227**, 61.

Schad, A., Timmer, J., and Roth, M. (2012) *Astron. Nachr.*, **333**, 991.

Schaeffer, N. and Cardin, P. (2005) *Phys. Fluids*, **17**, 104111.

Schaeffer, N. and Cardin, P. (2006) *Earth Planet. Sci. Lett.*, **245**, 595.

Schaeffer, N., Jault, D., Cardin, P., et al., (2012) *Geophys. J. Inter.*, **191**, 508.

Schartman, E., Ji, H., and Burin, M.J. (2009) *Rev. Sci. Instrum.*, **80**, 024501.

Schartman, E., Ji, H., Burin, M.J., et al. (2012) *Astron. Astrophys.*, **543**, A94.

Schmitt, D., Alboussière, T., Brito, D., et al. (2008) *J. Fluid Mech.*, **604**, 175.

Schmitt, D. (2010) *Geophys. Astrophys. Fluid Dyn.*, **104**, 135.

Schmitt D., Cardin P., La Rizza P., et al., 2013, *Eur. J. Mech. B*, **37**, 10.

Schou, J., Antia, S.M., Basu, S., et al. (1998) *Astrophys. J.*, **505**, 390.

Schou, J. (2001) in *Recent Insights into the Physics of the Sun and Heliosphere: Highlights from SOHO and other Space Missions* (eds P. Brekke et al.), ASP, p. 21.

Schrijver, C.J. and Martin, S.F. (1990) *Sol. Phys.*, **129**, 95.

Schrijver, C.J. and Zwaan, C. (2000) *Solar and Stellar Magnetic Activity*, Cambridge University Press.

Schrinner, M., Rädler, K.H., Schmitt, D., et al. (2005) *Astron. Nachr.*, **326**, 245.

Schrinner, M., Rädler, K.H., Schmitt, D., et al. (2007) *Geophys. Astrophys. Fluid Dyn.*, **101**, 81.

Schüssler, M. and Pähler, A. (1978) *Astron. Astrophys.*, **68**, 57.

Schüssler, M. (1981) *Astron. Astrophys.* 94, L17.

Schultz-Grunow, F. (1959) *Z. Angew. Math. Mech.*, **39**, 101.

Schwope, A.D., Hambaryan, V., Haberl, F., et al. (2005) *Astron. Astrophys.*, **441**, 597.

Seehafer, N. (1990) *Sol. Phys.*, **125**, 219.

Seilmayer, M., Stefani, F., Gundrum, T., et al. (2012) *Phys. Rev. Lett.*, **108**, 244501.

Sellwood, J.A. and Balbus, S.A. (1999) *Astrophys. J.*, **511**, 660.

Shakura, N.I. and Sunyaev, R.A. (1973) *Astron. Astrophys.*, **24**, 337.

Shalybkov, D., Rüdiger, G., and Schultz, M. (2002) *Astron. Astrophys.*, **395**, 339.

Shalybkov, D. and Rüdiger, G. (2005) *Astron. Astrophys.*, **438**, 411.

Shalybkov, D. (2006) *Phys. Rev. E*, **73**, 016302.

Shtemler, Y.M., Mond, M., Rüdiger, G., et al. (2010) *Mon. Not. R. Astron. Soc.*, **406**, 517.

Shtemler Y.M., Mond, M., and Liverts, E., 2011, *Mon. Not. R. Astron. Soc.*, **413**, 2957.

Sigl, G., Olinto, A.V., and Jedamzik, K. (1997) *Phys. Rev. D*, **55**, 4582.

Simon, G.W., Title, A.M., Topka, K.P., et al. (1988) *Astrophys. J.*, **327**, 964.

Simon, G.W., Brandt, P.N., November, L.J., et al. (1994) in *Solar Surface Magnetism* (eds R.J. Rutten et al.), Kluwer, p. 261.

Simon, G.W. and Weiss, N.O. (1997) *Astrophys. J.*, **489**, 960.

Sisan, D.R. (2004) *Hydromagnetic Turbulent Instability in Liquid Sodium Experiments*, PhD Thesis, University of Maryland.

Sisan, D.R., Mujica, N., Tillotson, W.A., et al. (2004) *Phys. Rev. Lett.*, **93**, 114502.

Siwak, M., Rucinski, S., Matthews, J.M., et al. (2010) *Mon. Not. R. Astron. Soc.*, **408**, 314.

Siwak, M., Rucinski, S., Matthews, J.M., et al. (2011) *Mon. Not. R. Astron. Soc.*, **415**, 111.

Skumanich, A. (1972) *Astrophys. J.*, **171**, 565.
Snellman, J.E., Käpylä, P.J., Korpi, M.J., et al. (2009) *Astron. Astrophys.*, **505**, 955.
Snodgrass, H.B. (1984) *Sol. Phys.*, 94, 13.
Soderblom, D.R. and Däppen, W. (1989) *Astrophys. J.*, **342**, 945.
Soward, A.M. and Dormy, E. (2010) *J. Fluid Mech.*, **645**, 145.
Spence, E.J., Roach, A.H., Edlund, E.M., et al. (2012) *Phys. Plasmas*, **19**, 056502.
Spiegel, E.A. and Zahn, J.P. (1992) *Astron. Astrophys.*, **265**, 106.
Spies, G.O. (1988) *Plasma Phys. Control. Fusion*, **30**, 1025.
Spitzer, L. (1962) *Physics of Fully Ionized Gases*, John Wiley & Sons, Inc., New York.
Spruit, H.C. (1977) *Mem. Soc. Astron. Ital.*, **68**, 392.
Spruit, H.C. (1987) in *The Internal Solar Angular Velocity* B. (eds R. Durney et al.), D. Reidel, p. 185.
Spruit, H.C. (1999) *Astron. Astrophys.*, **349**, 189.
Spruit, H.C. (2002) *Astron. Astrophys.*, **381**, 923.
Spruit, H.C. (2003) *Sol. Phys.*, **213**, 1.
Sridhar, S. and Subramanian, K. (2009) *Phys. Rev. E*, **80**, 066315.
Starchenko, S.V. (1997) *J. Exp. Theor. Phys.*, **85**, 1125.
Starchenko, S.V. (1998a) *Phys. Fluids*, **10**, 2412.
Starchenko, S.V. (1998b) *Stud. Geophys. Geod.*, **42**, 314.
Steenbeck, M. and Krause, F. (1965) *Monatsber. Dtsch. Akad. Wiss. Berl.*, **7**, 335.
Stefani, F., Gundrum, T., Gerbeth, G., et al. (2006) *Phys. Rev. Lett.*, **97**, 184502.
Stefani, F., Gundrum, T., Gerbeth, G., et al. (2007) *New J. Phys.*, **9**, 295.
Stefani F., Gerbeth G., Gundrum Th, et al., 2008, *Astron. Nachr.* 329, 652.
Stefani F., Gerbeth G., Gundrum Th, et al., 2009, *Phys. Rev. E*, **80**, 066303.
Stefani F., Eckert S., Gerbeth G, et al., 2012, *Magnetohydrodynamics*, **48**, 103.
Stępień, K. (2000) *Astron. Astrophys.*, **353**, 227.
Stergioulas, N. and Font, J.A. (2001) *Phys. Rev. Lett.*, **86**, 1148.
Stewartson, K. (1966) *J. Fluid Mech.*, **26**, 131.
Stieglitz, R. and Müller, U. (2001) *Phys. Fluids*, **13**, 561.
Stix, M. (1981) *Astron. Astrophys.*, **93**, 339.
Stix, M. (1989) *Rev. Mod. Astron.*, **2**, 248.
Stix, M. and Skaley, D. (1990) *Astron. Astrophys.*, **232**, 234.
Stix, M. (2002) *The Sun: An Introduction*, Springer.
Strassmeier, K.G., Pichler, T., Weber, M., et al. (2003) *Astron. Astrophys.*, **411**, 595.
Strassmeier, K.G. (2004) in *Stars as Suns: Activity, Evolution and Planets*, (eds A.K. Dupree et al.), IAU Symp. 219, p. 11.
Strassmeier, K.G., Weber, M., Granzer, T., et al. (2012) *Astron. Nachr.*, **333**, 663.
Strugarek, A., Brun, A.S., and Zahn, J.P. (2011) *Astron. Nachr.*, **332**, 891.
Suijs, M.P.L., Langer, N., Poelarends, A.J., et al. (2008) *Astron. Astrophys.*, **481**, L87.
Szklarski, J. (2007) *Astron. Nachr.*, **328**, 499.
Szklarski, J. and Rüdiger, G. (2007) *Phys. Rev. E*, **76**, 066308.
Tabatabaei, F.S., Krause, M., Fletcher, A., et al. (2008) *Astron. Astrophys.*, **490**, 1005.
Tataronis, J.A. and Mond, M. (1987) *Phys. Fluids*, **30**, 84.
Tayler, R.J. (1957) *Proc. Phys. Soc. B*, **70**, 31.
Tayler, R.J. (1973) *Mon. Not. R. Astron. Soc.*, **161**, 365.
Tayler, R.J. (1980) *Mon. Not. R. Astron. Soc.*, **191**, 151.
Taylor, G.I. (1923) *Philos. Trans. R. Soc. A*, **233**, 289.
Thierbach, M., Klein, U., and Wielebinski, R. (2003) *Astron. Astrophys.*, **397**, 53.
Thompson, M.J., Christensen-Dalsgaard, J., Miesch, M.S., et al. (2003) *Annu. Rev. Astron. Astrophys.*, **41**, 599.
Tilgner, A. (2004) *Geophys. Astrophys. Fluid Dyn.*, **98**, 225.
Toqué, N., Lignières, F., and Vincent, A. (2006) *Geophys. Astrophys. Fluid Dyn.*, **100**, 85.
Tout, C.A. and Pringle, J.E. (1992) *Mon. Not. R. Astron. Soc.*, **259**, 604.
Travnikov, V., Eckert, K., and Odenbach, S. (2011) *Acta Mech.*, **219**, 255.
Tuominen, I., Brandenburg, A., Moss, D., et al. (1994) *Astron. Astrophys.*, **284**, 259.
Ulrich, R.K. (1993) in *Inside the Stars* (eds W. Weiss et al.), p. 25.
Umurhan, O.M. (2006) *Mon. Not. R. Astron. Soc.*, **365**, 85.
Urpin, V.A. and Rüdiger, G. (2005) *Astron. Astrophys.*, **437**, 23.
Vainshtein, S.I. and Kichatinov, L.L. (1983) *Geophys. Astrophys. Fluid Dyn.*, **24**, 273.
Vandakurov, Y.V. (1972) *Sov. Astron.*, **16**, 265.

Velikhov, E.P. (1959) *Sov. Phys. JETP*, **9**, 995.

Vempaty, S. and Loper, D.E. (1975) *Phys. Fluids*, **18**, 1678.

Vempaty, S. and Loper, D.E. (1978) *Z. Angew. Math. Phys.*, **29**, 450.

Vincent, A., Michaud, G., and Meneguzzi, M. (1996) *Phys. Fluids*, **8**, 1312.

Vishniac, E.T. and Brandenburg, A. (1997) *Astrophys. J.* 475, 263.

Vorontsov, S.V., Christensen-Dalsgaard, J., Schou, J., et al. (2002) *Science*, **296**, 101.

Vogt, S.S. and Penrod, G.D. (1983) *Publ. Astron. Soc. Pac.* 95, 565.

Wade, G.A., Silvester, J., Bale, K., et al. (2009) *ASP Conf. Ser.*, **405**, 499.

Walker, G.A.H., Croll, B., Kuschnig, R., et al. (2007) *Astrophys. J.* 659, 1611.

Wang, Y., Noyes, R.W., Tarbell, T.D., et al. (1995) *Astrophys. J.*, **447**, 419.

Ward, F. (1965) *Astrophys. J.*, **141**, 534.

Wardle, M. (1999) *Mon. Not. R. Astron. Soc.* 307, 849.

Wareing, C.J. and Hollerbach, R. (2009) *Astron. Astrophys.*, **508**, 39.

Wasiutynski, J. (1946) *Astrophys. Nor.*, **4**, 1.

Watson, M. (1981) *Geophys. Astrophys. Fluid Dyn.*, **16**, 285.

Watts, A.L., Andersson, N., Beyer, H., et al. (2003) *Mon. Not. R. Astron. Soc.*, **342**, 1156.

Weber, M. and Strassmeier, K.G. (2005) in *Cambridge Workshop on Cool Stars, Stellar Systems and the Sun* (eds F. Favata et al.), p. 1029.

Weber, M., Strassmeier, K.G., and Washuettl, A. (2005) *Astron. Nachr.*, **326**, 287.

Wei, X. and Hollerbach, R. (2008) *Phys. Rev. E*, **78**, 026309.

Wei, X. and Hollerbach, R. (2010) *Acta Mechanica*, **215**, 1.

Weiss, N.O. (1965) *Observatory*, **85**, 37.

Weiss, N.O. (1966) *Proc. R. Soc. Lond. A*, **293**, 310.

von Weizsäcker, C.F., 1943, *Z. Astrophys.*, **22**, 319.

von Weizsäcker, C.F., 1948, *Z. Naturforsch.*, 3a, 524.

Wendt, F. (1933) *Ingenieur-Archiv*, 4, 577.

Widrow, L.M. (2002) *Rev. Mod. Phys.*, **74**, 775.

Wielebinski, R. and Krause, F. (1993) *Astron. Astrophys. Rev.*, **4**, 449.

Wilson, P.R., Burtonclay, D., and Li, Y. (1997) *Astrophys. J.*, **489**, 395.

Withjack, E.M. and Chen, C.F. (1974) *J. Fluid Mech.*, **66**, 725.

Wöhl, H., Brajša, R., and Hanslmeier, A., et al., 2010, *Astron. Astrophys.*, **520**, A29.

Woosley, S.E. and Heger, A. (2006) *Astrophys. J.*, **637**, 914.

Wright, G.A.E. (1973) *Mon. Not. R. Astron. Soc.*, **162**, 339.

Xing, L.F., Shi, J.R., and Wei, J.Y. (2007) *New Astron.*, **12**, 265.

Yavneh, I., McWilliams, J.C., and Molemaker, M.J. (2001) *J. Fluid Mech.*, **448**, 1.

Yoon, S.C., Langer, N., and Norman, C. (2006) *Astron. Astrophys.*, **460**, 199.

Yoshimura, H. (1981) *Astrophys. J.*, **247**, 1102.

Youd, A.J. and Barenghi, C.F. (2006) *J. Fluid Mech.*, **550**, 27.

Yousef, T., Brandenburg, A., and Rüdiger, G. (2003) *Astron. Astrophys.*, **411**, 321.

Yousef, T.A., Haugen, N.E.L., and Brandenburg, A. (2004) *Phys. Rev. E*, **69**, 056303.

Yousef, T.A., Heinemann, T, Schekochihin, A.A., et al. (2008) *Phys. Rev. Lett.*, **100**, 184501.

Zahn, J.P. (1989) in *Inside the Sun* (eds G. Berthomieu et al.), p. 425.

Zahn, J.P. (1992) *Astron. Astrophys.*, **265**, 115.

Zahn, J.P. (1993) in *Astrophysical Fluid Dynamics*, Les Houches 1987, p. 561.

Zahn, J.P., Brun, A.S., and Mathis, S. (2007) *Astron. Astrophys.*, **474**, 145.

Zeldovich, Y.B., Ruzmaikin, A.A., and Sokoloff, D.D. (1983) *Magnetic Fields in Astrophysics* Gordon and Breach Science Publishers.

Zhang, H., Sakurai, T., Pevtsov, A., et al. (2010) *Mon. Not. R. Astron. Soc.*, **402**, 30.

Zhang, H. (2012) *Mon. Not. R. Astron. Soc.*, **419**, 799.

Zhao, J. and Kosovichev, A.G. (2004) *Astrophys. J.*, **603**, 776.

Zhao, J., Wang, X.F., and Zhang, H. (2011) *Sol. Phys.*, **270**, 23.

Ziegler, U. (1996) *Astron. Astrophys.*, **313**, 448.

Ziegler, U., Yorke, H.W., and Kaisig, M. (1996) *Astron. Astrophys.*, **305**, 114.

Ziegler, U. (2002) *Astron. Astrophys.*, **386**, 331.

Ziegler, U. (2004) *Comput. Phys. Commun.*, **157**, 207.

Zikanov, O. and Thess, A. (1998) *J. Fluid Mech.*, **358**, 299.

Index

Symbols
1D approximation, 171
3D spectral MHD code, 208
3 m diameter spherical experiment, 325

A
A star, 252
AB Dor, 46
absolute instability, 244
accretion disks, 162, 317
accretion theory, 194
acoustic oscillations, 2, 65
activity belts, 7
advection problem, 169
advection velocity, 167
advection-dominated dynamo, 32
advection-dominated solar dynamo, 131
Alfvén crossing times, 184
Alfvén frequency, 85, 159, 200, 216
Alfvén radius, 9
Alfvén velocity, 74, 198
Alfvén waves, 118, 316
Alfvén–Coriolis wave, 216
α effect, 95, 144, 155, 269, 280
α experiment, 271
α^2-dynamo, 171
α-tensor, 144, 161, 168
AMRI experiment, 221, 230
analytical theory, 19
anelastic approximation, 131
angular momentum transport, 13, 20, 194, 262
angular velocity, 69
anisotropic diffusion, 126
anisotropic forcing, 15
antidiffusion, 129
antisolar rotation, 24
$\alpha\Omega$-dynamo, 32, 101, 168
Ap stars, 103

Arcturus, 56
azimuthal magnetorotational instability (AMRI), 186, 219, 223, 274

B
baroclinic flow, 26, 28, 37
baroclinic term, 37
battery effect, 176
beryllium, 63
β viscosity, 187, 213
blades, 324
Bochner theorem, 143
boron, 63
boundary condition, 76, 136
boundary layer, 324
boundary layer effect, 109
buoyancy, 66
buoyancy forces, 26
buoyancy frequency, 189
butterfly diagram, 32, 154

C
Ca II H&K, 11
cell size, 206
channel, 150
channel mode, 214
chemical mixing, 249, 266
clustered SN explosions, 164
collapsing core, 111
color index, 9
column density, 194
conducting cylinders, 224
conducting walls, 204
confinement parameter, 78
controlled endplates, 187
controlled rings, 305
convection velocity, 53
convection zones, 71
convective cores, 103

copper electrodes, 255
copper endplate, 240
core-envelope coupling, 64
Coriolis force, 14, 75, 313
Coriolis number, 14, 59
CoRoT data, 44
CoRoT-2a, 48
correlation function, 196
correlation tensor, 19, 262
correlation time, 263
Cottbus SRI experiment, 193
counterrotating jet, 298, 318
Cowling theorem, 109, 204
cross-correlation, 14
cross-helicity, 134
crossover point, 260
current helicity, 85, 96, 107, 129, 153, 280
current-free fields, 182
cycle time, 34
cylindrical rotation law, 27

D

decay of sunspots, 134
density scale height, 19
density stratification, 132, 150
density-stratified Taylor–Couette flows, 188
diamagnetic pumping, 79, 129
differential rotation, 277, 317
differential temperature, 27, 29
diffusion equation, 77, 266
dipolar magnetic field, 307
dipolar parity, 5, 33
disk dynamos, 168
dispersion relation, 178, 198, 304
DIV-CURL correlation, 133
Doppler imaging, 11
Doppler magnetograms, 137
Doppler sounding, 235
Doppler velocimetry, 303, 308
double-diffusive instability, 89
drift frequency, 237
drift rate, 90, 222
driven turbulence, 116, 150
DTS experiment, 307
dynamo equation, 142
dynamo excitation, 95
dynamo experiments, 203
dynamo number, 61, 280
dynamo regime, 48

E

early Universe, 176
Earth's core, 288

eddy diffusivity, 32, 126, 130, 168
eddy viscosity, 22, 56, 262, 264
eddy viscosity tensor, 22
eigenmodes, 84
Ekman circulation, 243
Ekman layer, 242, 288
Ekman number, 288
Ekman pumping, 230, 244
Ekman scale, 46, 75
Ekman vortices, 241, 246
Ekman–Hartmann layer, 243
electric currents, 251
electromagnets, 303
electromotive force (EMF), 124, 128, 165, 262, 263
Elsasser number, 211, 228, 288, 316
end-effects, 239
entropy equation, 28
entropy gradient, 22
ε Eri, 44
equatorial acceleration, 24
equatorial drift, 5
equatorial symmetries, 84
equator–pole difference, 2
equipartition field, 157
equipartition value, 120
η-quenching, 124
experiment GATE, 255
external coil, 240

F

F stars, 53, 61
Ferraro isorotation law, 291, 310
filling factors, 165
fixed wave number, 234
FK Coma, 57
flux of angular momentum, 72
flux tubes, 218
fluxgate sensors, 255
force-free, 155
fossil field, 4, 248
Fourier mode, 65, 225
Fourier spectrum, 179
Froude number, 190
fully convective stars, 126

G

galactic dynamo, 174
galactic rotation law, 172
galinstan (GaInSn), 198, 229, 252, 254
gallium, 198
GATE, 256
geodynamo simulations, 316

Grashof number, 27
gravitation waves, 65
gravity, 14
gravity darkening, 31
growth rate, 111, 171, 252, 256, 277
growth time, 101
gyrochronology, 9, 58

H
Hα emission, 162
Hall effect, 109, 216, 282, 284
Hall instability, 200
Hall parameter, 284
Hall probes, 303, 308
Hall time, 110, 283
Hall-MRI, 112, 218
Hall-TI, 283
halo gas, 160
Hartmann current, 243
Hartmann layer, 242, 293
Hartmann number, 74, 220, 254, 289
HD 141943, 49
helical background field, 129, 237, 273, 281
helical magnetorotational instability (HMRI), 232, 272
helioseismology, 63
Helmholtz-Zentrum Dresden-Rossendorf, 221
hemisphere rule, 153
high conductivity, 144, 252
high-conductivity limit, 120, 147
HINODE, 109, 137
horizontal temperature gradient, 28
H–R diagram, 249
hydrodynamic simulation, 107
hydromagnetic dynamos, 95
hydrostatic equilibrium, 166
HZDR, 255

I
II Peg, 13
induction equation, 109
inductionless nature, 306
inertial oscillations, 321
inner sphere rotation rate, 324
instability window, 106
insulating cylinders, 224
insulating walls, 201
intermediate-mass stars, 103
interstellar medium, 161
interstellar turbulence, 157
IT Com, 13

J
jet structures, 294
Jupiter-like profile, 325

K
κ^1 Ceti, 44
Kelvin–Helmholtz type, 314
Kelvin–Helmholtz type instabilities, 297
Kepler disk, 215
Kepler flow, 209
Kepler limit, 205
KEPLER mission, 248
Kepler rotation, 212
KIC 7341231, 57, 248
KIC 8366239, 57, 248
kinematic dynamo, 142, 322
kinetic helicity, 85, 95, 107, 131, 267, 280
kink mode, 88
kink-type instability, 100, 249, 261
Kolmogorov exponent, 161
Kolmogorov spectrum, 179

L
laboratory experiment, 228
Lalande 21185, 51
Λ effect, 14, 35
lapping time, 42
large-scale dynamos, 281
left-hand modes, 273
left-hand spirals, 221
Legendre spectral element method, 208
Lehnert number, 315
light element, 63
liquid metals, 148, 198
liquid sodium, 302
lithium, 63, 89, 228, 248, 265
local approximation, 237
longitudinal drift, 89
Lorentz force, 6, 91
low conductivity, 252
low-conductivity limit, 120, 147, 257
LQ Hya, 46
LQ Lup, 49
Lundquist number, 91, 117, 200, 233, 268, 303

M
M 81, 158
M dwarf rotation, 51
M dwarfs, 62
Mach number, 18, 162, 200
Madison Plasma Couette Experiment, 325
magnetic boundary conditions, 201
magnetic diffusivity, 61
magnetic eddy diffusivity, 136

magnetic energy, 279
magnetic field belts, 92
magnetic field modulus, 104
magnetic fluctuations, 264
magnetic Mach number, 223
magnetic Prandtl number, 34, 143, 200, 238, 277
magnetic quenching, 172
magnetic Reynolds number, 76, 111, 156, 200, 233, 303, 306
magnetic spherical Couette flow, 313
magnetic suppression, 121
magnetic suppression of SRI, 197
magnetic-diffusivity tensor, 124
magnetoconvection, 122, 137
magneto-Coriolis wave, 233
magneto-inertial waves, 314
magnetorotational instability (MRI), 176, 186
magnetostrophic MRI, 317
marginal stability, 93, 172
Maryland apparatus, 321
Maryland experiment, 302
Maryland group, 314
Maunder minimum, 4
Maxwell pressure, 120
Maxwell stress, 74, 120, 282
Maxwell tensor, 121
mean electromotive force, 140
mean-field electrodynamics, 195
mean-field hydrodynamics, 2
mean-field model, 35
mean-field theory, 131
mercury, 198
meridional circulation, 6, 109
meridional flow, 23, 24, 32
mesogranulation, 131
metallicity, 68
MHD equations, 73
MHD simulations, 166, 178
microscopic diffusivity, 176
microscopic viscosity, 63
mirror-symmetric turbulence, 143
mixing-length, 22
mode spectrum, 225
modified Reynolds number, 256
Most mission, 12, 44
MRI, 109, 317

N
Navier–Stokes equation, 91, 125
near-boundary jets, 47
near-equator belts, 55
near-surface shear layer, 4

negative-pressure phenomenon, 123
Neptune-like profile, 325
neutrino emission, 111
neutron star, 110, 247, 284
NGC 4414, 159, 177
NGC 4631, 158
NGC 4736, 158
NGC 5775, 159
NGC 6946, 157
Nirvana Code, 16, 122, 136, 165
nonaxisymmetric modes, 204, 236
nondiffusive transport, 14
nonlinear regime, 224, 259
no-slip conditions, 201

O
oblique rotator, 111
Ohmic decay, 78
Ohmic dissipation, 228
Ohmic heating, 256
$\Omega \times J$ effect, 97, 125
opacity, 80
overreflection theory, 313
overshoot region, 18, 30

P
parity breaking, 97, 267
Pencil Code, 15, 141, 148
penetration depth, 33
penetration region, 76
penetrative convection, 71
phase velocity, 221
pitch angle, 158, 173, 234
planet, 48
planetary cores, 152
plasma β, 216
plexiglass, 240, 244
plexiglass endplates, 230
polar branch, 6
pole–equator temperature difference, 54
poloidal field, 97
potential difference, 271
Potsdam ROssendorf Magnetic InStability Experiment (Promise), 240
primordial magnetic field, 64, 177
Princeton experiment, 187
Princeton MRI experiment, 305
Promise, 221, 230, 235
Promise 2, 244
protoneutron star (PNS), 282
protoplanetary disks, 114
Proxima Cen, 51
Pm $\rightarrow$ 0 approximation, 290

Index | 345

pseudo-Kepler flow, 217
pseudo-Kepler line, 190
pseudo-Kepler rotation, 185, 237
pseudoscalar, 97, 99, 132, 153, 269, 272
pseudovacuum, 212
pseudovector, 14
pumping term, 146

Q
quadrupolar fields, 291
quadrupolar geometry, 178
quasi-galactic rotation, 189, 228
quasi-galactic rotation law, 180
quasi-geostrophic approximation, 322
quasi-linear approximation, 115
quasi-linear theory, 143
quasi-static approximation, 204, 290

R
R58, 49
radial jet, 324
radial jet instability, 298
radiation zone dynamo, 101
radio-polarization data, 157
rapid rotation, 20
rapid rotators, 60
Rayleigh limit, 187, 202, 205, 233
RBS1223, 283
red giants, 55
regularity conditions, 79
resting outer cylinder, 202
return flow, 9
return flow instability, 299
return flow/Shercliff layer branch, 301
Reynolds number, 24, 194, 220, 303
Reynolds rules, 115
Reynolds stress, 120
Reynolds stress tensor, 15
right-hand modes, 273
right-hand spirals, 221
rigid rotation, 38
r-modes, 65, 282
Roberts flow, 126
Rossby number, 14
Rossby waves, 325
rotation-activity-age relation, 249
rotational history, 249
rotational quenching, 87, 256
rotational shear, 72
rotational velocity, 249
RS CVn binary systems, 13

S
saturation of TI, 184

Schmidt number, 228, 265
Schwarzschild criterion, 1
screening problem, 31
second-order correlation approximation (SOCA), 115
seed field, 175
shear flow, 138, 146
shear flow dynamos, 142
shear flow experiment, 139
shear-induced dissipation, 149
shearing box, 165
shearing box simulations, 216
Shear–Hall instability (SHI), 109, 111, 217
Shercliff layer, 289, 304
Shercliff shear layer instabilities, 299
short-wave approximation, 67
sidereal rotation, 2
Skumanich law, 9
small magnetic Prandtl number limit, 198
small-gap approximation, 198
SN distribution, 165
SN explosion, 110, 166
sodium, 198
solar convection zone, 1
solar radiation zone, 63, 68
solar-type rotation, 43
solar-type stars, 66
sound speed, 137
sound waves, 313
specific angular momentum, 247
spectral lines, 51
spectral MHD code, 91
spectral tensor, 115, 153
spectrum of the nonaxisymmetric modes, 106
speed of sound, 16
spherical Couette dynamos, 321, 326
spherical Couette flow, 287, 305, 310, 317
spherical model, 111
spin-down law, 9
spin-down of the stellar core, 247
split-ring endplates, 188
SRI, 190
SRI experiment, 193
SST, 137
stability lines, 207
stability map, 71, 261
stability problem, 65
stagnation point, 24, 33
stainless steel, 302
standard MRI, 208, 275
star formation rate (SFR), 121, 175
star-disk coupling, 104
steady quadrupoles, 172

stellar coronae, 9
stellar dynamo, 54
Stewartson layer, 287, 322
stratorotational instability, SRI, 186
strong-field branch, 177
strong-field regime, 270
Strouhal number, 115
sunspot cycle, 32
sunspot formation, 123
sunspot observations, 4
superbubble, 164
supergranulation layer, 18
supernova (SN) explosions, 161
superrotating regions, 292
superrotation, 39, 198
supersonic rotation, 195
surface temperature, 12

T

T Tau stars, 13
tachocline, 35, 64, 72
tachocline dynamo, 34
tachocline layer, 17
tachocline theory, 81
tachocline thickness, 74
tangent cylinder, 108, 293
Tayler generator, 271
Tayler instability (TI), 86, 99, 180, 186, 247
Taylor number, 27
Taylor vortices, 206, 259, 274
Taylor-number puzzle, 29, 40
Taylor–Couette (TC) flow, 185
Taylor–Couette container, 249
Taylor–Dean flow, 243
Taylor–Proudman, 17
Taylor–Proudman theorem, 188, 209, 287, 310
test-field method, 166
thermal convection, 16, 36
thermal diffusivity, 35
thermal diffusivity tensor, 61
thermal wind, 27, 36
torque measurements, 187
torsional oscillations, 5
total pressure, 122
transition zones, 91
travel frequency, 234
traveling wave, 240, 241, 246
turbulence frequency, 117

turbulence spectrum, 136
turbulent background flow, 304
turbulent diffusivity, 64, 263
turbulent dynamos, 95
turbulent EMF, 195
turbulent magnetic Prandtl number, 130
turbulent pressure, 119, 139
turbulent pumping, 164
turbulent viscosity, 14
turnover time, 52
two-dimensional α-tensor, 151
types of equatorial symmetry, 67

U

ultrasonic high-focus transducers, 240
ultrasound measurements, 241
uniform current, 275

V

V889 Her, 12, 49
V2253, 13
vacuum condition, 74
viscosity frequency, 237
viscous fluxes, 23

W

Watson's value, 93
wave number, 74, 84
white dwarfs, 247
white-noise spectra, 139
white-noise spectrum, 147
wide-gap container, 188
winds, 104

X

X-ray brightness, 283
X-ray emissions, 111
X-type neutral point, 318
XX Tri, 57

Y

young clusters, 249

Z

ZAMS stars, 58
ζ And, 13
ZZ Ceti, 248